Phase Transitions in
Soft Condensed Matter

NATO ASI Series

Advanced Science Institutes Series

A series presenting the results of activities sponsored by the NATO Science Committee, which aims at the dissemination of advanced scientific and technological knowledge, with a view to strengthening links between scientific communities.

The series is published by an international board of publishers in conjunction with the NATO Scientific Affairs Division

A	**Life Sciences**	Plenum Publishing Corporation
B	**Physics**	New York and London
C	**Mathematical and Physical Sciences**	Kluwer Academic Publishers Dordrecht, Boston, and London
D	**Behavioral and Social Sciences**	
E	**Applied Sciences**	
F	**Computer and Systems Sciences**	Springer-Verlag
G	**Ecological Sciences**	Berlin, Heidelberg, New York, London,
H	**Cell·Biology**	Paris, and Tokyo

Recent Volumes in this Series

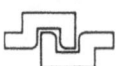

Series B: Physics

Phase Transitions in Soft Condensed Matter

Edited by

Tormod Riste

Institute for Energy Technology
Kjeller, Norway

and

David Sherrington

Imperial College of Science, Technology, and Medicine
London, United Kingdom

Plenum Press
New York and London
Published in cooperation with NATO Scientific Affairs Division

Proceedings of a NATO Advanced Study Institute on
Phase Transitions in Soft Condensed Matter,
held April 4-14, 1989,
in Geilo, Norway

Library of Congress Cataloging-in-Publication Data

Phase transitions in soft condensed matter / edited by Tormod Riste
and David Sherrington.
      p.   cm. -- (NATO ASI series. Series B, Physics ; v. 211)
   "Proceedings of a NATO Advanced Study Institute on Phase
Transitions in Soft Condensed Matter, held April 4-14, 1989, in
Geilo, Norway"--T.p. verso.
   "Published in cooperation with NATO Scientific Affairs Division."
   Includes bibliographical references.
   ISBN-13:978-1-4612-7862-7   e-ISBN-13:978-1-4613-0551-4
   DOI: 10.1007/978-1-4613-0551-4

   1. Condensed matter--Congresses. 2. Phase transformations
(Statistical physics)--Congresses.   I. Riste, Tormod, 1925-
II. Sherrington, D. C. III. NATO Advanced Study Institute on Phase
Transitions in Soft Condensed Matter (1989 : Geilo, Norway)
IV. North Atlantic Treaty Association. Scientific Affairs Division.
V. Series.
QC173.4.C65P442 1989
530.4'1--dc20   89-26644
                                                          CIP

© 1989 Plenum Press, New York
Softcover reprint of the hardcover 1st edition 1989

A Division of Plenum Publishing Corporation
233 Spring Street, New York, N.Y. 10013

PREFACE

   This volume comprises the proceedings of a NATO Advanced Study
Institute held in Geilo, Norway, between 4 - 14 April 1989. This
Institute was the tenth in a series held at Geilo on the subject of
phase transitions. It was the first to be concerned with the growing
area of soft condensed matter, which is neither ordinary solids nor
ordinary liquids, but somewhere in between.

   The Institute brought together many lecturers, students and active
researchers in the field from a wide range of NATO and some non-NATO
countries, with financial support principally from the NATO Scientific
Affairs Division but also from Institutt for energiteknikk, the Nor-
wegian Research Council for Science and the Humanities (NAVF), The
Nordic Institute for Theoretical Atomic Physics (NORDITA), the Norwegian
Physical Society and VISTA, a reserach cooperation between the Norwegian
Academy of Science and Letters and Den norske stats oljeselskap a.s
(STATOIL).

   The organizing committee would like to thank all these contributors
for their help in promoting an exciting and rewarding meeting, and in
doing so are confident that they echo the appreciation also of all the
participants.

   Soft condensed matter is characterized by weak interactions between
polyatomic constituents, by important·thermal fluctuations effects, by
mechanical softness and by a rich range of behaviours. The main emphasis
at this Institute was on the fundamental collective physics, but preparation techniques and industrial applications were also considered.

   The introductory lectures set the scene and introduced the nature of
the constitutent systems and their cooperative consequences and conceptual make-ups for the whole field, with later talks emphasizing sub-
fields in greater detail. The lectures were supplemented by research
seminars and poster sessions, all of which proved conductive of considerable discussion and continuing exchange. These proceedings include
write-ups of the majority of the lectures and seminars and a selection
of the posters.

   One important class of soft condensed matter comprises thermotropic
liquid crystals, which are composed of relatively small and simple molecular building blocks but produces a plethora of interesting phases
characterized by orientational and spatial order. These are the subject
of several articles.

   Polymers are composed of larger molecules. Many aspects of their
phase transitions, association and gelation were discussed and reported
here, as also other physical properties.

Amphiphilic molecules are soap molecules with a polar head and an alkyl tail, able to bind respectively to polar liquids and hydrocarbons. Structures formed by amphiphilic molecules in contact with water are two-dimensional films and membranes, which may further interact to form lyotropic liquid crystals with a wide range of polymorphism and interesting vibrational and defect properties, which were discussed, as also were lipid mono- and bi-layers and Langmuir-Blodgett films. In water-oil mixtures, surfactant amphiphilic molecules, binding to both constituents, lead to microemulsions and to micelles, of great interest to the petroleum and food industry, as well as possessing fascinating properties to excite the physicist.

Many interesting aspects are associated with the static and dynamic properties of liquid surfaces and interfaces and with the wetting of solid surfaces.

Fractals have been a subject of considerable interest in recent years and were the subject of the Institute in 1985. An ideal system to study their statics and dynamics is that of areogels and this study was a natural link between this Institute and the 1985 Institute. Other aspects of fractals and growth were also reported.

As noted earlier, although the emphasis was on fundamental microscopic physics, industrial applications were also discussed and illustrated: soft condensed matter is already a major part of our lifes and its use is growing. An interesting application involving modern polymeric processing to illustrate fundamental features is the use of perfectly engineered polystyrene spheres and related shapes immersed in water or ferrofluid to model microparticle interactions on a scale providing ready optical observations. In a sense this study (second lecture) represents a sort of "periodic boundary conditions" on the Institute, but in fact the subject is growing in many conceptual dimensions, as the discussions reported here amply demonstrate.

The subject of phase transitions in soft condensed matter is still in its infancy, compared with more conventional condensed matter physics, but it posseses a richness and a potential which are likely to provide excitement and challenges for many years to come. We hope that the papers presented in this volume will help germinate in others the euphoria which is already rapidly growing in the participants of the tenth Geilo Institute.

Finally, we would like to express our deep gratitude, and that of ten generations of participants, to Gerd Jarrett of the Institutt for energiteknikk, Kjeller, Norway who did all the practical organization of all ten of the Geilo Institutes, including the preparation of their proceedings, with incredible efficiency and smoothness. Without Gerd the series would not have been the success it seems to have been.

Tormod Riste   David Sherrington

CONTENTS

STATES AND PHASE TRANSITIONS IN SOFT CONDENSED MATTER:

AN INTRODUCTION TO THE INTERACTIONS

P. Pincus

Materials Department
University of California, Santa Barbara
Santa Barbara, California 93106

I.   INTRODUCTION

What is meant by "soft condensed matter?" In the context of this NATO Advanced Study Institute, soft systems are those where the relevant interactions are weak and thermal fluctuations play an important role. However, this is not a sufficiently sharp definition because all materials that have higher order or weakly first order phase transitions, e.g., magnetic materials, superconductors, etc., have this property. I believe that "softness," in addition, implies a relatively high bulk or osmotic compressibility. For a system with a transition temperature, T_c, the elastic modulus, G, scales as

$$G = T_c L^{-3} \tag{I.1}$$

where L is the characteristic spatial scale which is typically the size of the relevant building blocks (not too near a second order phase transition). Thus, softness suggests L >> a, where a is an atomic dimension. With this in mind, we are led to the consideration of correlated systems composed of superatomic objects. These may be fluids (vanishing static shear modulus, or weak solids, e.g., colloidal crystals, ordered micellar phases, etc. For the most part, then, this Advanced Study Institute will focus on the physics of complex fluids, where the complexity is associated with the multiple energy scales required to adequately describe the systems. In this introductory lecture, I will attempt to set the stage by reviewing the fundamental interactions which control the ultimate physical properties.

What are these relevant energy scales? We will require two or three, $E_1 > E_2 > E_3$. The largest of these, $E_1 >> T$ (T is the absolute temperature in energy units), provides the <u>intra</u>molecular binding and is generally associated with

1

covalent bonds, ionic bonds, metallic bonds, etc., the typical origins of the cohesive energy of all solids. We shall not discuss these further in the same sense that while nuclear forces are responsible for nuclear stability, atomic physics does not specifically discuss them. They provide the atomic building blocks which are then considered as essentially stable, unalterable moities in atomic processes. E_2 describes the <u>inter</u>molecular forces which are responsible for the correlations which dominate this conference. These are principally Van der Waals interactions, electrostatic interactions, hydrogen bonds. When these forces are <u>attractive</u>, they may provoke self-assembly of the molecules into supermolecular structures such as micelles. These weaker aggregates may, in turn, interact with one another through the final energy scale, E_3. For our purposes, E_2 is typically somewhat greater than T while E_3 is generally comparable or somewhat less than T.

For the organization of this presentation, we arbitrarily divide the systems under consideration into those involving non-polar solvents, which we denote as Van der Waals systems, and those where we are dealing with polar solvents and charge separation is allowed, the Coulomb systems. These are discussed in sections II and III, respectively.

What are some examples of soft condensed matter systems and how do we characterize the relevant numbers of degrees of freedom? As discussed previously, we focus on fluids or weak solids. Most fluids are relatively incompressible and a simple example of a soft system where the softness is related to anisotropic behavior is provided by thermotropic liquid crystals. These are often rod-like molecules which may have one or more orientationally ordered phases in between the solid and isotropic fluid. A nematic liquid crystal is described by an ordered phase of quadrapoles where the relevant degree of freedom may be the director, a unit vector parallel to the molecular axis. More complexity is introduced, for example, when the molecules have a chirality, such as for the polypeptides in their helical conformations. Then, the order parameter may have two components, describing, respectively, the amplitude and wavevector of a possible cholosteric phase. A simple polymer solution is also a two component system, the solvent and the polymer. However, quasi-incompressibility allows description of a solution of perfectly flexible polymers in terms of a single variable, the concentration of polymer or, alternatively, the volume fraction of the sample occupied by the monomers which are linked together to form the chains. similarly, a melt of block copolymers (which may be visualized as two homopolymers, A and B, which are covalently linked together to form a single A-B chain) might be considered as a single component system where the variable is the local A (B) concentration. Obviously, a macroscopic A-B phase separation is impossible but more local microphase separation may lead to a rich variety of ordered and disordered structures.

Block copolymers are one example of what are called amphiphiles. These are molecules which generally have two, more or less, dissimilar parts which may be not soluble in the same solvent. Other amphiphiles are soaps which are molecules with a smallish polar or zwitterionic head group and one or two alkyl tails of moderate molecular weight (ten to twenty carbons). Amphiphiles are generally surface active in that they favor segregation to interfaces where both their tendencies may be accommodated at the cost of loss of entropy of mixing. A solution of amphiphiles should be considered as a two--component system where the variables are the concentration and orientation of the head-to-tail axis. Under certain circumstances, the amphiphiles may aggregate into clusters called micelles (Fig. 1) where the interior is composed of that part of the molecule which is unhappy in the given solvent and protected from it by the outer corona. In some cases, these micelles are quite stable with a rather well-defined aggregation number. Then, the micellar solution may be viewed as an effective single component system. (The behavior of colloidal particles with their stabilizing coronae are quite similar in this regard.) However, as a function of some external control parameter such as temperature, the micelles may spontaneously reorganize, e.g., into cylindrical or lamellar structures (Fig. 1); then, other degrees of freedom such as shape variables or interfacial curvatures must be introduced. Even more complex situations arise by adding another solvent. For example, typical microemulsions are single phase systems composed of oil, water, amphiphiles, and often salts, alcohols, etc.

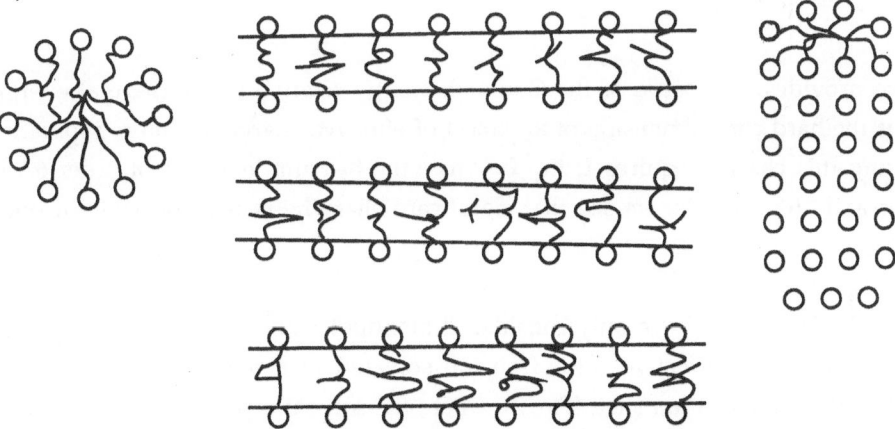

Figure 1.   Sketch of spherical, cylindrical and lamellar structures in an aqueous environment

3

Thus, we are generally dealing with complex fluids. It is clearly important for understanding the underlying physical principals to carefully make sure that we are well aware of those degrees of freedom which are relevant in any particular case. The constraints which simplify the problems by reducing the number of variables are determined by the interaction energies that are operative.

As in all of condensed matter physics, the fundamental force which determines all of the physical phenomena is the electromagnetic interaction, which may arbitrarily be divided into static and dynamical components. The dominant electrostatic interaction is Coulomb's law between monopoles of charges q_1 and q_2,

$$V_c(r) = q_1 q_2 / \varepsilon r \qquad (I.2)$$

in cgs units, where e is the dielectric constant of the medium in which the charges are imbedded. For two electronic charges (e) separated by an atomic distance ($\approx$ 1A), the corresponding energy is of order 10dV or approximately 10^5 K. This implies an extremely strong tendency for the formation of electrically neutral objects, atoms. Maintenance of electrical neutrality is often a dominant constraint in condensed matter systems. Furthermore, when two neutral "atoms" approach one another, at atomic separations the overlap of the electronic wave functions accompanied by the strong Coulomb force yields an extremely steep effective repulsive interaction, the "hard core." Thermally induced charge separation, ionization, occurs in the ambient temperature range only when the dielectric constant of the medium in which the charges are imbedded is sufficiently high. This is conveniently described in terms of the Bjerrum length,

$$\ell = \left(e^2 / \varepsilon T \right) \qquad (I.3)$$

which provides a measure of the Coulomb energy relative to the thermal energy. If L is the hard core dimension of an object of effective charge q, charge separation yielding this charge requires $L > q^2 \ell$. Since the Bjerrum length is of order 6Å in water and 10^3Å in oils, we see that significant charging is only possible in polar solvents.

Until now, we have only considered monopoles and their interactions with one another. The effective size of an ion solvated in a polar solvent is largely controlled by a hydration shell (Fig. 2), where the solvent dipoles are locally forced to point toward (or away) from the ion because of the dipole-monopole

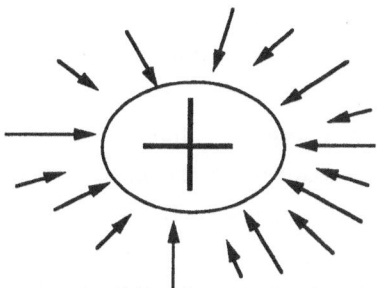

Figure 2. Sketch of a solvated ion with
its hydration shell. The arrows
represent water dipoles.

interaction. This deformation of the solvent will generally extend radially a distance determined by the bulk correlation length of the solvent, i.e., the distance over which the pair distribution decays toward zero.

There are several situations where electrical dipole-dipole interactions are important in determining structures. Ionomers[1] (Fig. 3) are functionalized

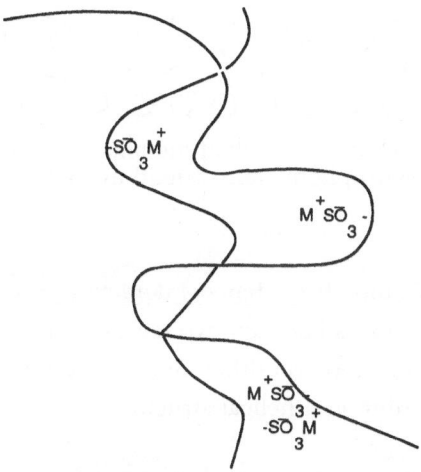

Figure 3. A schematic representation of
ionomers in solution aggregating
under the influence of dipolar
attractions between the polar groups

polymers where the substituents are polar or zwitterionic, i.e., they possess dipole moments in nonpolar solvents. The anisotropic dipolar interaction,

$$U_d = (\vec{\mu}_1 \bullet \vec{\mu}_2 - 3\vec{\mu}_1 \bullet \hat{r}\hat{r} \bullet \vec{\mu}_2)r^{-3} \qquad (I.4)$$

5

where $\vec{\mu}_1$ are two dipoles separated by the vector $\vec{r}$, provides an attractive coupling between the dipoles of magnitude μ^2/b^3 where b is their distance of closest approach; this may become as large as a fraction of an electron-vole and thus, greatly exceed thermal energies. Thus, the dipolar substituents provide a mechanism for physically cross-linking the polymers which ultimately leads to interesting physical phenomena such as gelation and shear thickening.

Langmuir monolayers are liquid-air interfaces (typically water/air) which are decorated with adsorbed amphiphiles (Fig. 4) which self-organize at the surface with their polar heads in the water and the aliphatic tails extending above the surface. The dipoles are more or less perpendicular to the surface and, in this case, their mutual interaction is repulsive. It has been suggested[2,3] that the dipolar interaction plays a central role in the organization of the amphiphiles at the interface.

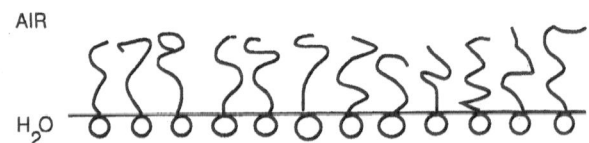

Figure 4.   A sketch of a surfactant monolayer on a water surface.
The hydrophobic tails extend away from the solvent.

A more specific Coulomb mediated interaction is the hydrogen bond that occurs when a proton resonates between two ions, e.g., in a polypeptide the proton links an oxygen and nitrogen ion on different repeat units (generally separated by 5–7 other monomers) leading to a helical structure.

Magnetic dipole-dipole interactions (also of the form of Eqn. (I.4) but with μ now standing for a magnetic dipole moment) are generally much weaker than the corresponding electrical interactions. Nevertheless, there exist a few situations in colloidal systems where they plan an important role in determining structures and dynamical responses. This principally involves magnetic particles suspended in a carrier solvent, ferrofluids. Under appropriate conditions of concentration, temperature, pH, etc., the magnetic particles may aggregate into chains[4] (Fig. 5), driven by the anisotropic nature of the dipolar interaction. Skeltorp[5] has used ferrofluids as a complex magnetic solvent in which he has studied larger latex particles which behave as larger magnetic "holes." These magnetic holes also interact with one another via magnetic dipolar forces.

Finally, quantum dipolar fluctuations of the atomic structure of neutral atoms and molecules lead to the ubiquitous Van der Waals interaction[6]

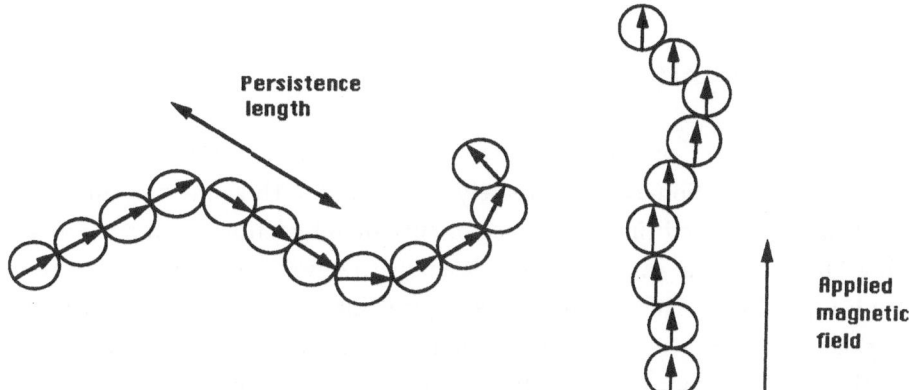

Figure 5.   Chaining of ferromagnetic (or superparamagnetic) colloidal particles as induced by magnetic dipolar interactions

between matter. For like species, this interaction is always attractive. The strength and range of this interaction may be estimated by the following simple argument. Consider two neutral atoms separated by a distance, r. Suppose atom (1) has an instantaneous fluctuating electric dipole moment of magnitude μ. This creates an electric field (Eqn. I.4) of magnitude $E \approx \mu r^{-3}$ at the site of atom (2), which induces a dipole moment on atom (2) of magnitude αE where α is the atomic polarizibility. The interaction between the two fluctuating dipoles is then

$$V \approx -\alpha <\mu^2>/r^6 \qquad (I.5)$$

where $< >$ denotes a quantum mechanical expectation value. The electronic polarizibility, α, is of order a^3, where a is the atomic radius. The mean square dipole moment of a spherically symmetric atomic shell is of order $(ea)^2$. This results in an attractive Van der Waals interaction which may be expressed as

$$V \approx -A(\alpha/r)^6 \qquad (I.6)$$

where the Hamaker constant $A \sim e^2/a \sim 1eV$. For macroscopic objects of linear dimension b, the form of the interaction remains the same at large separations[6] $(r >> b)$ with a replaced by b. The Hamaker constant is characteristic of the chemical nature of the material and is independent of macroscopic organization of the molecules. However, when the particles are immersed in a solvent, the effective Hamaker constant depends on the contrast in polarizibilities between the solvent and the particle, leading typically to a reduction of A by one or two orders of magnitude. In fact, a good "rule of thumb" is that Hamaker constants between particles in a solvent are within an order of magnitude of thermal energies.

II   VAN DER WAALS FLUIDS

Let us consider organic solvents with low dielectric constants, ($\varepsilon \approx 1$), where charge separation is an extremely rare event. Then the intermolecular potentials are dominated (in the absence of specific interactions such as hydrogen bonding) by the long range Van der Waals attractions and the short range Coulomb repulsions. This results in the typical intermolecular potential energy curve sketched in Fig. 6, which is often modelled as a Lennard-Jones (6-12) potential. The important feature is that there is an attractive minimum of depth, Δ, located at a typical molecular radius, a. Roughly speaking, the binding energy Δ scales with the relevant Hamaker constant, A. If the attractive

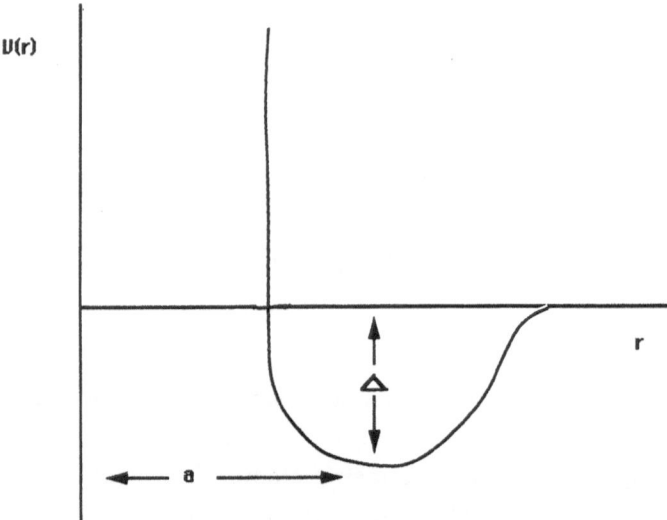

Figure 6.   A typical intermolecular potential energy
            curve

minimum dominates the situation, we expect some type of phase separation into an ordered solid, disordered glass, or a distribution of aggregates. If entropy of mixing is more important, then a solution or stable suspension (colloidal particles) obtains. This competition is most easily understood in the case of a dilute array of the interacting moities where a virial expansion[7] may be used to describe the effects of interactions. then, the contribution of the interparticle interactions to the free energy density F_{int}, is given by

$$F_{int}/T = (1/2)B_2c^2 + (1/6)B_3c^3 + \text{.......} \qquad \text{(II.1)}$$

where the effective two-body coupling constant, B_2, is given by

$$B_2 = \int\left[1 - e^{-V(r)/T}\right]d^3r. \qquad \text{(II.2)}$$

Then, with the potential (Fig. 6) in mind, if $T \gg \Delta$, the second virial coefficient is dominated by the hard care and $B_2 \approx a^3$. On the other hand, if $t \ll \Delta$, $B_2 \approx -a^2(\delta a)\, e^{\Delta/T}$, and there is a strong negative second virial coefficient. here, δa is a measure of the width of the attractive well. Thus, if $V(r)$ is assumed to be temperature independent (which may not be the case because there is solvent entropy included therein), a sketch of B_2 might look like Fig. 7. Generally, there exists a temperature, Θ, where the second virial coefficient vanishes, i.e., a cancellation between attractive Van der Waals interaction and entropy of mixing, and effective three body repulsions control the thermodynamics. The range $B_2 > 0$ is characteristic of "good" solvents while $B_2 < 0$ denotes "poor" solvent conditions.

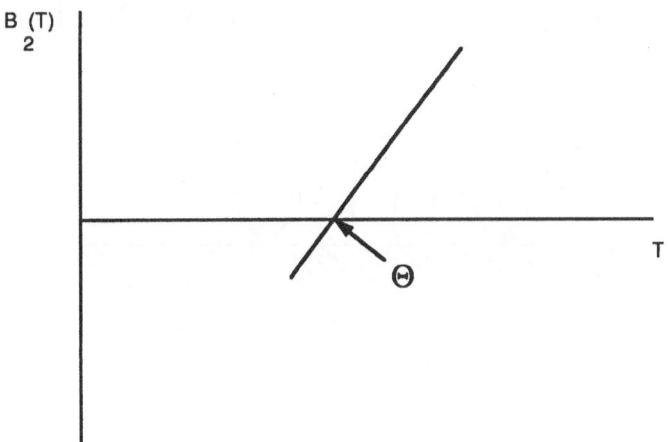

Figure 7.   A sketch of the temperature dependence of
             the second virial coefficient in the
             neighborhood of the Flory Temperature

Let us apply these ideas to the case of flexible polymers in solution. We assume the chains to be composed of N monomers of dimension a, giving a contour length, $L_0 = Na$. In the absence of interactions, the polymer takes on a random walk, or Gaussian, conformation in order to maximize its entropy. This results in a random coil of projected size, $R_0 \approx N^{1/2} a$, (N >> 1). Let us consider a dilute solution of these chains with separation much greater than R_0. Then, using Eqn. (II.1), the interaction free energy of the coil is given by

$$f_{int} \approx (1/2) T B_2 c^2 R_0^3 \approx T N^{1/2} (B_2/a^3) \qquad (II.3)$$

where c is now the concentration of monomeric units. Notice that so long as $B_2 > 0$, this interaction diverges with the molecular weight of the polymer. This leads to a <u>non-perturbative</u> swelling[8] of the chain, i.e., c decreases to reduce this excluded volume energy. This results in a coil size $R > R_0$, with $R \alpha N^\nu$, ($\nu \approx 0.6$ in three dimensions). On the other hand, under poor solvent conditions, $B_2 < 0$, the polymer tends to strongly collapse toward a compact structure when $|B_2/a^3| > N^{-(1/2)}$. Thus, only very weak attractive interactions are sufficient to make high molecular weight polymers insoluble.

Under good solvent conditions, high molecular weight polymers may, themselves, induce long range interactions between, e.g., colloidal particles. This can be illustrated by considering the confinement of a polymer coil between two inpenetrable walls[8] (Fig. 8), separated by a distance, h. If $h < R$, the coil size in the solvent far from any wall, this requires deforming the polymer from its preferred

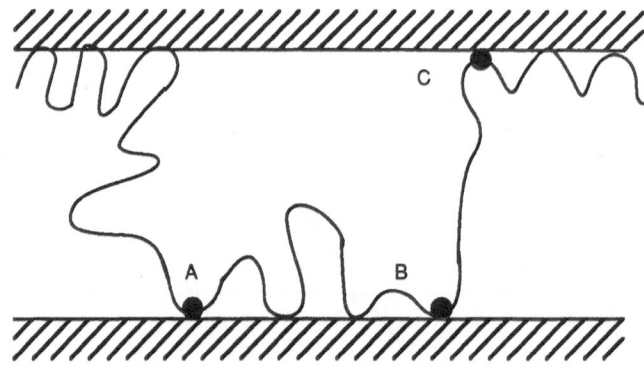

Figure 8. A polymer chain confined between inpenetrable surfaces

conformation, decreasing its entropy, and demanding that work be done by the confining walls on the chain. The corresponding free energy cost, $\Delta F \approx (R/h)^2 T$, is straightforwardly determined by a linear response analysis. If a dilute solution of n coils per unit area is trapped between the walls, this leads to an effective repulsive interaction energy per unit area between the two walls which is given by $U_{eff} = n\Delta F$ with an equivalent disjoining pressure, $\Pi \alpha\ h^{-3}$, between the interfaces. Therefore, in a rather trivial manner, the polymers induce a fairly long range ($\approx R$) interaction between two surfaces. [It should, however, be noted that increasing the concentration of polymers beyond their overlap concentration, i.e., the concentration where the chains interpenetrate to form a transient network[8], decreases the effective range to ξ, the typical network dimension (Fig. 9), which scales from R at the overlap threshold to a monomer dimension, a, in the limit of vanishing solvent.]

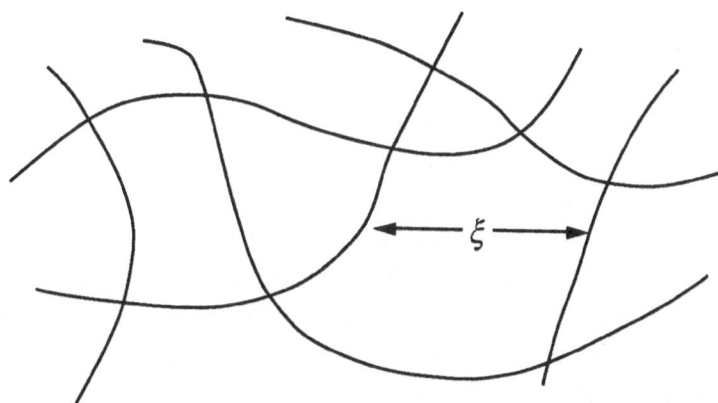

Figure 9.   The transient network of a semi-dilute
polymer solution

Until now, we have only considered flexible polymers where there are no preferred orientations of the bonds connecting adjoining monomers along the chain.   Generally, however, there does exist some chain stiffness which is described by a persistence (or Kuhn) length, L, which is the distance along the chain required for a finite change in orientation of a unit vector tangent to the backbone of the polymer.   Then, we may think of the polymer as a renormalized random walk chain (called a worm-like chain) of L_0/L steps of length L, leading to an expanded coil of dimension (for $L_0/L > 1$).

$$R \approx \left(L_0/L\right)^{1/2} L = \left(L_0/L\right)^{1/2}. \qquad (II.4)$$

If $L > L_0$, the polymer may be considered as a rigid rod. Rigid polymers are of importance for the manufacture of ultra-high modulus fibers[9] and therefore have attracted a great deal of attention in recent years.

Processing and characterizing work-like and rigid chain polymers is generally difficult because they have limited ranges of solubility. This may be understood by considering the second virial coefficient, B_2, (Eqn. II.2) between two rigid segments of length L. In Eqn. (II.2), the potential V(r) should be considered to be a function of both the relative positions and <u>orientations</u> of the two interacting rods. Onsager[10] has shown that purely repulsively interacting hard core cylinders of length L and radius, a, have an effective excluded volume of order L^2a, for isotropic suspensions. When the hard core is augmented with attractive dispersion forces, simple geometric estimates[11] yield

$$B_2 \cong L^2a\left[1 - (a/L)^\sigma \exp\left(LA/aT\right)\right] \tag{II.5}$$

where the exponent σ is of order unity. Notice that the exponential term dominates and leads to phase separation for sufficiently long persistence lengths. This tendency for long rods to be insoluble is easily understood in terms of the binding energy between two rods which are parallel and at a separation placing them in the Van der Waals minimum. This energy clearly scales as $|(L/a)A|$ and thus diverges as L grows. Thus, in contrast to flexible polymers, rigid and worm-like polymers that have long enough Kuhn lengths can never be thermodynamically stable in solution.

What is the origin of this polymer rigidity? In fact, there are several reasons why chains are not perfectly flexible. The most common and classical mechanism involves steric hindrance (excluded volume effects) associated with more-or-less bulky side groups which locally inhibit chains to have small radii of curvature. This effect is often discussed in terms of the rotational isomer model[12] and we will not consider it further. For polymers which are electrically charged (see Section III), the intrachain Coulomb repulsions lead to chain stiffening. A more novel driving force for rigidity occurs for conjugated polymers such as the polyalkylthiophenes which may be doped with electrons in Π orbitals, which is discussed below.

As a simple model for conjugated polymers in solution, we may consider an inherently flexible chain[13] which may be doped with a few electrons which occupy an empty Π^* band. The asymmetric nature of the Π orbitals will engender a resonance integral, $t(\theta)$, for electronic transfer from one monomer to an adjacent one which depends on the relative orientation of the two monomeric units. The magnitude of this matrix element will be optimized for a locally rod-like structure, implying that lowering of the electronic kinetic energy favors chain rigidity. This tendency is opposed by the loss of orientational entropy as the polymer straightens. If we assume that the electron is self-trapped on a rigid segment of length, L, the free energy associated with the Kuhn length is estimated by

$$\Delta F \cong t \left(a/L\right)^2 + T(L/a) \qquad \text{(II.6)}$$

where t is the optimum of $t(\theta)$ and the first term is the zero point energy of an electron confined to a box of length L; the second term is the loss of entropy associated with the rigidification, T per monomer. Minimization with respect to L, gives an electronic self-localization length, $L^* \approx (t/T)^{1/3} a$. These self-trapped electrons are called conformons[14] in analogy to polarons in solids. For typical resonance integrals of order 1eV, $L^* \approx 50$Å. Thus, a few percent doping leads to rigid chain polymers which should tend to aggregate as discussed above. Such behavior is indeed observed[15] for the soluble polythiophenes.

Let us turn now to the consideration of surfactants in mixtures of oil and water. With proper choice of the amphiphiles, the surfactants may solubilize the oil and water into single thermodynamic transparent, low viscosity phases called microemulsions. Microemulsions and their phase behavior may be understood in terms of surfactant decorated interfaces[16] separating microphase separated oil and water domains. Locally, these interfaces may resemble the monolayers sketched in Fig. 4. The localization of the surfactants at the interfaces is driven by the amphiphilic character of the molecules, i.e., hydrophilic (often polar) head groups and hydrophobic (alkyl) tails. Indeed, Schulman[17] has suggested that the total interfacial area is related to the number of surfactant molecules by the following argument.

Let us idealize the situation to imagine zero solubility of the surfactants in either the oil or water with all of the surfactant molecules residing at the oil-water interface. Thus, the total interfacial area, S, will be controlled by the number of surfactant molecules, n. The part of the free energy associated with the oil-water interfaces may be written, to lowest order in spatial gradients of the interface, as

$$f \cong \gamma_0 S + nG(\Sigma) \qquad \text{(II.7)}$$

where $\Sigma = S/n$ is the interfacial area per surfactant, γ_0 is the bare oil-water interfacial energy (≈ 50 dynes/cm), and G represents the Gibbs energy of the interactions of the surfactants with the interface and among themselves. Minimization with respect to Σ gives

$$\gamma_0 + \partial G/\partial \Sigma | \Sigma^* = 0 \qquad\qquad (II.8)$$

and an optimum area per surfactant Σ^*. Then, the total interfacial area is $S^* = n\Sigma^*$. If G and γ_0 represent the dominant energies, then given h, the total interfacial area is fixed at S^*. The next contribution to the interfacial energy arises from the local curvatures[18],

$$f_c \cong (K/2)\int (R_1^{-1} + R_2^{-1} - 2R_0^{-1})^2 dS + (K'/2)\int (R_1^{-1} - R_2^{-1})^2 dS \qquad (II.9)$$

where $R_{1,2}$ are the two local radii of curvature, R_0 is the spontaneous curvature which represents the allowed tendency for the interface to prefer a curved shape, and K, K' are respectively the bending constant and saddle splay. For typical microemulsions, the bending constants are of order T and consequently, play integral roles in the determination of the geometrical nature of the phases. Thus, the interplay between K, K', and entropy dictate whether the microemulsions should be viewed as suspensions of swollen micelles, cylindrical micelles, bicontinuous structures, etc. The physical origin of the bending energies resides in the geometric shape of the surfactants and their resistance to increased tail or head packing engendered by bending of the interfaces at fixed area. For larger polymeric surfactants, the bending constants increase rapidly with molecular weight and thermal fluctuations become less important, resulting in high viscosity more or less ordered phases.

Let us consider an example of thermal interfacial fluctuations at the expense of bending energy. We take as an example a lamellar phase (Fig. 10), where each layer (separation h) may fluctuate about its means position with an amplitude $\xi(x,y)$ where (x, y) denote the position in the horizontal plane. Assuming no tendency for spontaneous curvature ($R_0 \approx \infty$),

$$R_1^{-1} = R_2^{-1} = \nabla^2 \zeta(x,y) \qquad\qquad (II.10)$$

Fourier transformation of $\zeta = \Sigma_q \zeta_q e^{iq \cdot r}$, and substitution into Eqn. (II.9), and application of the equipartition theorem yields

$$<|\zeta_q|^2> = (T/KSq^4) \qquad\qquad (II.11)$$

and an rms amplitude $<\zeta^2> = (T/K)S$. As $S \to \infty$, the rms amplitude diverges because of the long wavelength fluctuations. However, $<\zeta^2>$ cannot greatly

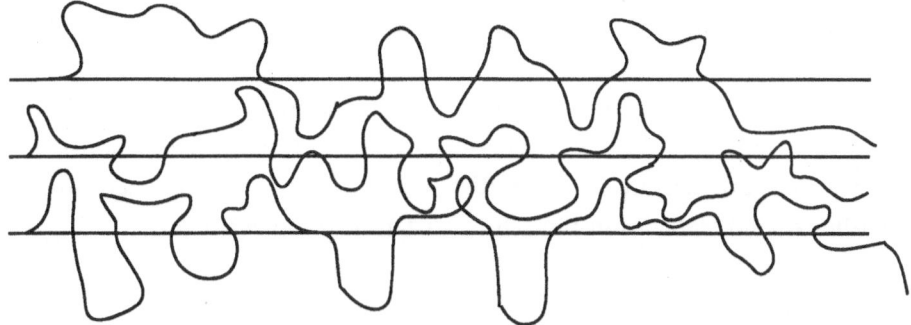

Figure 10.   A fluctuating lamellar phase. The horizontal lines represent the mean positions of the lamellae.

exceed h^2 because one layer is sandwiched between its two nearest neighbors (and interpenetration is forbidden). Just as for the previously discussed case of confined polymers, there is a free energy penalty for confinement[19],

$$\Delta F = T<\zeta^2>/h^2 \tag{II.12}$$

and a corresponding disjoining pressure

$$\Pi \cong \left(T^2/Kh^3\right). \tag{II.13}$$

This interlayer interaction is known as the Helfrich undulation interaction. In a beautiful series of X-ray diffraction experiments, Safinya[20] and his co-workers have demonstrated that the undulation force is dominant where any electrostatic interactions are screened out, but are ineffective in the presence of unscreened double layer interactions.

III   COULOMBIC FLUIDS

Let us now turn to a consideration of polar molecules in an aqueous environment where ionization exists and local charge densities are allowed. The lamellar surfactant phases (discussed above) provide a nice example where electrostatic interactions play an important role. Indeed, the problem of the structure of a charged surface in contact with a fluid which contains the neutralizing counterions (and possibly dissolved salts, acids, bases, etc.) has become classic in electro- and interfacial chemistry. A standard theoretical approach is based on the Poisson-Boltzmann[21] approximation for solving the electrostatic Maxwell equation,

$$- \varepsilon \nabla^2 \phi = 4\pi\rho \qquad (\text{III}.1)$$

where ε is the solvent dielectric constant, ϕ is the electrostatic potential (in cgs units) and ρ is the charge density. The essence of the P-B approximation is that the counterions are assumed to be distributed in the solvent according to Maxwell-Boltzmann statistics in the self-consistant *local* potential $\phi(r)$,

$$\rho = \rho_0 e^{-e\phi/T}, \qquad (\text{III}.2)$$

where ρ_0 is determined by overall charge neutrality. The approximation neglects finite counterion size effects (since only classical electrostatic interactions are included) and correlations among the counterions. If there are other ions, as with added salt, they are included in a manner analogous to Eqn. (III.2). Linearization of Eqn. (III.2) and substitution into Eqn. (III.1) yields Debye-Huckel approximation. However for the relatively high membrane surface charge densities ($\approx e/50\text{Å}^{-2}$) which are typical, this linearization is not appropriate and the full B-P equation is used. For the one dimensional geometry of a single surface or a lamellar stack, the B-P equation is exactly soluble[22] in a manner identical to classical one dimensional mechanics problems. The exact solution for the counterion charge density $\rho(x)$ at a distance x from the interface (in the absence of added electrolyte), is

$$\rho(x) = (2\pi\ell)^{-1}(x + \lambda)^{-2} \qquad (\text{III}.3)$$

where ℓ is the Bjerrum length and $\lambda\ (= \Sigma/2\pi\ell)$; Σ is the surface area per unit charge of the membrane. The counterions are then localized to a sheath of thickness λ and, for many purposes may be considered to be a two-dimensional ideal gas of ΣS counterions confined to the sheath. (S is the total surface area of the membrane.) Thus, to an excellent approximation, the electrostatic contribution, f_e, to the free energy is given by

$$f_e = nT\ln(a^3/\Sigma\lambda). \qquad (\text{III}.4)$$

This result may be used to estimate the Coulombic contributions to such parameters as the area per surfactant, curvature energy, spontaneous curvature,

16

etc. For example, using the Schulman argument, (Eqns. III.7, 8), the charges contribute to the area per surfactant an amount

$$\Delta\Sigma/\Sigma \;=\; T/(\Sigma\gamma_0) \;\cong\; a^2/\Sigma \;\cong\; 10^{-1} - 10^{-2}. \tag{III.5}$$

This is typically a modest contribution. On the other hand, the electrostatics is more significant for the curvature parameters. The effect on the bending energy, K, may be estimated by noting that, for a bent interface of local radius of curvature, R, the change in free energy is of order $f_e(\lambda\eta/R^2)$ which leads to

$$\Delta K/K \;=\; (T/K)(h/\ell) \tag{III.6}$$

which may become large. Therefore, for example, changing the pH of the water has a significant effect on the bending energy which, in turn, is central for the determination of phase boundaries. More precise calculations[23] substantiate this result. In a similar manner, the electrostatic contribution to the spontaneous curvature, ΔR_0, is given by

$$\Delta R/R_0 \;\cong\; \left|(R_0/\ell)(T/K)\right| \tag{III.7}$$

which may again be quite important. The sign of the effect on R_0 is such to have the counterions in the external phase. Their entropy of mixing is thus enhanced.

Another important electrostatic effect associated with finite sized objects such as micelles or polymer lattices is charge renormalization[24] which for linear assemblies of charges, e.g., charged polymers or rod-like viruses, is known as Manning condensation[25]. For example, consider a spherical particle of radius, b, carrying Z ionizable surface groups immersed in an aqueous solvent. If the groups are strong electrolytes, we might naively expect that all the groups ionize leading to a sphere of charge Z. Let us investigate whether or not this is self consistent. Suppose that we have a dilute solution of c identical spheres per unit volume, each with an effective charge Z^*.d Then, using Gauss theorem, the electrostatic potential at the surface of a sphere is $Z^*\ell/b$ in units of T, where $\pi\ell$ is the Bjerrum length. Since the chemical potential for the counterions must be spatially constant, we may approximately equate this surface potential to the entropy of mixing (for monovalent counterions) to give

$$Z^*\ell/b \;\cong\; -\ln\phi \tag{III.8}$$

where ϕ is the volume fraction of counterions. This leads to a maximum charge per sphere of order $(b/\ell)\ln\phi$ which is typically 10–100 charges for a 100 angstrom particle. This is generally equivalent to a rather weak degree of ionization,

$$Z^{\cdot}/Z \;\cong\; \Sigma/\ell b \tag{III.9}$$

which is only about ten percent for 100 angstrom micelles. This charge renormalization is now well established for micellar[26] and latex[27] systems.

In the absence of salt, charged lattices and micelles may order under the influence of the repulsive screened Coulomb interaction, leading to colloidal crystals[28]. However, because of the charge renormalization which limits the maximum charge per sphere to scale with its radius, there is a minimal size for crystallization (at a given temperature) which is approximately given by $b_{min} \cong 10\,\ell$. This predicts that charged particles smaller than about 50 Angstroms should not order. This seems to be consistent with existing data.

When does the Boltzmann-Poisson approximation break down? Wennerstrom and coworkers[29] have carried out simulations which demonstrate that the B-P approximation is insufficient when we are dealing with high surface charge densities (somewhat larger than is typical in micellar systems) and multivalent counterions. Subsequently, several groups[30,31] have shown that these simulations could be understood in terms of correlations between the counterions, finite counterion size effects, and counterion fluctuations—all effects which are neglected in the B-P approximation. In particular, the Swedish group[29] found that under extreme conditions, the Coulombic contribution to the interaction between two parallel surfaces could even become <u>attractive</u>. Indeed, this result is in agreement with Earnshaw's Theorem that there is no mechanical equilibrium possible with only Coulomb forces.

A nice example of double layer effects occurs with the lamellar phases discussed in Section II. We already mentioned the beautiful experiments of Safinya et al.[20] where the Helfrich undulation force was clearly demonstrated, using electrostatic interactions between the layers as a control parameter. Let us try to understand how the electrostatic interlayer forces have impact upon the undulation interaction[32]. Recall (Eqn. III.3) that the counterion distribution in the neighborhood of a single charged surface falls off as x^{-2} for $x \gg \lambda$. Since the counterions may be approximately considered as an ideal gas, the double layer contribution to the disjoining pressure between two lamellae separated by a distance, h, is roughly

$$\Pi(h) \cong T\rho(h) \cong T/(2\pi\ell h^2). \tag{III.10}$$

The corresponding interfacial energy per unit area is

$$\gamma(h)/T \cong (2\pi\ell h)^{-1}. \tag{III.11}$$

As at the end of the last section, suppose that the interfaces are fluctuating with a local displacement from equilibrium, $\zeta(x, y)$. For long wavelength, weak fluctuations, we may express the cost in free energy per unit area in the spirit of Eqns. (II.9–10) as

$$\Delta F = K(\nabla^2\zeta)^2 + T/(\pi\ell h^3)\zeta^2 \tag{III.12}$$

for $< \zeta^2 > \ll h^2$. If we Fourier transform and use the equipartition theorem to compute the mean square amplitude of the fluctuations, the electrostatic term in Eqn. (III.12 acts as a "mass" and cuts off the infra-red divergence. In contrast to the situation where the electrostatic interactions are, for example, screened by added electrolytes, the mean square amplitude is finite and approximately given by

$$< \zeta^2 > \cong (T/K)\xi^2 \tag{III.13}$$

where ζ is

$$\xi \cong (\ell K/hT)h. \tag{III.14}$$

If $K \cong T, h \gg 1$, which gives $\zeta/h < 1$. Thus, the mean square amplitude is much less than the interplane separation, h. The double layer intéraction has completely screened out the undulation force! In this case, the Helfrich interaction is not only much weaker than the double layer disjoining pressure (Eqn. III.11), but is exponentially screened as well. However, with added electrolyte, as soon as the Debye screening length becomes significantly smaller than the interplanar spacing, the Helfrich interaction switches back on. It is precisely this effect which Safinya et al. used to pin down the Helfrich force.

Finally, I should like to point out an area of polymer physics where Coulombic forces play an important but poorly understood role. Simple polyelectrolytes are polymers where each repeat unit carries an ionizable group. Manning condensation limits the linear charge density to approximately one charge per Bjerrum length. Isolated polyelectrolyte chains in polar solvents swell tremendously under the influence of the repulsive intramolecular electrostatic forces into a nearly rod-like conformation. With added salt, such chains shrink toward a Gaussian configuration. Odijk[33] has shown that in the presence of salt, the polymers are worm-like with a persistence (or Odijk) length which scales as $(\kappa^2 a)^{-1}$. At higher concentrations, in semi-dilute solutions, Witten and Pincus[34] have speculated that there is strong screening of the intramolecular forces by the intervening chains leading to a persistence length, L, given by

$$L \cong \kappa/\kappa_0^2 \qquad\qquad\qquad (III.15)$$

where κ_0 is the inverse Debye length solely arising from the counterions. Notice that in the absence of salt, L scales as κ_0^{-1} which is much shorter than the Odijk length. While these results are consistent with neutron scattering experiments, it is apparently much more difficult to understand dynamic phenomena. In particular, such elementary properties as viscosities of polyelectrolyte solutions remain to be explained. This is likely to become an active area of research in the coming years.

This work was partially supported by the US DOE 3DE–FG03–87ER45288 and the NSF #DMR58703399.

REFERENCES

1. For a general review, see W. J. MacKnight and T. R. Earnest, Jr., *Journal of Polymer Science: Macromolecular Reviews* 16:41 (1981).
2. D. Andelman, F. Brochard, P. G. de Gennes and J. F. Joanny, *C. R. Acad. Sci. Paris*, 301:675 (1985).
3. F. Brochard, J. F. Joanny and D. Andelman in "Physics of Amphiphilic Layers," J. Meunier, D. Langevin and N. Boccara, Eds., Springer-Verlag, Berlin (1987).
4. P. G. de Gennes and P. Pincus, *Phys. of Cond. Matter*, 11:189 (1970).
5. A. Skeltorp, This Conference.
6. J. Mohanty and B. W. Ninham, "Dispersion Forces," Academic Press, London (1976).
7. See for example: L. D. Landau and E. M. Lifshitz, "Statistical Physics," Pergamon Press Ltd., London (1958).
8. See: P. G. de Gennes, "Scaling Concepts in Polymer Physics," Cornell University Press, Ithaca (1979).
9. P. Smith, This Conference.
10. L. Onsager, *Ann. N. Y. Acad. Sci.* 51:627 (1949) and *in*: P. G. de Gennes, "The Physics of Liquid Crystals," Oxford Univ. Press, Oxford (1974).
11. P. Pincus, K. Kremer, J. Batoulis, to be published.
12. See: P. J. Flory, "Principles of Polymer Chemistry," Cornell University Press, Ithaca (1953).

13. See: Proceedings of the International Conference on Science and Technology of Synthetic Metals, Santa Fe, NM, June 1988 in: *Synthetic Metals* 29 (1989).
14. P. Pincus, G. Rossi, M. Cates, *Europhys. Lett.* 4:41 (1987).
15. J. P. Aime, F. Bargain, M. Schott, H. Edkhardt, R. L. Elsenbaumer, G. G. Miller, M. E. McDonnell and K. Zero, *Synthetic Metals* 28:C407 (1989).
16. For an excellent introduction to microemulsions, see: P. G. de Gennes and C. Taupin, *J. Phys. Chem.* 86:2294 (1982).
17. J. H. Schulman and J. B. Montagne, *Ann. N. Y. Acad. Sci.* 92:366 (1961).
18. W. Helfrich, *Z. Naturforsch* 28:6693 (1973).
19. W. Helfrich, *Z. Naturforsch* 33a:305 (1978).
20. C. Safinya, This Conference.
21. E. J. W. Verwey and J. Th. G. Overbeek, "Theory of the Stability of Lyophobic Colloids," Elsevier, Amsterdam (1948).
22. D. L. Chapman, *Phil. Mag.* 25:475 (1913); G. J. Gouy, *J. Phys.* 9:457 (1910); *Ann. Phys.*, 7:129 (1917).
23. P. Pincus, S. Safran, S. Alexander and D. Hone, *in:* "Physics of Finely Divided Matter," N. Boccara and M. Daoud, Eds., Springer-Verlag, Berlin (1985).
24. S. Alexander, P. M. Chaikin, P. Grant, G. J. Morales, P. Pincus and D. Hone, *J. Chem. Phys.* 80:5776 (1984).
25. G. S. Manning, *J. Chem. Phys.* 5:924, 934, 3249 (1969).
26. S. H. Chen, E. Y. Sheu, J. Kalus and H. Hoffmann, *J. Appl. Cryst.* 21:751 (1988).
27. Y. Monovoukas and A. P. Gast, *J. Colloid and Interf. Sci.*, to be published.
28. P. M. Chaikin, J. M. di Meglio, W. D. Dozier, H. M. Lindsay and D. A. Weitz, *in:* "Physics of Complex and Supermolecular Fluids," John Wiley & Sons, New York (1987).
29. J. Wennerstrom, *J. Chem. Phys.* 79:2221 (1984).
30. R. Kjellender and S. Marcelja, *J. Phys. Chem.* 90:1230 (1986).
31. M. Cates and J. F. Joanny, Unpublished.
32. P. Pincus, D. Andelman and J. F. Joanny, to be published.
33. T. Odijk, *J. Polym. Sci. Phys. Ed.* 15:477 (1977).
34. T. A. Witten and P. Pincus, *Europhy. Lett.* 3:315 (1987).

QUALITATIVE ASPECTS OF CONDENSATION, ORDERING AND AGGREGATION

A. T. Skjeltorp[1] and G. Helgesen[2]

[1] Institute for Energy Technolgy, N-2007 Kjeller, Norway
[2] Dept. of Physics, University of Oslo, Blindern, N-0316
Oslo 3, Norway

1. INTRODUCTION

This lecture is intended to demonstrate qualitative aspects of
various many-body phenomena like deposition, fracturing, aggregation,
crystallization, melting and vortices using model systems of uniform
microparticles dispersed in water or ferrofluid. The particles are
confined to monolayers between glass plates allowing direct microscopic
observations of local structure and movement of individual particles.

The particles which are used in the experiments are of uniform size
and shape (spherical, anisotropic) producing good manybody systems. The
interactions may also be varied using "tailor-made" particles (charged,
neutral, magnetic, antigen-antibody on surfaces). For small particles
(diam. $\leq$ 5 μm), the interaction may be made comparable to the disrupting
thermal energy induced by the Brownian motion.

2. EXPERIMENTAL

Although uniform latex particles were discovered already in 1947 at
Dow Chemical Company in the US[1], a major breakthrough came around 1980
when John Ugelstad at the University of Trondheim, Norway, discovered
the so called swollen emulsion polymerization technique[2]. This allows
production of very uniform particles in the size range from about 1 μm
to more than 100 μm with a coefficient of variance in diameter less than
1%.

Ugelstad et al. have also developed methods for preparation of
porous, magnetic and nonspherical particles[3,4] with a broad variation in
chemical structures and miscibilities (hydrophobic, hydrophilic). The
particles may thus be "tailor-made" and filled or coated with various
compounds and dispersed in various carrier fluids to suit a wide range
of important applications in industry, medicine and research[1,5]. Some of
these applications are illustrated schematically in Fig. 1.

A major extension to model physical processes is to disperse monosized particles in ferrofluid[6]. A typical ferrofluid is a colloidal sus-

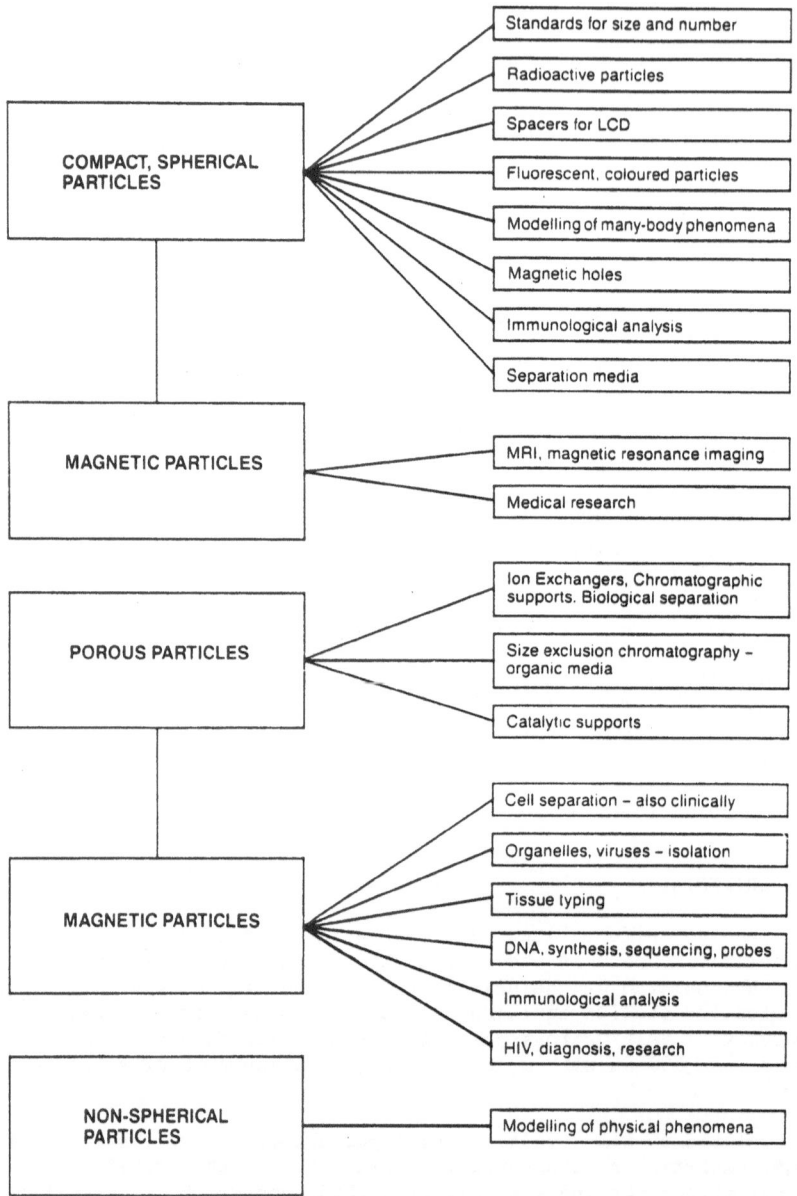

Fig. 1. Schematic illustration of various applications of monodisperse polymer particles.

pension of monodomain magnetic particles like magnetite of diameter
about 0.01 μm in a nonmagnetic carrier fluid like water or kerosene. A
surfactant covering the particles prevents agglomeration, and because of
the small size, Brownian motion prevents sedimentation. The ferrofluid
particles behave as dipoles and the fluid is ideally paramagnetic in
that it only becomes magnetic in an external field. The monosized
spheres are typically 100 - 1000 times larger than the ferrofluid particles, and the spheres will therefore move around in an approximately
uniform magnetic background. The holes created by the spheres appear to
possess magnetic moments antiparallel to an external field. This magnetic analogue of Archimedes prinsiple thus creates magnetic holes with
dipolar interactions controlled by external fields.

In most of the experiments to be discussed the models are essentially two-dimensional. This was realized by confining a monolayer of
dispersed spheres between glass plates, Fig. 2. The separation between
the plates could be adjusted evenly by using a small fraction of larger
spheres as spacers.

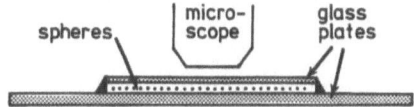

Fig. 2. Schematic experimental setup.

The interactions between the particles (Coulomb repulsion, van der
Waals attraction, steric, magnetic, antibody-antigen on surfaces) could
be controlled via the particle production and the type of carrier fluids
used.

For the magnetic measurements, uniform fields could be produced
both normal ($H_\perp$) and parallel ($H_{||}$) to the layer using Helmholtz coils.

The particles were observed directly in light microscopes with
video camera attrachment for recordings and digital analysis with
512x512 pixels resolution.

3. PARTICLE CONDENSATION AND FRACTURING

Under this heading we will demonstrate deposition of spheres onto a
surface as well as fracturing of the resulting packed structures. For
this, water-dispersed spheres are confined to a monolayer between two
glass plates and evaporation is allowed to take place along one edge of
the gap between the plates. During this process, the fluid is drawn towards the open edge of the cell where the spheres are deposited. It is
possible to vary the model parameters such as the rate of evaporation
and particle interactions. Two examples are shown in Fig. 3. In Fig.
3(a) the deposition process is rapid, producing a large number of lattice defects like vacancies, grain boundaries and dislocations. If the
deposition takes place much slower, more perfectly ordered structures
are formed as shown in Fig. 3(b). Various simulations have been performed[7,8] showing some resemblance with the present observations. However,
it is clear that cooperative restructuring processes involving the correlated motion of many particles takes place. This would require much
more extensive simulations than have been done so far.

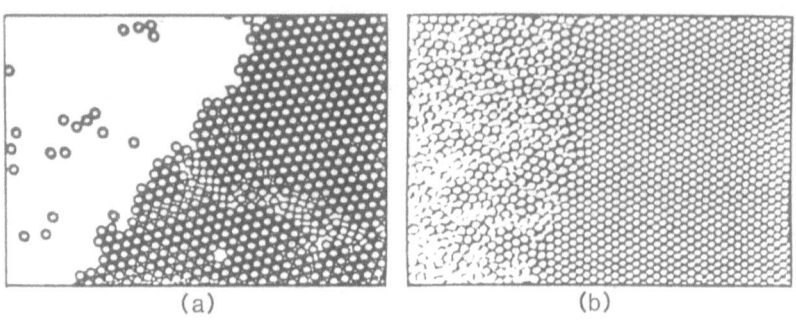

Fig. 3. Packing of polymer spheres between two parallel glass plates.
Fig. 3(a) shows the packing of 5.3 μm spheres under fast fluid
evaporation conditions to produce a structure with a large defect concentration. Fig. 3(b) shows the structure produced by
the relatively slow depositions of 3.4 μm spheres from a more
concentrated dispersion at a much lower evaporation rate. In
both cases the dispersed particles in the upper left part of the
figures are being deposited onto the deposit in the lower right
region.

Regular, closely packed structures like the one under formation in
Fig. 3(b) may provide interesting model systems for studies like flow in
porous media and other processes. In particular, it has been possible to
study the crack growth a plane[8]. For this, monosized sulfonated polystyrene spheres (d_1 = 3.4 μm) dispersed in water were packed to a dense
monolayer forming a polycrystal with relatively large crystalline grains
containing typically 10^5-10^6 spheres. The fracturing was realized as a
result of particle shrinking during a slow drying process reducing the
sphere diameter to d_2 = 2.7 μm. This produced breaking of the bonds between the spheres but not of the weaker bonds between the spheres and
the glass surface (no buckling).

The evolution of fracture in the whole sample showed many characteristic features. The first cracks were formed along the grain boundaries where fewer bonds had to be broken than inside the grains. Fig. 4
shows a typical initial crack propagation inside a grain from an originally regular lattice. The initial cracks propagated rapidly more or
less linearly. As more and more cracks were formed, they became more and
more irregular, reflecting the irregular strain field in the sample.
Figs. 5 and 6 show the final fracture pattern for a small and large area
of the sample, respectively. The visual appearance of this suggests
that the cracking patterns might be described in terms of the concepts
of fractal geometry. In fact digital analysis of the cracks for a wide
range of length scales produced an effective fractal dimensionality D =
1.68 ± 0.06[10].

Recently, a simple computer model has been developed[9,11] to represent
the present model system and good agreement has been obtained.

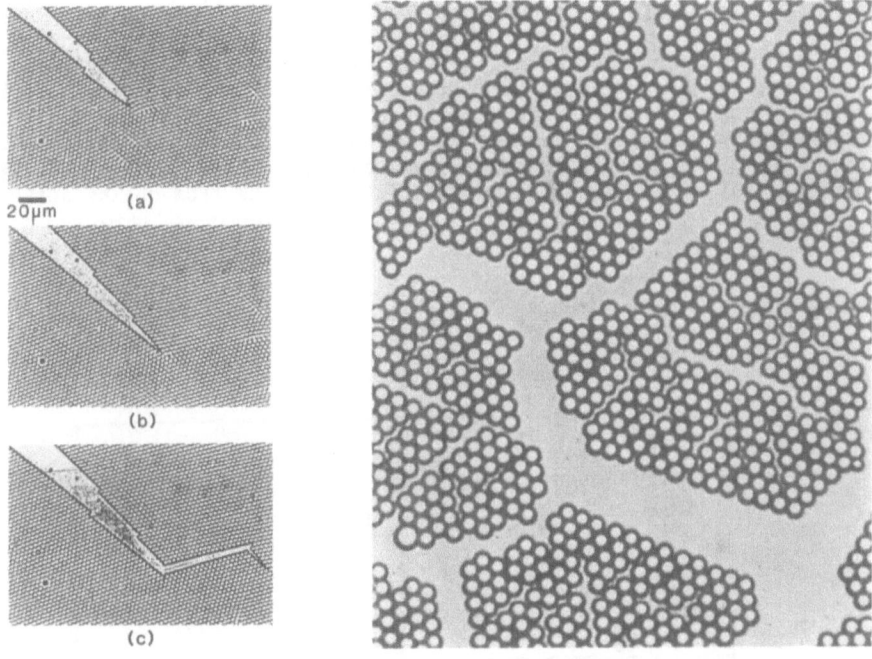

Fig. 4.   Fig. 5.

Fig. 4. Propagation of one crack (a) - (c), inside a regular lattice of
3.4 μm spheres subjected to uniform contraction.

Fig. 5. Final fracture pattern in a small area of the sample.

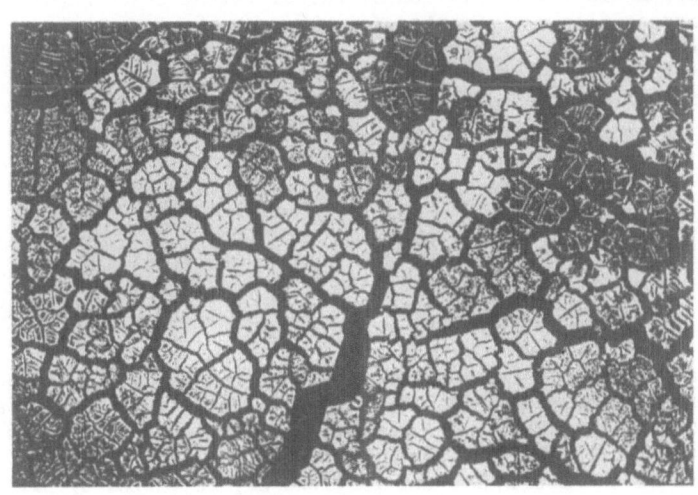

Fig. 6. Final fracture pattern in a monolayer of about 10^6 spheres.

27

4. AGGREGATION

The combination of small subunits like atoms or colloidal particles into large structures like aggregates, dendrites and crystals is a
common process in many areas of science and technology. The use of the
present monosized particles has proved to be very suitable for controlled experimental realizations to study such processes. It has thus been
possible to obtain a wide range of growth patterns from chainy struc-

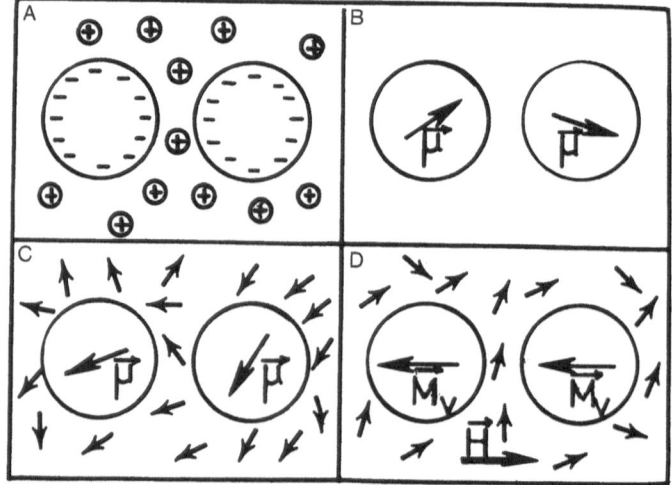

Fig. 7. Schematic representation of the different cases studied in the
aggregation experiments: (a) charged spheres with counter-ions;
(b) magnetized spheres with moment μ; (c) magnetized spheres in
ferrofluid (arrows represent moments on ferrofluid
particles); (d) nonmagnetic spheres in ferrofluid with an
external field producing a magnetic moment M_v (magnetic
hole).

tures to faceted single crystals by varying the type, strength and range
of the attractive interactions[12,13]. The different cases studied are
shown schematically in Fig. 7. A typical sample contains 10^4 - 10^6
spheres diffusing in a plane (Fig. 2). For the situation in Fig. 7(a),
the strength of the attractive interactions may be varied by balancing
the short-range attractive van der Waals forces against the net repulsive electrostatic forces between the charged spheres by varying the
counter-ion concentration[14]. The interaction potential for two spheres
may be estimated using the Derjaguin-Landau-Verwey-Overbeck (DLVO)
model[15].

It is customary to introduce the "sticking probability" between spheres as a parameter. High sticking probability for predominantly short-range attractive van der Waals forces leads to ramified or fractal structures whereas faceted single crystals are formed in the limit of extremely low sticking probability for weakly interacting particles (secondary minimum in the potential). These two extreme cases are shown in Fig. 8. It should be noted that for the fast growth case in Fig. 8(a), particle-cluster as well as cluster-cluster aggregation take place. This socalled diffusion-limited cluster aggregation (DLCA) process produces a fractal dimension D = 1.49 ± 0.06[12] in close agreement with simulations (D = 1.485 ± 0.015)[16]. To our knowledge there has been no physical realization of the diffusion-limited aggregation (DLA) model[17] of microparticles in a plane, producing a fractal dimension D ≃ 1.7.

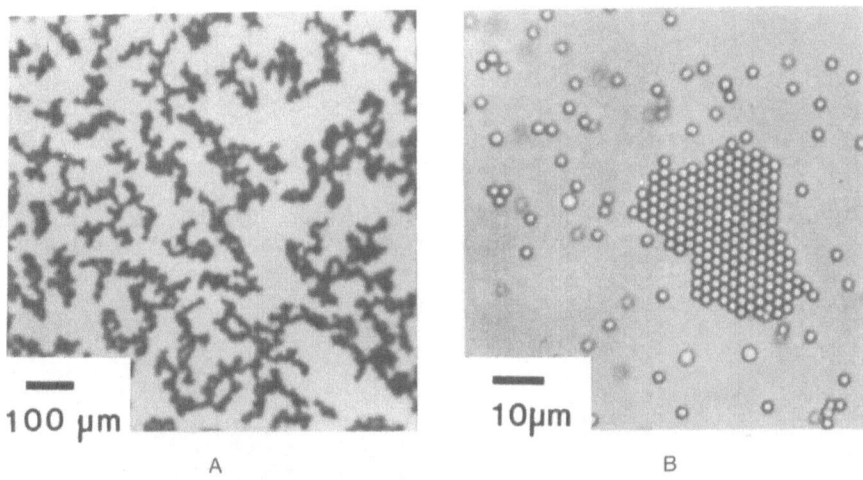

A   B

Fig. 8. Fast (a) and slow (b) cluster formation for the case in Fig. 7(a) using spheres of diamter 4.7 µm and 1.9 µm, respectively.

As will be shown below, diffusing microspheres interacting via long-range dipolar magnetic forces (the case in Fig. 7(b)) produces aggregates which are quite different from those resulting from short-range particle forces[13]. For these experiments two types of polystyrene spheres labeled M1 and M2 were used. The M1 spheres (d = 4.5 µm) contained iron oxide (30 % by weight) evenly distributed in pores. These spheres were to a very good approximation "superparamagnetic" as seen from the magnetization measurements in Fig. 9(a)[18].

The M2 spheres (d = 3.6 µm) contained iron oxide (30% by weight) in the form of evenly distributed grains in a thin shell (~ 0.2 µm) near the surface. These spheres possessed a maximum remanent magnetization M_r = 1.4 emu/g as seen from the magnetization curve in Fig. 10(a)[19].

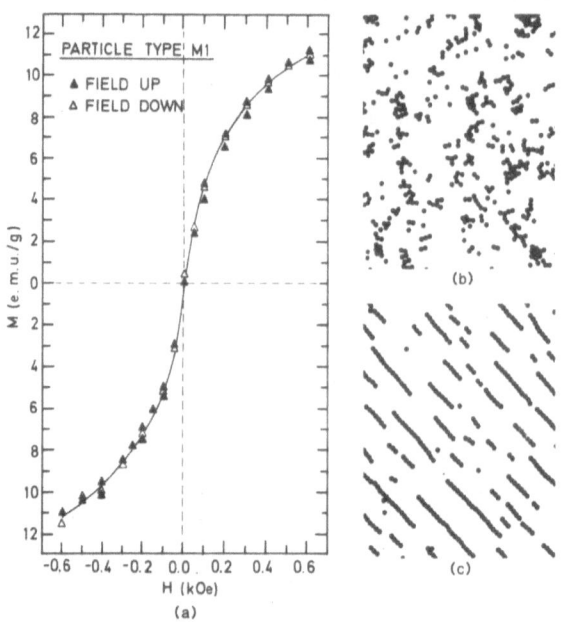

Fig. 9. (a) Magnetization curve for M1 spheres; (b) limited aggregation in zero field; (c) chain formation in an external field $H_{||}$ = 1 Oe.

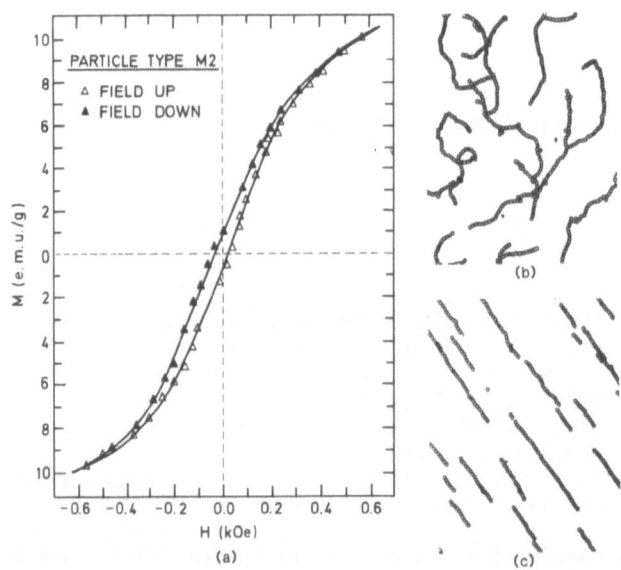

Fig. 10. Same legend as in Fig. 9 for M2 spheres.

The M2 spheres could also be magnetized to various levels of remanent magnetization M_r = 0 - 1.40 $\pm$ 0.13 emu/g, Fig. 11(A). The dispersion was stabilized so that the electrostatic and van der Waals interactions between the spheres were negligible compared to the magnetic forces. The magnetized spheres may be considered as point dipoles with magnetic moment μ = M_r ($\pi d^3/6$). The dimensionless parameter which determines the effective strength of the dipole-dipole interaction relative to the disruptive thermal energy is

$$K_{dd} = \mu^2/d^3 k_B T. \qquad (1)$$

and that which characterizes the strength of the dipole-field interactions is

$$K_{df} = \mu H/k_B T. \qquad (2)$$

The experiments were performed using samples with concentrations less than 10% with random initial distributions of spheres confined to a monolayer. To avoid the influence of the earth magnetic field and stray fields, the samples were enclosed in mumetal.

Fig. 9(b) shows that diffusing particles of type M1 formed small clusters which dispersed easily. Although these particles apparently do not have remanent magnetization, the weak attractive forces between the spheres could be due to local interactions between weakly magnetized grains near the surface. Fig. 9(c) shows the effect of a small external field $H_{||}$ = 1 Oe along the layer. As may be seen, this produced pronounced chaining. The effective dipole-field interaction parameter was estimated from Eq. (2) to be $K_{df} \approx 70$ for an induced moment M = 0.07 emu/g.

Fig. 10(b) shows that the M2 spheres aggregated readily into a random chainy structure. This is consistent with a large dipole-dipole coupling constant $K_{dd} \approx 1360$ (Eq.(1)). Fig. 10(c) shows the alignment of the same spheres along an external field of $H_{||}$ = 1 Oe corresponding to a dipole-field constant K_{df} = 1250.

Figs. 11(a) - 11(c) show a series of aggregated structures formed after a few hours for different K_{dd} for the M2 spheres. It is seen that there is a increasing tendency to form chains and open loops as K_{dd} increases, reflecting the preference of alignment of the dipoles.

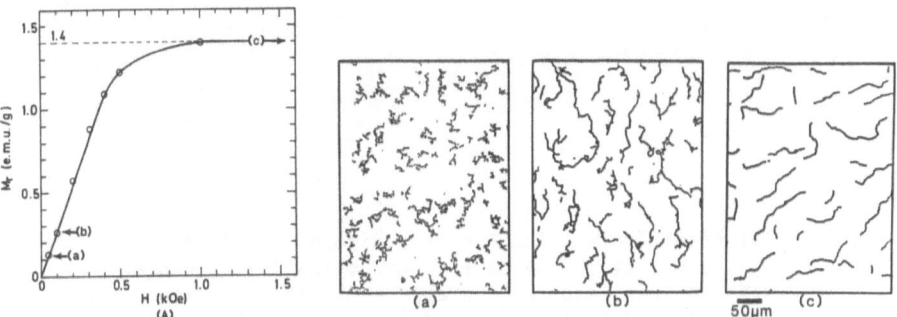

Fig. 11. (A) Remanent magnetization M_r versus external applied field H. Aggregates for increasingly reduced magnetization: (a) K_{dd} = 16; (b) K_{dd} = 100; (c) K_{dd} = 1360.

The fractal dimensions of these aggregates were determined from the usual log-log plot of radius of gyration versus the number of particles in each cluster. Fig. 12(a) shows the variation of D versus K_{dd} for no external fields. As may be seen, D becomes significantly lower as K_{dd} increases and agrees within experimental error with the simulated value $D = 1.23 \pm 0.12$ for large K_{dd}[19]. Also included in Fig. 12(a) is the value $D = 1.49 \pm 0.06$ found for cluster aggregation of nonmagnetic spheres. As may be seen, this result appears to be close to the limiting value also for magnetized spheres as $K_{dd} \rightarrow 0$.

The temporal evolution of the cluster size distribution was analyzed using the following proposed scaling relation for cluster-cluster aggregation[20]:

$$n_s(t) \simeq s^{-2}(t)f(s/S(t)) \qquad (3)$$

Here, $n_s(t) = N_s(t)/N_0$ with $N_s(t)$ the number of clusters with s particles and N_0 the total number of particles. The mean cluster size is given by

$$S(t) = \Sigma n_s(t)s^2 / \Sigma n_s(t)s \qquad (4)$$

where the sums are taken over all clusters. It is expected that

$$S(t) \sim t^z \qquad (5)$$

for $t \rightarrow \infty$ with z a critical exponent.

Fig. 12(b) shows the experimental results for the clustering of the $K_{dd} = 1360$ particles for a time span of $t = 8 - 165$ min. As may be seen, there is fair data collapse with an exponent $z = 1.7 \pm 0.2$. For the less magnetized spheres, similar results were obtained. The fitted exponent z was thus found to decrease slightly with reduced K_{dd}, reaching a limiting value $z = 1.4 \pm 0.2$ for low values of K_{dd} which is the same as the earlier simulated result for nonmagnetic DLCA[20].

Another type of aggregation occurs for magnetic spheres in ferrofluid[21] (Fig. 7(c)) as illustrated in Fig. 13. As may be seen,

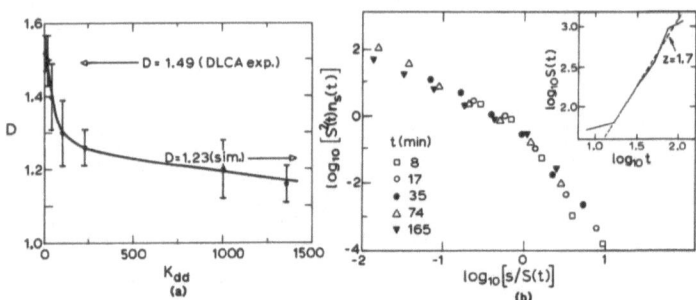

Fig. 12. (a) Fractal dimension D versus dipolar coupling constant K_{dd}. $D = 1.49 \pm 0.06$ is the DLCA value for nonmangetic spheres. For large K_{dd}, D approaches the simulated value $D = 1.23 \pm 0.12$ as discussed in the text. The solid line is only a guide to the eye. (b) Scaling of the temporal evolution of the cluster size distribution ($K_{dd} = 1360$). The inset shows the fit of $z = 1.7 \pm 0.2$ to the log-log plot of average cluster size vs time (Eq. (5)).

there is a more pronounced chaining and looping than in a nonmagnetic fluid. The reason for this may be that the spheres polarize the surrounding ferrofluid particles and with this "cloud" other spheres in the neighbourhood are more easily aligned and attracted, enhancing the chaining. So far there are no detailed calculations to describe this type of enhanced aggregation. Some form of molecular dynamics simulations seems to be required.

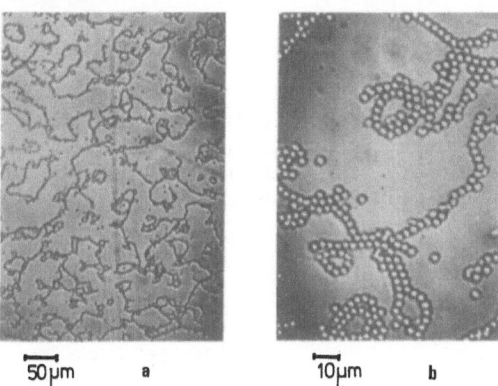

Fig. 13. Aggregation of 3 μm magnetic spheres in ferrofluid for small (a) and large (b) magnification.

The last form of aggregation to be demonstrated is the case in Fig. 7(d) with nonmagnetic spheres in ferrofluid[22] subjected to external fields. This produces magnetic holes (Sec. 5) with magnetic moments associated with each sphere antiparallel to the field. Fig. 14 shows typical chain build-up along the field for d = 1.1 μm diffusing spheres. It may be argued that the average number of spheres, N, in the chains is $N \simeq \Gamma_{||} = \pi^2 d^3 \chi_{eff} H^2_{||} / 36 k_B T$ (see Sec. 5, Eq. (10) with $H_{||}$ replacing

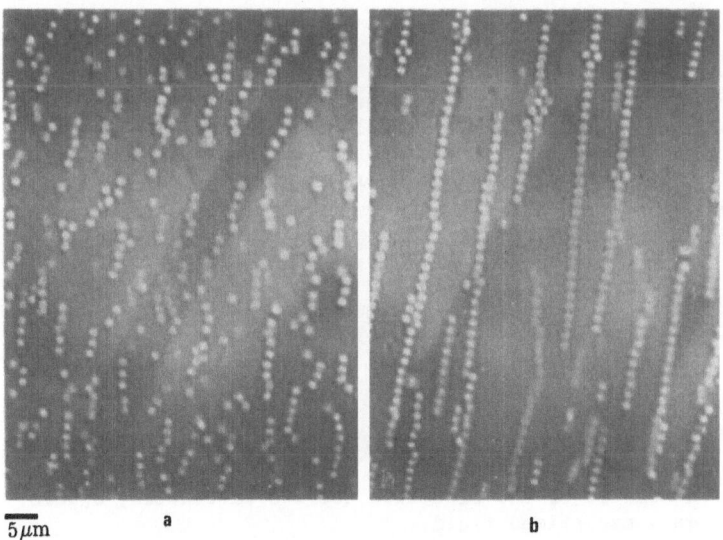

Fig. 14. Chain formation of D = 1.1 μm spheres with increasing parallel field: (a) $H_{||}$ = 5 Oe; (b) $H_{||}$ = 30 Oe.

$H_\perp$) with the effective volume susceptibility of the present ferrofluid[22] $\chi_{eff} = 0.13$, k_B Boltzmann's constant and T absolute temperature. For the cases in Fig. 14, with $H_{||} = 5$ Oe and 30 Oe, the corresponding values for N are 3 and 90, respectively, in fair agreement with observations.

5. LATTICES OF MAGNETIC HOLES

Some years ago it was discovered[6] that by dispersing monosized nonmagnetic spheres/particles in a ferrofluid, it was possible to produce a system of interacting magnetic dipoles which has been called "magnetic holes". The concept is very simple and corresponds to the magnetic analogue of Archimedes' principle as shown schematically in Fig. 15.

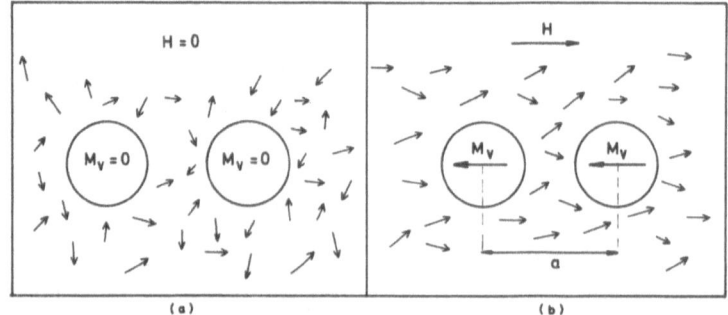

Fig. 15. Magnetic hole effect. Nonmagnetic spheres dispersed in ferrofluid without (a) and with (b) an external magnetic field. The latter case produces apparent magnetic moments, M_v, related to the spheres and dipolar interactions as discussed in the text. The arrows indicate the directions of the magnetic moments of the much smaller ferrofluid particles.

Due to the paramagnetic property of the ferrofluid, it is possible to turn on and off the effective magnetic interactions between the spheres which is generally not possible for interacting magnetic spheres. If the spheres are confined to a monolayer between solid boundaries (Fig. 16), the interaction energy between two spheres is approximately the dipolar interaction[23]:

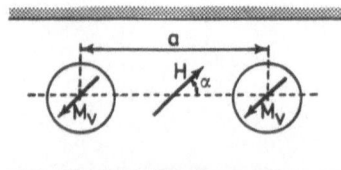

Fig. 16. Schematic of the situation for two interacting spheres (holes) in a magnetized fluid.

$$E_{dd}(\alpha) = E_0(1-3\cos^2\alpha),\qquad\qquad(6)$$

with $E_0 \equiv [M_v(\alpha)]^2/a^3$ for an external field $H(\alpha)$ making an angle α relative to the layer, and with separation a between the spheres. The effective magnetic moment of the sphere, $M_v(\alpha)$, corresponds to the magnetic moment of the displaced fluid with volume $V = \pi d^3/6$ for a sphere of diameter d. For low fields

$$M_v(\alpha) = - V\chi_{eff}(\alpha)H(\alpha), \qquad (7)$$

where $\chi_{eff}(\alpha)$ is the effective volume susceptibility of the ferro-fluid corrected for demagnetizing factors[6].

Eq. (6) shows that for a field $H_\perp$ normal to the layer, there will be repulsive interactions between the spheres given by

$$E_{dd}^\perp = M_v^2/a^3. \qquad (8)$$

This situation is shown in Fig. 17 with the formation of a triangular lattice.

Likewise, for a field $H_{||}$ parallel to the layer, Fig. 18(a), there will be attractive interactions between the spheres given approximately by

$$E_{dd}^{||} = - 2M_v^2/a^3. \qquad (9)$$

The resulting chain structure is shown in Fig. 18(b).

There may also be other possibilities like a combination of $H_{||}$ and $H_\perp$ producing repulsive pairs of spheres. If $H_{||}$ is rotating in the plane while $H_\perp$ is stationary, the whole triangular lattice will consist of spinning pairs.

Another interesting aspect of magnetic holes is that for small spheres ($\leq 2\mu m$) the Brownian motion effectively introduces a thermodynamic temperature in the system. It is thus possible to study crystallization and melting in two dimensions by varying the external field $H_\perp$. The controlling parameter for the stability of structure formation is

$$\Gamma_\perp = M_v^2/a^3 k_B T. \qquad (10)$$

An example of this is shown in Figs. 19 and 20 for the direct lattice as well as the diffraction pattern obtained using a collimated laser beam as the illumination source in combination with a Bertrand lens in the microscope. As may be seen, the system is near the melting point in Fig. 20. If one turns the page and views along the paper one sees parallel rugged lines. This signifies long-range orientational order but only short-range positional order typical for a hexatic phase. (see Aharony et al., these proceedings). The diffraction pattern (6-sided, see Fig. 20(b)) and calculated correlation functions[7] for the present system also have the signatures of such a phase. However, it is practically impossible to distinguish this from the results obtained for a polycrystalline sample with low-angle grain boundaries or a system with coexisting liquid-like and solid-like phases. So far it appears that the nature of the two-dimensional melting for isotropic systems remains unresolved.

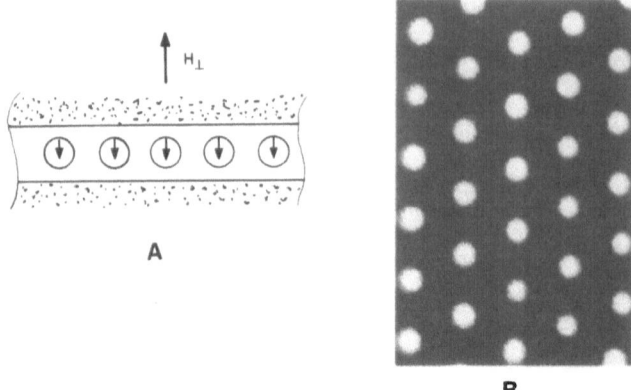

A

B

Fig. 17. Repulsive dipolar interactions between magnetic holes for a field normal to a monolayer. (a) Schematic cross-section of the cell with external field $H_\perp$ and corresponding effective magnetic moments (arrows). (b) Correspondng ligh microscope picture of 10 µm spheres.

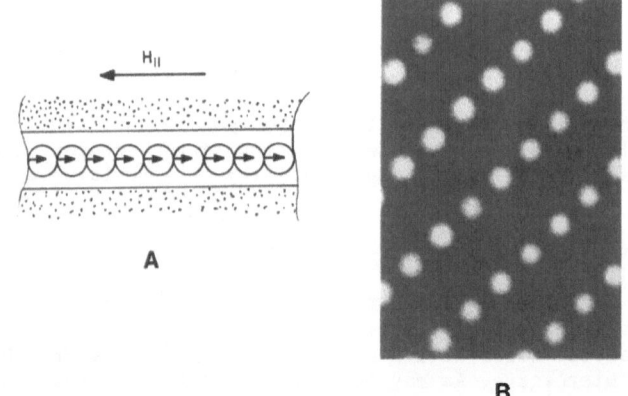

A

B

Fig. 18. Same legend as in Fig. 17, but now for an external field $H_{||}$ parallel to the layer producing attractive dipolar interactions. The spheres in the chains (b) are actually touching each other although they appear to be separated. This is due to the fact that the ferrofluid absorbs light and only the top part of the spheres closest to the glass boundary is visible.

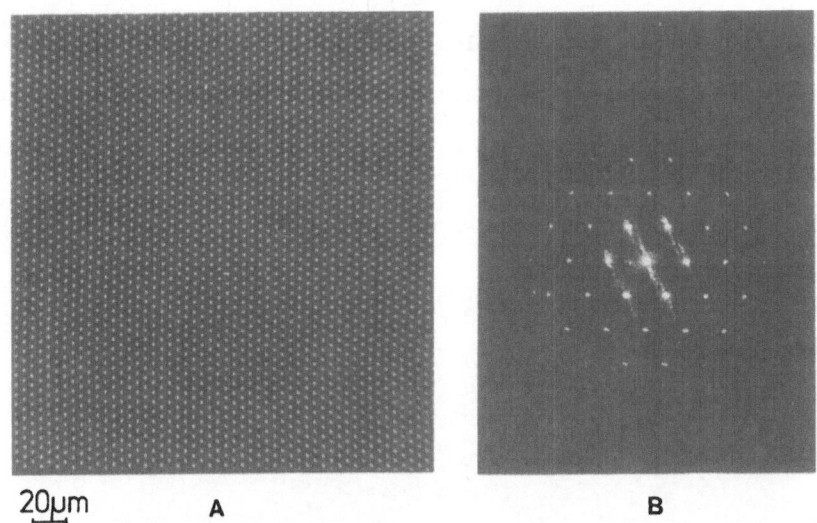

Fig. 19. Direct picture (a) of 1.9 µm spheres in ferrofluid with $H_\perp$ ≈ 200 Oe and diffraction picture (b).

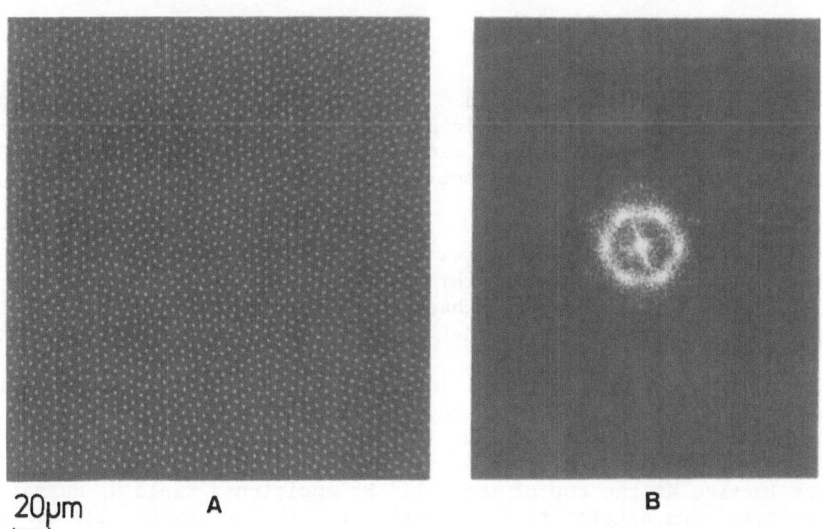

Fig. 20. Same situation as in Fig. 19, but with $H_\perp$ = 13 Oe.

Nonspherical magnetic holes offer many interesting model systems. As an example, Fig. 21 shows pear-shaped particles approximately 2.5 μm long and 1.7 μm across confined to a monolayer in ferrofluid. In Fig. 21(a), there is no external field and there is a random position and orientation of the "pears". In Fig. 21(b) a field $H_\perp$ = 110 Oe normal to the layer produces a triangular structure as for spheres. The added feature is that the particles behave as if they possessed polar heads preferring an up-down arrangement. For a triangular structure this introduces frustration which is comparable to a two-dimensional triangular antiferromagnet which have no long-range order. However, for the present system the positions are not necessarily fixed as in a crystal. In Fig. 21(b) the field is sufficiently high to produce positional order and a frustrated up-down lattice. Below a critical field the positional order is lost which relaxes the up-down frustration.

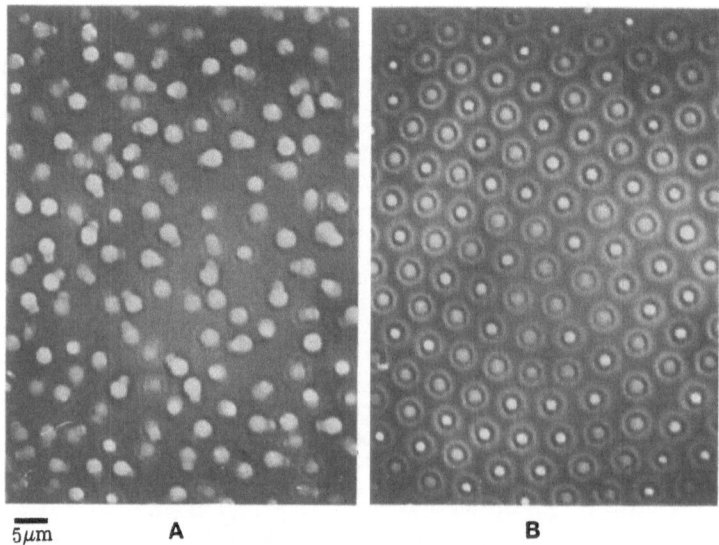

5μm   A   B

Fig. 21. Pear-shaped particles in a magnetic fluid with: (a) no external
         field; (b) a normal field $H_\perp$ = 110 Oe, where the up and
         down orientation may be identified from small white discs and
         larger gray discs, respectively.

        Preliminary experiments indicate that for this case a network of "up-down" particles percolating through the system may be formed. However, it appears that the system has to be "heat treated" by going up and down in field and given adequate time to sort out the best network.

        Another unusual phase transition takes place for magnetic holes in a gravitational field. For this, the sample cell is placed vertically using spheres with specific weight 1.04 g/cm^3 slightly less than that for ferrofluid (1.15 g/cm^3)[22]. The spheres will thus move up and form a compact lattice at the top of the cell. By applying a field $H_\perp$ normal to the layer, the spheres tend to repel each other producing the curved lattice illustrated in Fig. 22. This satisfies the transition between a dense and dilute triangular lattice. At each field $H_\perp$, there is a stable structure formed as the result of the competing compressive bouyancy forces depending on the height and the repulsive dipolar forces depending on the separation. The same type of curved structures are seen for a cell placed horizontally when a field normal to the layer pushes spheres out from a close packed region. In contrast to the case for the vertical cell, the structures here are not stabilized before the spheres are distributed evenly throughout the cell.

Aperiodic structures of this type are also seen in the florets of daisies, sunflowers etc., with grain boundaries forming one-dimensional quasicrystals[25].

Curved structures are also observed in a system of mm-sized magnetic spheres vibrating on a slanted plate. Reference is made to the presentation by Piotr Pieranski in these proceedings discussing the analytical solution to this problem.

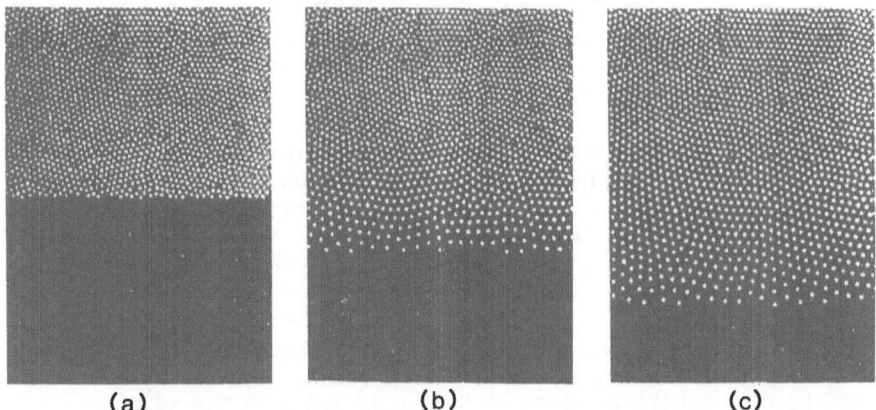

| (a) | (b) | (c) |

Fig. 22. Phase transition of magnetic holes in a gravitational field:
(a) initial packed structure; (b) normal field $H_\perp$ = 100 Oe after 1 min., and (c) 5 min.

6. ROTATING PAIR OF MAGNETIC HOLES

The last part of this lecture will present some recent experimental and numerical studies[26] of the response of pairs of magnetic holes confined between two parallel glass plates when subjected to rotating magnetic fields in the plane (Fig. 23). This produces an average attractive force between the holes forming a bound pair and a torque which tends to align the pair axis along the instantaneous direction of the field. Due to the viscous drag on the spheres, there is a balance between the magnetic driving torque and this viscous drag torque. This produces various modes of motion, (Fig. 24), depending on the frequency of the rotating magnetic field.

The experimental setup is shown schematically in Fig. 23(a). Uniformly sized microspheres with diameters either 10, 25 or 96 µm, were dispersed in kerosene-based ferrofluid[22,27] and confined between two glass plates. The spacing between the plates were several times the diameter of the spheres. Two pairs of Helmholtz coils were used to produce two sinusoidal fields, $H_0 \sin \omega_H t$ and $H_0 \sin(\omega_H t + \pi/2)$ in the plane with amplitude H_0. The sample cell (20x20 mm^2) contained a very dilute dispersion of polystyrene spheres. This produces only a few pairs of spheres which were far apart and thus not interacting. The frequencies of the various modes were low ($\lesssim$ 1 Hz) and could easily be measured manually using a stop-watch.

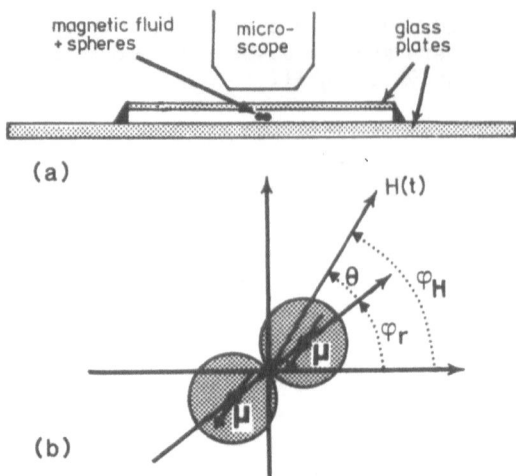

(a)

(b)

Fig. 23. (a) Side-view of experimental setup. (b) Top-view for
coordinate system of two magnetic holes rotating in the plane
between the two glass plates in (a) and driven by the planar
field H(t) as discussed in the text.

The coordinate system for a bound pair of magnetic holes (spheres)
in the plane is shown in Fig. 23(b). The apparent magnetic moment
carried by each sphere is given by $M_v = - V\chi_{eff}H$ with $V = \pi d^3/6 =$
volume of the sphere with diameter d and χ_{eff} the effective volume
susceptibility of the ferrofluid[6]. The interaction energy between
the spheres is given to first order by the dipolar energy:

$$U(\theta) = M_v^2(1-3\cos^2\theta)/d^3 \qquad (11)$$

with $\theta = \varphi_H - \varphi_r$ the angle between the pair axis and the direction of the
field[2,8].

The torque acting on the pair is given by $T_H = - dU(\theta)/d\theta$

$$T_H = - \varepsilon \sin(2\theta) \qquad (12)$$

with $\varepsilon = \pi^2\chi_{eff}^2 H_x^2 d^3/12$. The rotating pair is subject to a viscous
torque acting opposite to the direction of motion and given by

$$T_\eta = -K\eta \, d\varphi_r/dt \qquad (13)$$

where $\varphi_r = \varphi_H - \theta$. Here, η is the viscosity of the fluid and K a geometrical
factor related to the sphere diameter[26].

The two torques in Eqs. (12) and (13) must be equal at equilibrium
resulting in the following basic equation of motion:

$$\frac{d\theta}{dt} = \omega_H - \omega_c \sin(2\theta) \qquad (14)$$

and where $\omega_c = (\varepsilon/K\eta)$.

As will be discussed below, ω_c represents a critical angular frequency for which the mode of rotation changes character. In the reasoning given above, we have neglected the inertia term $Id^2\varphi_r/dt^2$. This can be done as the moment of inertia is small and the viscosity is high (overdamped system).

Fig. 24 shows polar (t,θ)-diagrams of some of the observed modes of rotation which agrees very well with the simulations using Eq. (14). For frequencies ω_H less than a certain critical frequency ω_c the spheres rotate with a constant phase lag θ behind the field, Fig. 24(a). For ω_H slightly larger than ω_c as shown in Fig. 24(b), the rotation is steady from the start of the experiment until point (1) in the figure, where the particles rotates in the opposite direction until point (2) where the direction of rotation is once more with the rotating field. The motion is then regular for a long time (from one to several hundred cycles depending on ω_H) before a new instability occurs. For higher ω_H this may happen several times per full rotation of the particle axis as in Fig. 24(c), and for still higher ω_H the motion becomes even more "jerky", Fig. 24(d).

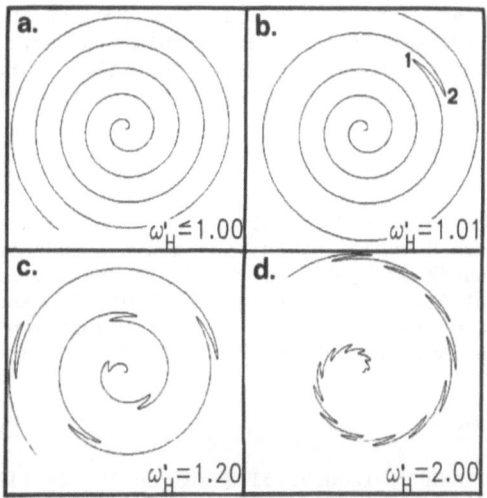

Fig. 24. Polar plot of the angle of the pair axis, φ_r vs. time (radially) for different values of the normalized angular frequency $\omega'_H = \omega_H/\omega_c$:
(a) below or just at the transition ω_c; (b) - (d) at the various angular frequencies above transition.

These observations can easily be explained from the equation of motion. Let us assume that, as observed in experiments, for ω_H sufficiently small, the pair follows the field, i.e. rotates with the same angular velocity but lagging behind by a constant angle $\theta(\omega_H)$. Since in this case $d\theta(\omega_H)/dt = 0$, direct integration of Eq. (14) gives:

$$\theta(\omega_H) = \frac{1}{2}\arcsin(\omega_H/\omega_c). \qquad (15)$$

This solution holds up to $\omega_H = \omega_c$ at which point the lag angle θ reaches $\pi/4$ and the magnetic torque T_H^c reaches its first maximum, Eq. (12). The system is then unstable because a small increase in θ will reduce the torque. It is then favourable to reduce the torque by rapidly increasing the lag angle θ until it reaches a value $\pi/2$ at which point T_H reverses sign forcing the pair to rotate backwards. This continues until the axis of the magnetic field again becomes parallel to the pair axis. By integrating Eq. (14) from the end of one backward rotation to the end of the next, one finds that the angular frequency of the backward rotations is:

$$\omega_b = 2\omega_H[1 - (\omega_c/\omega_H)^2]^{1/2} . \tag{16}$$

For high frequencies ω_H of the magnetic field ($\omega_H \gg \omega_c$), $\omega_b \simeq 2\omega_H$. This means that the pair makes a stop and moves backwards twice during the period of the rotating field. On the other hand, as ω_H decreases towards ω_c, ω_b drops to zero as for a "phase transition".

The agreement between the experiments and eq. (16) is rather good[26] except perhaps for ω_H very close to ω_c where we observed increasing fluctutions in ω_b which are not included in a "mean-field" type of description like Eq. (14).

Obvious extensions of these model studies are the introduction of anisotropy of the driving H-field[26], or allowing the particles to have radial motions or the use of more than two particles. This produces an even wider range of nonlinear dynamic processes, including transition to chaos[26].

ACKNOWLEDGEMENTS

The research was supported in part by Dyno Industrier A/S and by the Norwegian Research Council for Science and the Humanities (NAVF). We would also like to thank J. Ugelstand, T. Ellingsen and A. Berge, SINTEF, Trondheim for providing samples, and Piotr Pieranski for close collaboration on the gravitation induced phase transition and rotating pair, as well as J.L. McCauley, P. Meakin, and J.P. Hansen for valuable discussions.

REFERENCES

1. L.B. Bangs, M.T. Kenny, Industrial Research 18, 46 (1976).
2. J. Ugelstad et al., Adv. Colloid Interface Sci. 13, 101 (1980). Produced by Dyno Particles, N-2001 Lillestrøm, Norway.
3. J. Ugelstad, A. Berge, R. Schmid, T. Ellingsen, P. Stenstad, and A. Skjeltorp, in Polymer Reaction Engineering, edited by K.H. Reichert and W. Geiseler (Hüthig & Wepf, Heidelberg, 1986) p. 77.
4. A.T. Skjeltorp, J. Ugelstad, and T. Ellingsen, J. Coll. Interface Sci. 113, 577 (1986).
5. A. Berge et al. in "Polymer Colloids", edited by R.H. Ottewil (Plenum, 1989, in press).
6. A.T. Skjeltorp, Phys. Rev. Lett. 51, 2306 (1983).
7. P. Meakin et al., Phys. Rev. A34, 5091 (1986).
8. P. Meakin, Phys. Rev. A38, 994 (1988).
9. A.T. Skjeltorp and P. Meakin, Nature 335, No. 6189, 424 (1988).
10. A.T. Skjeltorp in "Random Fluctuations and Pattern Growth: Experiments and Models", edited by H.E. Stanley and N. Ostrowsky. (Kluwer Academic Publ., Dordrecht, 1988), p. 170.
11. P. Meakin, Thin solid Films 151, 165, (1987).
12. A.T. Skjeltorp, Phys. Rev. Let. 58, 1444 (1987).
13. G. Helgesen, A.T. Skjeltorp, P.M. Mors, R. Botet, and R. Jullien, Phys. Rev. Lett. 61, 1735 (1988).
14. Pawel Pieranski, Contemp. Phys. 24, 25 (1983).

15. B.V. Derjaguin and L. Landau, _Acta Phys. Chim._ 14, 633 (1941);
    E.J. Verwey and J. Th. G. Overbeck, _Theory of the Stability of Lyophobic Colloids_ (Elsevier, Amsterdam, 1948).
16. P. Meakin, Phys. Rev. Lett. 51, 1119 (1983);
    K. Kolb, R. Botet, and R. Jullien, Phys. Rev. Lett. 51, 1123 (1983).
17. T.A. Witten, Jr. and L.M. Sander, Phys. Rev. Lett. 47, 1400 (1981).
18. H. Fjellvåg and A.T. Skjeltorp (unpublished report, 1986).
19. H. Fjellvåg (unpublished report, 1987).
20. T. Vicsek and F. Family, Phys. Rev. Lett. 52, 1669 (1984).
21. A.T. Skjeltorp, J. Appl. Phys. 57, 3285 (1985);
    A.T. Skjeltorp, Physica 127B, 411 (1984),
22. Type EMG 909 from Ferrofluidics Corp.
23. In more accurate calculations one has to consider magnetic imaging effects both across the boundaries and on opposite spheres. See also M. Warner and R.M. Hornreich, J. Phys. A18, 2325 (1985).
24. A.T. Skjeltorp, J. Magn.Magn. Mat. 65, 195 (1987).
25. N. Rivier, J. de Physique, 47, C3-299 (1986).
26. Geir Helgesen, Piotr Pieranski, and Arne T. Skjeltorp, unpublished.
27. Ferrofluid with saturation magnetization $M_s = 200$ G, initial susceptibility, $\chi = 0.0829$ and viscosity $\eta^s = 5$ cP.
28. Eq. (11) shows that whenever $\cos^2 \theta < 1/3$ the spheres repel each other which would lead to separation. However, in order to simplify the experimental conditions for the present studies, the spheres were prepared in such a way that once in contact they stayed together.

GRAVITY'S RAINBOW -

STRUCTURE OF A 2D CRYSTAL GROWN IN A STRONG GRAVITATIONAL FIELD

Piotr Pierański

IFM PAN
Smoluchowskiego 17
Poznań, Poland

INTRODUCTION

A 2D system of particles interacting through long range repulsive forces

$$F(r)=Ar^{-q}, \quad q=2,3,4..., \tag{1}$$

where r is the distance between interacting particles, makes the simplest example of what at low temperatures can be aptly described as "soft condensed matter". At $T=0^{\circ}K$ the interval of densities at which the system displays crystalline order becomes infinite. In 2D, in the absence of any external fields, structure of the crystalline state is hexagonal (Crandall and Williams, 1974; Pieranski, 1980; Skjeltorp, 1983).

What would the structure be if the 2D crystal was grown in a strong gravitational field ?

Finding an answer to this question with but analytical tools seems to be a difficult task. It is the aim of this report to present an experiment in which the nature itself was asked to provide the answer.

EXPERIMENTAL

$N \simeq 10^{3}$ steel spheres of diameter d=1mm were placed within a flat, rectangular box whose upper and lower walls were made of glass. See Fig.1. N was a few times smaller than the number for which the spheres would be close packed within the box. The box was located between a pair of Helmholtz coils. When the supply current is switched on, the spheres are seen to escape from each other. This effect is due to the interaction between magnetic dipole moments induced in each sphere. The moments are parallel to each other and normal to the plane of the box, thus forces between them are of form (1) with q=4. A is here proportional to the mean square of the magnetic field produced by the coils system. (To avoid permanent magnetization of the spheres 50 Hz current was used.) Escaping from each other the spheres try to minimize energy of their magnetic interaction. Unlike in 1D systems, where the path to the ground state is smooth and straight, here, energy barriers often lock the magnetic spheres system within a local minimum. A delicate shaking of the box helps to overcome the obstacle. As a result, after a few minutes of such a shake-and-watch procedure, a stable

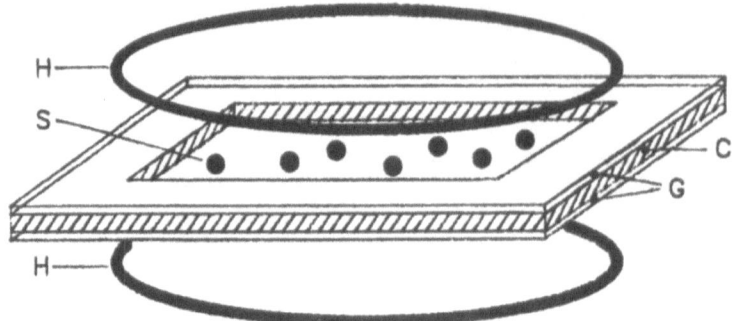

Fig.1. Experimental set-up. G – glass plates; C – cardboard spacer;
        S – steel spheres; H – Helmholtz coils.

configuration appears in the box. Analyzing it, one should be able to draw
some conclusions concerning structure of the global ground state of an infinite system.

    Two series of experiments aimed at finding structures of the ground state
were performed : with (i) horizontal and (ii) tilted box. Obviously, in experiments with the tilted box, the results depend on the tilt angle. In this
report we present but one, most typical case.

    In experiments performed with horizontally oriented box a typical final
configuration was a patchwork of a few grains of the hexagonal lattice. That
this is but an artefact induced by awkward boundary conditions, one can conclude repeating the experiment with a hexagonally shaped box. Here, a single
almost perfect grain can be easily achieved. All in all, experiments performed with horizontal box confirm that in absence of external fields the
ground state of the magnetic sphere system is hexagonal.

    This simple solution breaks down when gravitational field enters the box
plane. What the new solution will be, is difficult to guess. One can expect
that, for stability reasons, its structure should still be locally hexagonal
On the other hand, due to the density gradient induced by the field, cells
of the new configuration must be small at the bottom and large at the top of
the tilted box. Putting such a collection of differently sized cells together into a consistent structure is a formidable task. Fig.2 shows how
nature does the job. Similar results were obtained by Skjeltorp (1989) in
his experiments on systems of magnetic holes.

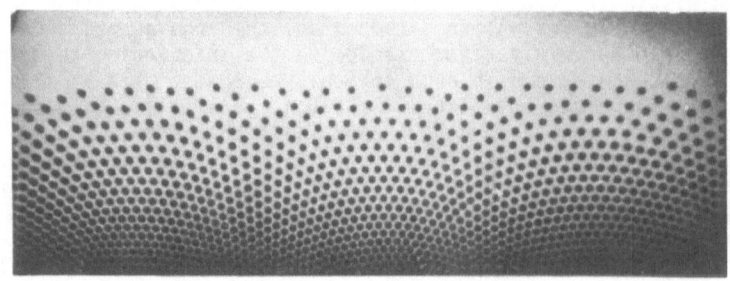

Fig.2. Gravity's rainbow.

As seen in the figure, the trick used here is building the 2D structure on a 1D lattice of pi/3 disclinations. Usually, "...isolated disclinations are energetically so costly as to be of little physical interest." (Mermin, 1979). The experiment we performed indicates a particular physical context within which it is a whole set of disclinations that allows the system to minimize its energy.

Looking at Fig.2 at low angles, one can easily distiguish three sets of basic atomic lines. In this gravity conditioned construction they are bent into arches remaining rainbows; hence the name, gravity's rainbow (GR), we used in the title. Analysing the angles at which the lines cross each other, one can note that they stay always close to pi/3. It was this angle conservation property that lead us to search for a formal description of the GR configuration among conformal transformations of the uniform hexagonal lattice. Preliminary analysis indicates that transformation (in complex plane)

$$w = \log z \qquad\qquad (2)$$

suits quite well the purpose. Fig.3 presents a piece of such an analytical construction. In accordance with what we see in experiment, the structure is periodic in the horizontal direction. On the other hand, its unit cells are infinite in the vertical direction while those of the GR structure are limited both from below (by presence of the hard core) and from above. The latter limit seems to stem from stability conditions. Numerical simulations we performed prove that, although in the considered here q=4 case conformally transformed hexagonal lattice presented in Fig.3 is not a ready for use solution, its pieces can serve well as initial configurations for a ground state searching procedure.

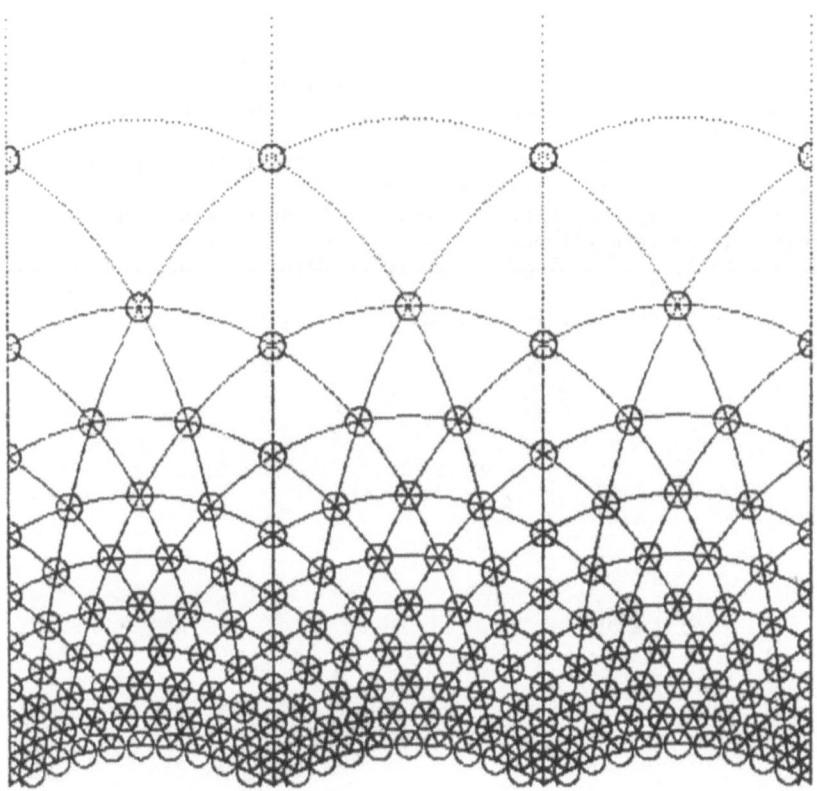

Fig.3 Hexagonal lattice transformed conformally according to Eq.2.

CONCLUSIONS

Experiment described above indicates that structure of the ground state of a 2D crystal grown in a strong external field is quite different from what one may find in textbooks of classical crystallography; its description needs more than the notion of a Bravais lattice. Large density gradient induced by such a field destroys translational symmetry, at least in the direction in which the field is applied. Structure which appears instead is, hoewever, not amorphous; a long range regularity of a new type is born. The problem we are faced with is to find the rules that would describe it. That classical crystallography is too rigid to copy with regular structures which are not infinite crystals has been argued in the past (Mackay, 1975). It seems that the gravity's rainbow structure is one of them.

ACKNOWLEDGEMENT

The idea of this work was born during my stay at the laboratory of A.Skjeltorp, Institute for Energy Technology, Kjeller. I would like to thank him for introducing me to the wonderful world of magnetic holes. Hospitality and kind help of G.Jarret, T.Riste and K.Otnes are gratefully acknowledged. That the GR structure can be approximated by a conformal mapping was pointed to me by K.W.Wojciechowski. Financial support was provided by NTNSF. This work has been sponsored in part by Polish Academy of Science under Project CPBP 01.12.1.4.1.

REFERENCES

Crandall, R. S. and Williams, R., 1971, Crystallization of Electrons on the Surface of Liquid Helium, Phys. Lett., 34A:404.

Pieranski, Pawel, 1980, Two-Dimensional Interfacial Crystals, Phys. Rev. Lett., 45:569.

Skjeltorp, A. T., 1983, One- and Two-Dimensional Crystallization of Magnetic Holes, Phys. Rev. Lett., 51:2306.

Skjeltorp, A. T., 1989, Qualitative Aspects of Condensation, Ordering and Aggregation, in this volume.

Mermin, N. D., 1979, The Topological Theory of Defects in Ordered Media, Rev. Mod. Phys., 51:591.

Mackay, A. L., 1975, Generalised Crystallography, Izvj. Jugosl. Centr. Krist.,10:15.

STRUCTURES AND PHASE TRANSITIONS IN THERMOTROPIC LIQUID CRYSTALS *

Tormod Riste

Institute for Energy Technology, POB 40,
N-2007 Kjeller, Norway

1. CLASSIFICATION

Liquid crystals[1] have structural order intermediate between conventional liquids and solids. Thermotropic liquid crystal phases form in pure compounds or homogeneous mixtures as the temperature is changed. Lyotropic liquid crystals form when amphiphilic molecules are dissolved in water, or another suitable solvent, and concentration is the main physical variable. Polymeric liquid crystalline order occurs in fluid polymer melts and solutions. We shall limit our discussion to thermotropic liquid crystals, for which the knowledge is most complete.

The structure of liquid crystals can broadly be classified as nematic, cholesteric and smectic, see Fig. 1. None of them have full three-dimensional (3-D) positional order, but some degree of orientational order. Most often the constituent molecules are elongated, as indicated in Fig. 1, but distinctly flat molecules make up the socalled discotic liquid crystals. The nematic phase has only orientational ordering of the molecules. The collection of molecules have one symmetry axis called the director n. The cholesteric phase has only orientational order, formed by the constituent chiral molecules. The director twists with a pitch comparable to the wavelength of light.

At least ten different smectic phases are known. In addition to orientational order they have some degree of positional order. Smectic-A(S_A) is a 1-D solid along the director and a 2-D liquid normal to it. In S_C the molecular axes are tilted with respect to the normal of the smectic layers. The remaining smectic phases, as e.g. S_B, see Fig. 1, all have more order than S_A and S_C. In order to discuss the different smectic phases it is necessary to distinguish molecular orientational order and bond orientational order. The latter may occur also in a substance composed of isotropic molecules, and is defined as

$$O(r) = \langle e^{i[\theta(0) - \theta(r)]}\rangle.$$

---

*This brief overview has been written for the proceedings. A talk on the same topic was given by Pawel Pieranski at this Advanced Study Institute.

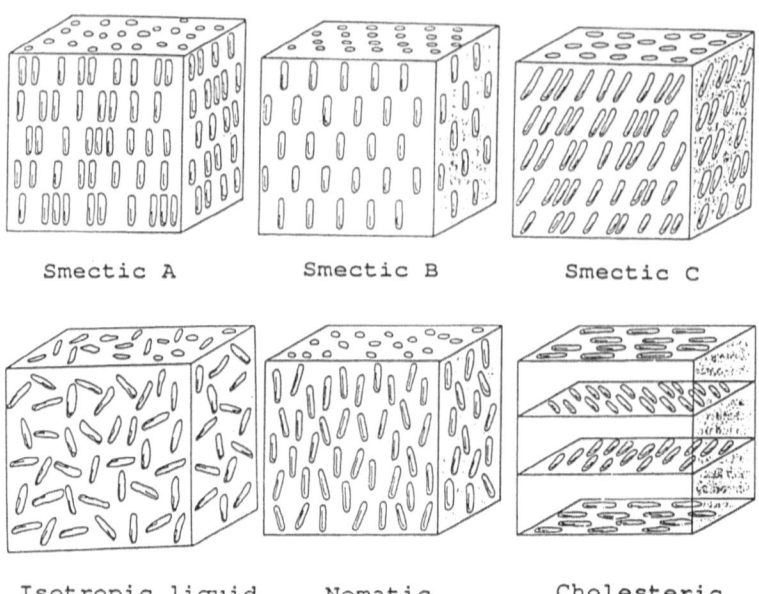

Smectic A   Smectic B   Smectic C

Isotropic liquid   Nematic   Cholesteric

Fig. 1. Most common types of liquid crystal ordering.

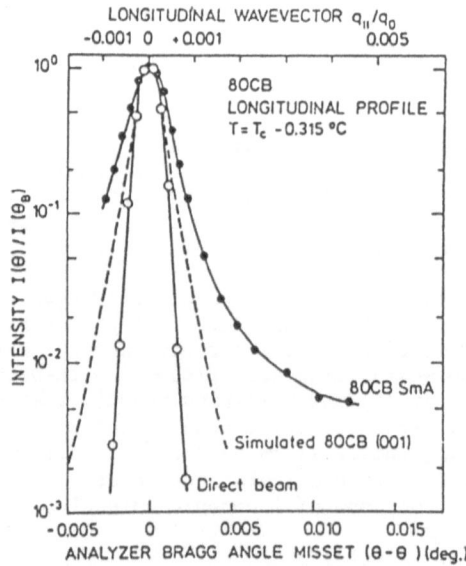

Fig. 2. Intensity profile for X-ray scattering from a smectic A
substance, 8OCB. The simulated profile is the one expected if
the order were truly long-range.

(Als-Nielsen et al[4])

Here $\theta(r)$ is the angle between the bond direction at position r and some reference axis. The conventional positional order G(r) and O(r) may now decay over short, long and infinite distances. Of special interest is the hexatic phase which plays an important role in the discussion of 2-D crystals. The distribution between a thick, hexatic sample and a 3-D crystal is rather subtle, and experimentally very difficult.

The blue phase[2] was earlier classified as belonging to the cholesteric phase, but is in fact a separate phase. It is in a sense a crystalline liquid. The molecules are positionally disordered and have no fixed mean positions, but the spatial pattern of molecular orientations can be assigned to a translation group. Usually the blue phase falls between the cholesteric and the isotropic phase, but direct transition to the smectic phase has been observed. Three blue phases have been found, one of them is amorphous.

PHASE TANSITIONS[3]

Some substances, as para-azoxy-anisole (PAA), has only one liquid crystalline phase, the nematic phase, between the melting point and the socalled clearing point. The behaviour of PAA with increasing temperature is

$$\text{crystal} \xrightarrow{118^0\text{C}} \text{nematic} \xrightarrow{135.5^0\text{C}} \text{isotropic liquid}$$

On cooling the nematic phase can be supercooled more than 25^0. Smectogenic phases may exhibit a whole sequence of liquid crystal phases. An example of that is terephtal-dibutyl-aniline (TBBA). In addition to two crystalline modifications and the nematic (N) and isotropic phase (I) there are five different smectic phases.

$$Cr_{II} \xrightarrow{-33^0C} Cr_I \xrightarrow{113^0C} S_G \xleftarrow{144^0C} S_C \xleftarrow{172^0C} S_A \xleftarrow{200^0C} N \xleftarrow{236^0C} I.$$

$$\Big\downarrow^{84^0C}$$

$$S_{VII} \xleftarrow{68^0C} S_H$$

Returning to the case of PAA, the continuous rotational symmetry of the isotropic liquid phase is broken when cooling from the isotropic liquid to the nematic phase. The translational invariance is broken when the sample is further cooled to the crystalline phase. In the nematic phase the order parameter S measures the average deviation (β) of the long molecular axes from n, and is in its simplest form written as $S = 1/2 \langle 3 \cos^2\beta - 1\rangle$. In the isotropic phase $S = 0$, in the nematic phase $S \neq 0$. Positive and negative values of S are allowed, corresponding to positive and negative birefringence. The Landau expansion for the free energy must therefore contain a cubic term, which means that the nematic/isotropic transition is of the first-order type.

The most studied phase transition in liquid crystals is the N/S_A transition, which was thoroughly covered in an earlier ASI at Geilo[4]. The S_A state consists of a one-dimensional density wave along the average direction of the molecular axes. The order parameter contains the magnitude and phase of this density wave, and is a complex quantity. S_A has strongly anisotropic elastic properties and fluctuations in layer positions diverge logarithmically. Thus G(r) has no true long-range order normal to the layers, but decays algebraically as $r^{-\eta}$. The associated X-ray scattering, given by its Fourier transform, gives peaks that are sharp, but wider than a Bragg peak, see Fig. 2. In the nematic phase, close to the N/S_A-transition, the smectic density wave exists, but with a relatively short decay length. The transition is observed to be second order or weakly first order. The true nature of the transition is still not known, a puzzling result is the existence of apparently two diverging correlation lengths. The S_{C-N}-transition is also a challenge. It is always observed to be of first order, although the symmetry of the order parameter permits a second-order change.

The hexatic, smectic phases have quasi-long-range orientational order, and are discussed in Aharony's lecture at this school, together with a thorough discussion of associated phase transitions. Hexatic and other smectic phases offer a unique possibility for studying the strong fluctuations of low-dimensional systems, even in 3-D samples, and represent condensed matter at the lower marginal dimensionality.

The cholesteric-nematic phase transtion induced by a magnetic field is, according to de Gennes[3], a second order transition of the nucleation type. This type of transition has hysteresis but no critical fluctuations.

The transitions between the blue phase region and its neighbouring phases, the cholesteric and the isotropic phase, are both of first order[2]. The transitions between the various blue phases are weakly first order.

In addition to transitions between different phases of liquid crystals these are also orientational phase transitions within a single phase. A wellknown example is the Fredericksz transition in nematics[1]. Less well known are the anchoring transitions discussed by Pieranski in this volume.

REFERNCES

1. Extensive reviews are given in P.G. de Gennes, The Physics of Liquid Crystals, Oxford U.P., London (1974), and in S. Chandrasekhar, Liquid Crystals, Cambridge U.P., Cambridge (1977).
2. For a recent review, see D.C. Wright and N.D. Mermin, Rev. Mod. Phys. 61, 385 (1989).
3. J.D. Litster and R.J. Birgeneau, Phys. Today 35, 26 (1982).
4. J. Als-Nielsen, J.D. Litster, R.J. Birgeneau, M. Kaplan and C.R. Safinya, Ordering in Strongly Fluctuating Condensed Matter Systems, T. Riste, ed., Plenum, New York (1980), pp 57 and 357.
5. Fluctuations, Instabilities and Phase Transitions, T. Riste, ed., Plenum, New York and London (1975), p. 1.

MULTICRITICALITY IN HEXATIC LIQUID CRYSTALS

Amnon Aharony

School of Physics and Astronomy
Raymond and Beverly Sackler Faculty of Exact Sciences
Tel Aviv University, Tel Aviv 69978, Israel

and

Department of Physics
Massachusetts Institute of Technology
Cambridge, Massachusetts, 02139, U.S.A.

INTRODUCTION

The elegant theories of defect mediated transitions[1] provide an interesting scenario for melting of two dimensional crystals. In particular they predict an intermediate phase with quasi-long-range orientational order: the hexatic[2]. However, observation of the crystal-hexatic-liquid sequence in either experiments or simulations has proved controversial[1] - the hexatic phase can be pre-empted by a direct discontinuous melting transition. Birgeneau and Litster[3] observed that certain liquid crystal phases can be regarded as three dimensional stacked hexatic layers. The coupling of layers in the third dimension, endows them with true long-range orientational order, and changes the nature of the corresponding transitions[4]. However, by looking at successively thinner films of such liquid crystals one may hope to approach the two dimensional limit, and indeed recently several groups have been engaged in such a pursuit, using diffraction studies of freely suspended liquid crystal films[5-7]. These studies may provide the most accurate and cleanest probe[1] of the two dimensional melting sequence.

The above developments had two major consequences. First, the identification of the various hexatic phases led to a detailed classification of the large variety of thermotropic liquid crystal phases. The essential structural details of these phases are illustrated in the Appendix[8,9]. Second, the diffraction patterns yield novel crossover critical phenomena. In three dimensions, this presented single experiments which yield a simultaneous measurement of many anisotropy crossover exponents,[10,11] previously obtained from separate complex experiments. For very thin layers, the patterns allow detailed studies of the crossover from two to three dimensions, and provide new quantitative information on the two dimensional hexatic phase.[12] The present paper aims at a brief review of these developments.

BOND ORIENTATIONAL HEXATIC ORDER

Let $\theta(\vec{r})$ be the angle between the bond direction at position $\vec{r}$ and some reference axis (see Fig.1). The bond orientational (BO) order parameter is then defined as

$\Psi_6(\vec{r}) = \langle e^{i6\theta(\vec{r})} \rangle$ where the brackets indicate a coarse grain average. The BO correlation function, $G_6(\vec{r})$, does not decay to zero, i.e.,

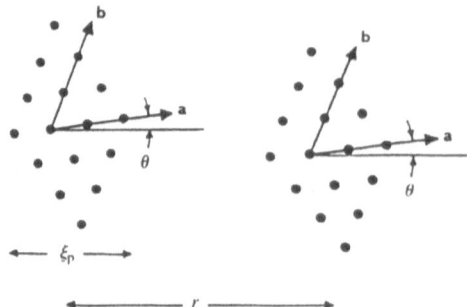

Fig. 1.   Bond orientational order in a triangular two-dimensional lattice.   Molecular positions are correlated only over a short distance ξ_p, while the orientation of the lattice vectors $\vec{a}$ and $\vec{b}$ is preserved throughout the sample.

$$\lim_{|r|\to\infty} G_6(r) = \lim_{|r|\to\infty} \langle e^{i6[\theta(\vec{0})-\theta(\vec{r})]} \rangle = \text{const} \neq 0, \tag{1}$$

in a two-dimensional crystal with sixfold rotation symmetry.

Naively, one would suspect that introducing an enormous number of dislocations would melt the crystal.   However, unbinding the dislocation pairs does not completely melt the crystal,[2] that is, the system does not become an isotropic fluid when the dislocations unbind. The details of the ordering and the functional form of the correlation function are model dependent.   In the Halperin-Nelson theory,[2] after the dislocations unbind, the positional correlations are short-ranged and the BO correlations are quasi long-ranged, that is, the positional correlations decay exponentially and the BO correlations decay algebraically:

$$\lim_{|r|\to\infty} G_\rho(r) = \lim_{|r|\to\infty} \left[\langle \rho(\vec{r})\rho(0)\rangle - \rho_0^2 \right] \sim e^{-r/\xi_p}, \text{ but } \lim_{|r|\to\infty} G_6(r) \sim r^{-\eta} . \tag{2}$$

A second disclination unbinding transition is necessary to achieve an isotropic, two-dimensional fluid,   Thus, in the Halperin-Nelson theory, melting (freezing) is a continuous, two step process in two dimensions.   The strange intermediate phase with short-range positional but quasi long-range BO order was dubbed <u>hexatic</u> for two-dimensional systems with hexagonal symmetry.

As explained in the Appendix, one now identifies several smectic liquid crystal phases as having three dimensional (3D) BO long range order.[3] Without coupling between the smectic layers, one would expect an algebraic decay of $G_6(r)$.   We return to this two dimensional (2D) behavior below, when we discuss very thin layers.   As soon as there is any coupling between the layers, the quasi long range BO order turns into real long range order, with a finite value of the average BO order parameter $C_6 = \text{Re} \int d\vec{r}\, \psi_6(\vec{r})/V$. Since $\psi_6(\vec{r})$ is a complex order parameter, its critical phenomena are in the same universality class as the magnetic XY model, or any other model with a two-component order parameter.[4,13]

ANISOTROPY CROSSOVER CRITICAL PHENOMENA

Using the coarse grained order parameter $\psi_6(\vec{r})$, the critical behavior of the hexatic phase transition in three (and more) dimensions may be studied starting from a Ginzburg-Landau-Wilson Hamiltonian,[10]

$$H = \int d\vec{r} \left\{ \frac{1}{2} \left[|\nabla \Psi_6|^2 + a|\Psi_6|^2 \right] + u_4 |\Psi_6|^4 + u_6 |\Psi_6|^6 + h \operatorname{Re} \Psi_6 \right\},$$ (3)

where a is linear in the temperature T, and h is the ordering field. Near the transition temperature T_c, we expect[14] the free energy to scale as $F(t,h) = |t|^{2-\alpha} f\left[h/|t|^\Delta \right]$, with $t = (T - T_c)/T_c$ and with α and $\Delta = \beta + \gamma = \beta\delta = (d+2-\eta)/2$ being the XY-model specific heat and "gap" exponents. Here, $C_6 \sim h^\delta$ when t=0, while $C_6 \sim |t|^\beta$, the correlation length diverges as $\xi \sim |t|^{-\nu}$ and the order parameter susceptibility diverges as $|t|^{-\gamma}$ for h=0.

Writing $\psi_6 = x + iy$, we have $|\psi_6|^2 = x^2 + y^2$, and the Hamiltonian (3) is equivalent to that of a general two-component order parameter, (x,y). For h=0, this Hamiltonian is isotropic. There are many ways to break this isotropy. In general, this is done by adding to (3) a term $H_n = g_n \int d\vec{r} \operatorname{Re}(\psi_6^n)$. Close to the isotropic XY model critical point (small g_n), the free energy scales as[14]

$$F(t,g_n) = |t|^{2-\alpha} f_n \left[g_n / |t|^{\phi_n} \right]$$ (4)

where ϕ_n is an appropriate crossover exponent.

For n=2, $H_2 = g_2 \int d\vec{r} (x^2 - y^2)$ implies a higher transition temperature for the x (if $g_2 < 0$) or the y component, i.e. uniaxial symmetry. This yields crossover to Ising model behavior. The XY behavior occurs only at the bicritical point, defined by $g_2 = 0$. The two Ising critical lines approach the bicritical point as $|g_2| \sim |T - T_B|^{\phi_2}$ This was first confirmed experimentally,[15] with $\phi_2 = 1.17 \pm 0.02$, for the uniaxial antiferromagnet Gd AlO$_3$, where a uniform field represents g_2.

For n=3, $H_3 = g_3 \int d\vec{r} (x^3 - 3xy^2)$ prefers ordering along three directions at 120° from each other, reflecting the symmetry of the 3-state Potts model. This can be achieved e.g., by a uniaxial stress along [111], in a system which otherwise prefers to order along any one of the cubic axes. Indeed, experiments[16] near the structural transition in SrTiO$_3$ yielded $\phi_3 \approx 0.5 \pm 0.1$.

Similarly, n=4,5 and 6 represent cubic, pentagonal and hexagonal symmetries. The tetracritical phase diagram[17] in Gd LaO$_3$ yielded[15] $\phi_4 = 0. \pm 0.03$. We are not aware of direct experimental measurements of ϕ_n for n > 4.

Given Eq. (4), we note that[10]

$$C_{6n} = \int d\vec{r} \operatorname{Re} \left[\psi_6^n \right] / V = (\partial F / \partial g_n)_{g_n = 0} \sim |t|^{2-\alpha-\phi_n} \sim C_6^{\sigma_n},$$ (5)

with

$$\sigma_n = \frac{2-\alpha-\phi_n}{2-\alpha-\Delta} = \frac{2(d-\lambda_n)}{d-2+\eta},$$ (6)

where $\lambda_n = \phi_n / \nu$.

A detailed renormalization group calculation of the exponent λ_n is given in Refs. 7 and 10. Briefly, this calculation is based on a perturbative expansion in both u_4 and g_n. To leading order in both, the term of order $u_4 g_n$ in the recursion relation for g_n has a combinatorial factor $n(n-1)$ (related to the number of ways to connect two factors ψ_6 in $u_4 |\psi_6|^4$ with two such factors in $g_n \, \text{Re}(\psi_6^n)$). This results in the form

$$\sigma_n = n + x_n \, n(n-1) ,\qquad (7)$$

with

$$x_n \simeq 0.3 - 0.008n. \qquad (8)$$

EXPERIMENTS OF 3D FREELY SUSPENDED FILMS

Details of the experimental techniques are given elsewhere.[6,7,8,10,11] The basic idea is to create a freely suspended film, by drawing the liquid crystal material across a hole in a thin plate, using a wiper blade.[18] One then uses X-ray or electron diffraction to study the structure factor in the plane of the film.

Fig. 2 shows the expected diffraction patterns in the normal isotropic fluid, i.e. the S_A

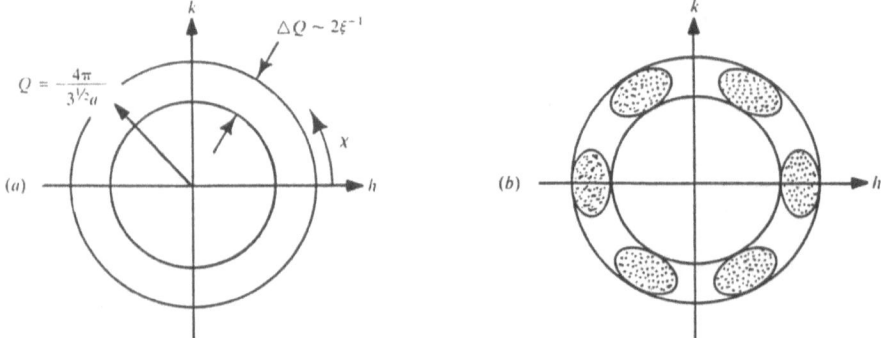

Fig. 2. Schematic illustration of the in-plane diffraction patterns for (a) an isotopic and (b) an orientationally ordered hexatic smectic liquid crystal.

phase (Fig. 2a) and in the hexatic (S_{BH}) phase (Fig. 2b).[19] For the former, one   expects a peak at a momentum transfer $Q \sim 4\pi/\sqrt{3}\,a$, where a is the average molecular separation.

If the fluid is isotropic, the intensity and shape of this peak must be independent of direction. Therefore, if one scans the angular variable χ, one should not see any variation in the strength of the scattering. The width of the ring, ΔQ, should be $\sim 2/\xi_p$ where ξ_p is the length scale on which positional correlations between the molecules decay. Such diffraction patterns are seen ubiquitously in fluids. The hexatic diffraction pattern is illustrated in Fig. 2b; the position and widths ΔQ are as for the fluid phase described above. However, in the BO ordered hexatic phase, the fluid develops a sixfold modulation in the angular variable χ, that is, the angular isotropy is broken in the hexatic phase. In spite of this angular anisotropy, the system is still a fluid, due to the short-range positional order. At a lower temperature the material may freeze, condensing the modulated ring of fluid scattering into sharp spots. Sharp diffraction (Bragg) peaks are the signature of a crystal.

Although hexatic BO order in the S_{BH} phase was demonstrated some time ago,[20] quantitative measurements were difficult due to a variable number of domains in the area probed. This practical problem was resolved by Brock et al[11] by creating single domain samples. This was achieved by looking at the S_I phase of the liquid crystal 8OSI, in a small magnetic field. The magnetic field orients the tilt direction of the S_C phase. The weak coupling between that tilt and the BO order then determines a unique orientation of all the bonds, i.e. a single domain.

Both the X-ray scattering,[7,10,11] and the electron diffraction [6,21] experiments scan the scattering intensity as function of the angle of rotation in the plane, $S(\chi)$. Long range BO order implies a six-fold modulation in $S(\chi)$. Indeed, both types of experiments exhibit such a modulation, on both thick and thin films. Quantitatively, this modulation is analyzed using the Fourier cosine series,[11]

$$S(\chi) = I_0 \left\{ \frac{1}{2} + \sum_{n=1}^{\infty} C_{6n} \cos 6n \, (90^0 - \chi) \right\} + I_{BG} ,\qquad (9)$$

where χ is the angle between the in-plane component of $\vec{q}$ and $\vec{H}$. The coefficients C_{6n}, defined in Eq. (5), measure the amount of 6n-fold ordering in the sample. With the constant term chosen as 1/2 in Eq. (9), the C_{6n} approach 1 for perfect BO order. Each of the C_{6n} is an independent BO order parameter. Both for thick and thin films, the measured values of C_{6n} grow gradually as the temperature is lowered.

Fig. 3 reproduces the temperature dependence of the first seven Fourier coefficients near the $S_C \rightarrow S_I$ transition in a thick film of 8OSI.[7,10,11] The small tails at high temperatures reflect the fact that the coupling to the tilt acts like an ordering field on the BO order parameter (i.e. $h \neq 0$ in Eq. (3)), destroying the sharp phase boundary. However, since this

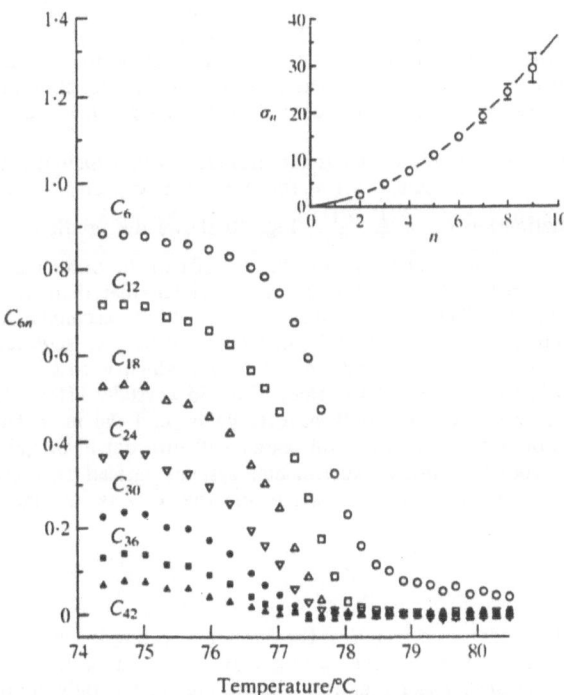

Fig. 3.   First seven Fourier coefficients describing the hexatic ordering in a thick film of 8OSI. The inset shows the exponents σ_n versus n, together with a fit to $\sigma_n = n + 0.295n \, (n-1)$.

field is small, the C_{6n}'s still obey the scaling form (5). Indeed, the data gave excellent fits to $C_{6n} = C_6{}^{\sigma_n}$, and the inset in Fig. 3 shows that the σ_n's agree beautifully with the theoretical prediction (7). Thus, a single experiment on 8OSI yielded eight crossover exponents, which otherwise required many separate complicated experiments.

MEAN FIELD THEORY FOR WEAKLY COUPLED PLANES

Even though the three-dimensional XY-model accurately predicts the experimentally observed hexatic scattering, the model is unsatisfying for two reasons. First, heat capacity data taken on the 8OSI system suggest that the heat capacity exponent, α, is approximately 0.5.[22] Second, the original motivation for the stacked hexatic model of well-ordered smectic liquid crystal phases was the intuitive idea that the weakly coupled smectic layers should display some of the novel physics predicted to occur in two dimensions. It is unsatisfying not to have any remnants of that two-dimensional physics survive in the three-dimensional model.

By applying ideas that were first developed for linear-chain compounds,[23] and more recently applied to the coupling of the CuO_2 planes in high T_c superconductors,[24] both of these objects can be addressed. We begin with the solution for the two-dimensional problem, $C_6 = \chi_{2D} h$ for $T > T_{2D}$. The susceptibility, χ_{2D}, is given by the Kosterlitz-Thouless[25] form $\chi_{2D} \sim \frac{1}{T} \exp[A(T - T_{2D})^{-1/2}]$. The effect of the weak interlayer coupling can be included in mean field theory (MFT) by letting $h_{eff} = h + JC_6$. This implies the equation of state

$$bC_6^3 + a(T)C_6 - h = 0 \qquad (10)$$

where $a(t) = (\chi_{2D}^{-1} - J)$. Both b and h are assumed to be independent of temperature. This equation of state treats the two-dimensional fluctuations precisely; only the three-dimensional aspects are treated in MFT. The free energy, F, is obtained by integrating the equation of state with respect to C_6. C_6 is then the root of the equation of state which minimizes F. The best fit of the order parameter data to the equation of state is shown in Fig. 4a. Clearly, these simple MFT ideas describe the order parameter data quite well.

A more stringent test of this model is the accuracy with which it predicts the form of the heat capacity. A functional form for the heat capacity is obtained from the thermodynamic relationship $C_v = \frac{1}{T} \frac{\partial^2 F}{\partial T^2}$. Fig. 4b shows the predicted form for the heat capacity using the parameters obtained from the best fit to the order parameter data plotted along with the measured heat capacity data. It is evident that there is qualitative but not quantitative agreement. Thus as in other systems, MFT accounts adequately for the temperature dependence of the order parameter but it is less satisfactory for the heat capacity. We believe, nevertheless, that this approach, which is so successful for the CuO_2 planar magnets, will also work well for smectic liquid crystals. However, to describe the heat capacity properly one will certainly have to go beyond the mean field approximation. it seems quite possible that the true critical region will turn out to be quite narrow and that the value $\alpha \simeq 1/2$ found in many experimental systems instead represents the effects of Gaussian pre-critical fluctuations and some of the effects of the two dimensional fluctuations.

THIN FILMS

Although the thermodynamic argument for harmonic scaling in the critical region is valid for all dimensions d, the form of the scaling may vary. In particular, the scaling form $\sigma_n = n + x_n n(n-1)$ becomes exact near four dimensions, and is only approximately valid in three dimensions. As the number of spatial dimensions is reduced, the perturbation series fails to converge and a more complicated expression for σ_n may be expected. However, something can be said about what one expects to see as the system approaches two

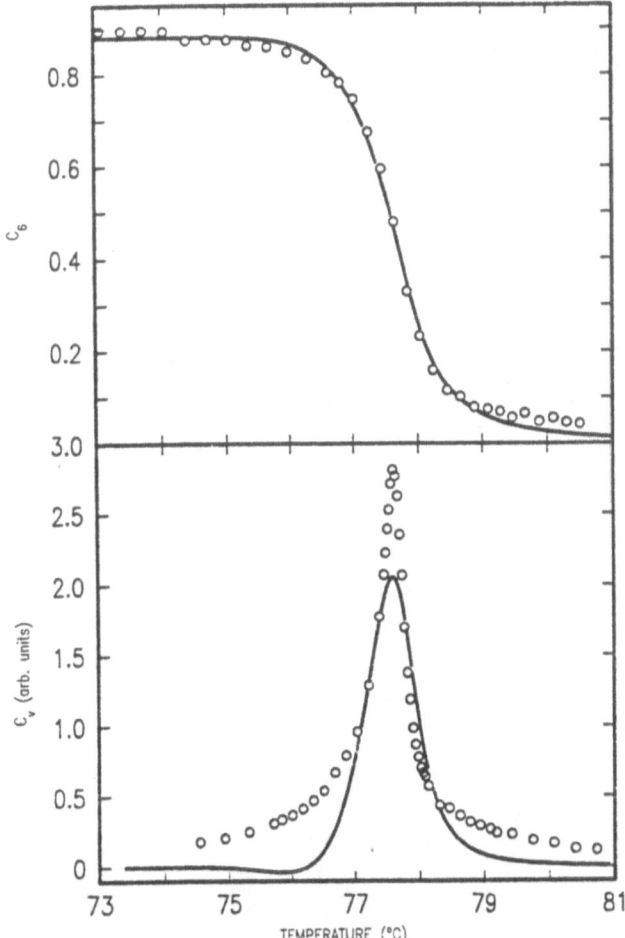

Fig. 4.   (a) Open circles are the experimentally determined values of $C_6(T)$. The solid line is the best fit of the data to a mean field coupled stack of hexatic layers. (b) Open circles are the experimental heat capacity data of reference 22. The solid line is the predicted form for $C_v(T)$ using the parameters obtained by the best fit to the $C_6(T)$ data above.

dimensions. As the volume of phase space is reduced, the amplitude of fluctuations grows. In two dimensions, these fluctuations are so large, they produce an algebraic decay of the (formerly long-range ordered) correlation functions (see Eq. (2)). An algebraically decaying correlation function, in turn, implies that the susceptibility to an ordering field has the functional form, $\chi_{6n} \sim L^{2-\eta n^2}$, where L is the sample size and $\eta_{6n} = \eta n^2$ is a simple consequence of a harmonic theory for the phase fluctuations of Ψ_6 ($\eta = \frac{1}{4}$ in the Kosterlitz–Thouless theory.) This implies that for $n \geq 3$, χ_{6n} approaches zero and we do not get long-range, 6n-fold BO order. Therefore, in an X-ray scattering experiment on an infinite system, in the presence of an ordering field, one should expect to see only the first few harmonics. In a finite sized system, with or without an applied field, one expects $C_{6n} \sim C_6^{n^2}$,[26] giving a similar n^2 scaling. Therefore, as the system crosses over from three dimensions to two dimensions, one expects, quite generally, to see the higher harmonics of the fundamental order parameter "turn off". Analysis of the thin film data in Ref. 7 indeed indicates the absence of all but the n=1 term.

Two-dimensional behavior was also seen by Cheng, et al.[6,21] In a beautiful transmission electron diffraction experiment studying very thin films, they were able to measure C_{6n} for n=1,2,3 and 4 samples of 12 wt.% of 4-proprionylphenyl-trans-(4-n-pentyl) cyclohexane carboxylate (PP5CC) in 650BC, with 2, 4 and 6 layers. This system exhibits S_A and S_{BH} phases. Cheng et al[21] fit their diffraction patterns to $C_{6n} = C_6^{\sigma_n}$, with $\sigma_n = n + \lambda n(n - 1)$. When forced to fit the data on thin films, λ becomes temperature dependent, and decreases to values smaller than one ($\sigma_n < n^2$) with increasing number of layers.[21] Thus it appears that even with four layers, describing the diffraction patterns requires a better understanding of the interlayer couplings.[12]

In two dimensions, the fluctuations in the amplitude Ψ_0 of Ψ_6 (which led to the result[9] $\sigma_n = n + \lambda n (n - 1)$) are negligible,[26] and we set $\Psi_0=1$. For a film of ℓ layers, we denote the angle at site $\vec{r}$ on the i'th layer by $\theta_i(\vec{r})$. Ignoring topological defects, the energy cost of slowly varying fluctuations is described by the effective (spin-wave) Hamiltonian[12]

$$H = \int d^2r \left[\frac{K_A}{2} \sum_{i=1}^{\ell} (\nabla\theta_i)^2 + \frac{J}{2} \sum_{i=1}^{(\ell-1)} (\theta_{i+1} - \theta_i)^2 \right]. \tag{11}$$

K_A is the effective Frank constant[2] in each of the layers, while J measures the interplane coupling. Since we are dealing with a Gaussian theory, the normalized Fourier coefficients of the diffraction intensity are given by

$$C_{6n} = \frac{1}{\ell} \sum_{i=1}^{\ell} < e^{6i\theta_i(\vec{r})} > = \frac{1}{\ell} \sum_{i=1}^{\ell} e^{-18n^2 \langle \theta_i^2 \rangle} =$$

$$= \frac{1}{\ell} \sum_{i=1}^{\ell} e^{-A_i^{(\ell)} n^2}, \tag{12}$$

where $\sum_{i=1}^{\ell} /\ell$ explicitly deals with the average over the layers. Due to the free boundary conditions, the fluctuations $\langle \theta_i^2 \rangle$ will differ from layer to layer ($\langle \theta_i^2 \rangle$ is expected to decrease from outside layers to inside layers). Clearly the scaling $C_{6n} = C_6^{\sigma_n}$ is no longer valid, and in fact $\left[\text{due to the symmetry } \langle \theta_i^2 \rangle = \langle \theta_{\ell-i+1}^2 \rangle \right] C_{6n}$ should be fitted to the average of $\ell/2$ (ℓ even) or $(\ell + 1)/2$ (ℓ odd) factors of $\exp(-A_i^{(\ell)} n^2)$.

To calculate $\langle \theta_i^2 \rangle$ from Eq. (11) we need to construct the eigenmodes. It is easy to check that the transformation to normal modes

$$\phi_0(\vec{r}) = \frac{1}{\sqrt{\ell}} \sum_{i=1}^{\ell} \theta_i(\vec{r}) \tag{13a}$$

and

$$\phi_m(\vec{r}) = \sqrt{\frac{2}{\ell}} \sum_{i=1}^{\ell} \cos\left[\frac{m\pi}{\ell}\left(i - \frac{1}{2}\right)\right] \theta_i(\vec{r}) \qquad (m=1,2...,\ell-1), \qquad (13b)$$

leads to a Hamiltonian

$$H = \int d^2 r \sum_{m=0}^{(\ell-1)} \left[\frac{K_A}{2} (\nabla\phi_m)^2 + 2J \sin^2\left(\frac{m\pi}{2\ell}\right)\phi_m^2\right]. \qquad (14)$$

Inverting the transformation in Eq. (13) leads to

$$A_i^{(\ell)} = 18 \langle\theta_i^2\rangle = \frac{18}{\ell}\left[\langle\phi_0^2\rangle + 2 \sum_{i=1}^{(\ell-1)} \cos^2\left[\frac{m\pi}{\ell}\left(i - \frac{1}{2}\right)\right]\langle\phi_m^2\rangle\right]. \qquad (15)$$

The averages of the normal mode amplitudes are easily calculated from Eq. (14). For the zeroth made

$$\langle\phi_0^2\rangle = \int \frac{d^2 q}{(2\pi)^2} \frac{1}{K_A q^2} = \frac{1}{4\pi K_A}\ln\left(\frac{L}{a}\right)^2, \qquad (16a)$$

where $\left(\frac{L}{a}\right)$ is the size of the system in units of molecular distance a. For the other modes

$$\langle\phi_m^2\rangle = \int \frac{d^2 q}{(2\pi)^2} \cdot \frac{1}{K_A q^2 + 4J \sin^2(m\pi/2\ell)} =$$

$$= \frac{1}{4\pi K_A}\ln\{[1+\tilde{J}(\ell,m)]/[(a/L)^2 + \tilde{J}(\ell,m)]\}, \qquad (16b)$$

when $\tilde{J}(\ell,m) = 4(J/K_A)(a/\pi)^2 \sin^2(m\pi/2\ell)$. Equations (12),(15) and (16) summarize our results[12] for the coefficients C_{6n} needed to fit the diffraction pattern, completely characterized by the three parameters K_A, Ja^2 and (L/a). For systems of five layers of more (with three or more $A_i^{(\ell)}$'s) it is therefore in principle possible to extract the effective Frank constant K_A, and to follow its temperature dependence. This information can then be used to test predictions of defect mediacted transition theories[2]. For example the exponent η for decay of correlations ($\langle\Psi_6(r)\Psi_6^*(0)\rangle \sim r^{-\eta}$) can be obtained from the L dependence of C_6 in Eqs. (15) and (16) and equals $\eta = 18/(\pi\ell K_A)$. Close to the hexatic to liquid transition it should have a universal drop[2] from $\eta^- = \left[1 - b\sqrt{T_c - T}\right]/4$ to $\eta^+ = 0$. Near the hexatic to solid transition η should vanish as ξ^{-2}, where ξ is the diverging correlation length at this transition[2]; $\xi \sim \exp[b'|T - T_m|^{-0.3699634}]$.

To obtain some insight into the behavior of $A_i^{(\ell)}$, we examine the limit of weak interplane coupling, $(a/L)^2 \ll Ja^2/K_A \ll 1$. In this limit

$$A_i^{(\ell)} \simeq \frac{9}{2\pi\ell K_A}\left[2\ln\left(\frac{L}{a}\right) + (\ell - 1)\ln\left(\frac{K_A \pi^2}{4Ja^2}\right) + a_i^{(\ell)}\right], \qquad (17a)$$

with

$$a_i^{(\ell)} = -4 \sum_{m=1}^{(\ell-1)} \cos^2\left[\frac{m\pi}{\ell}\left(i-\frac{1}{2}\right)\right] \ln\left[\sin\left(\frac{m\pi}{2\ell}\right)\right]. \qquad (17b)$$

From Eq. (17a) we see that on increasing the number of layers a crossover to three dimensional behavior $\left[A_i^{(\ell)}\right.$ independent of $\left.\ell\right]$ occurs for $\ell \gtrsim \ell_x \sim 2\ln\left[\frac{L}{a}\right]/\ln\left[\frac{K_A\pi^2}{4Ja^2}\right]$.

However, the coefficients $A_i^{(\ell)}$ in Eq. (17b) decay slowly as $1/i$ on moving away from the surface. This is the characteristic decay of correlation functions in three dimensional Gaussian models. In order to study the full crossover to the three dimensional result $C_{6n} \sim C_6^{n+\lambda n(n-1)}$ close to the transition, one must also include amplitude fluctuations.[10]

Another interesting crossover involves the variation of the transition temperature with J. For J=0, the layers are decoupled and topological defects (occurring independently in each layer) become relevant when $\eta = 18/(\pi K_A) = 1/4$, independent of ℓ. For $J \to \infty$, all layers behave as one and $\eta = 18/(\pi\ell K_A)$; the defects now extend over all layers and the critical value of $K_A = 72/(\pi\ell)$ decreases with ℓ. Note, however, that K_A is the effective rigidity after renormalization by topological defects. Since the experimental transition temperatures do not vary strongly with the number of layers ℓ, we expect that renormalization of K_A by defects is quite strong, and depends on ℓ. How does the crossover from independent defects in each layer at J=0, to a line of defects crossing through ℓ layers as $J \to \infty$ occur? For small J, the interplane energy cost of an isolated defect (on a single plane) increases roughly as $J\lambda^2$ over a distance λ. Equating this result with the energy cost of $2\pi K_A \ln(\lambda)$ for creating vortices in neighboring layers, leads to the conclusion that between two layers a "defect line" can move at most by $\lambda \sim (K_A/J)^{1/2}$. Naturally as $J \to 0, \lambda$ increases and eventually saturates to L. The partition function for a single defect now behaves as

$$Z_1 \sim (L/a)^{-\pi\ell K_A/36}\left[\frac{L}{a}\right]^2 \left[\frac{K_A}{J}\right]^{(\ell-1)}$$, where the first factor is the energy cost, the second

factor is a center of mass entropy , and the final term comes from the entropy of fluctuations of the defect line in between layers. The transition condition is obtained roughly by setting $Z_1 \sim O(1)$: For $L \gg \lambda$, $K_A^c = 72/(\pi\ell)$ at the transition, while for $\lambda \sim L$, $K_A^c = 72/\pi$; and in between the result is effected by the finite size L.

As the hexatic to liquid transition is approached, K_A is renormalized by the topological defects, in a way that may depend on ℓ. Therefore, it is not clear how to compare the values of $A_i^{(\ell)}$ for different ℓ's. Away from the transition, the spin wave approximation described above may give the correct answers for all ℓ, with the same values of the parameters K_A, Ja^2 and (L/a). This predicts strong relations between the C_{6n}'s for different layers thicknesses.

In Ref. 12, we refitted the data of Ref. 21 to our Eqs. (12) and (17), and found good consistency. It is thus possible to use such experiments to measure the temperature dependence of K_A.

CONCLUSION

The study of stacked hexatics has yielded a complete classification of the large variety of thermotropic liquid crystal phases. It has also presented a single experiment which yields a simultaneous measurement of many crossover anisotropy exponents, previously obtained from separate complex experiments. It is also an excellent system to reveal dimensional crossover and the interplay between two- and three-dimensional physics. For very thin layers, the stacked hexatic layers can be used to extract the hexatic stiffness, or Frank constant, and thus test predictions of defect mediated transition theories.

ACKNOWLEDGEMENTS

Most of the work reviewed here was done in collaboration with J. D. Brock, R. J. Brigeneau and J. D. Litster. The work on thin layers was done with M. Kardar.

The liquid crystal research at MIT has been supported by the National Science Foundation-Materials Research Laboratory under contract number DMR 84-18718 and by NSF grant DMR 86-19234; that at Tel Aviv by grants from the US - Israel Binational Science Foundation and the Israel Academy of Science and Humanities .

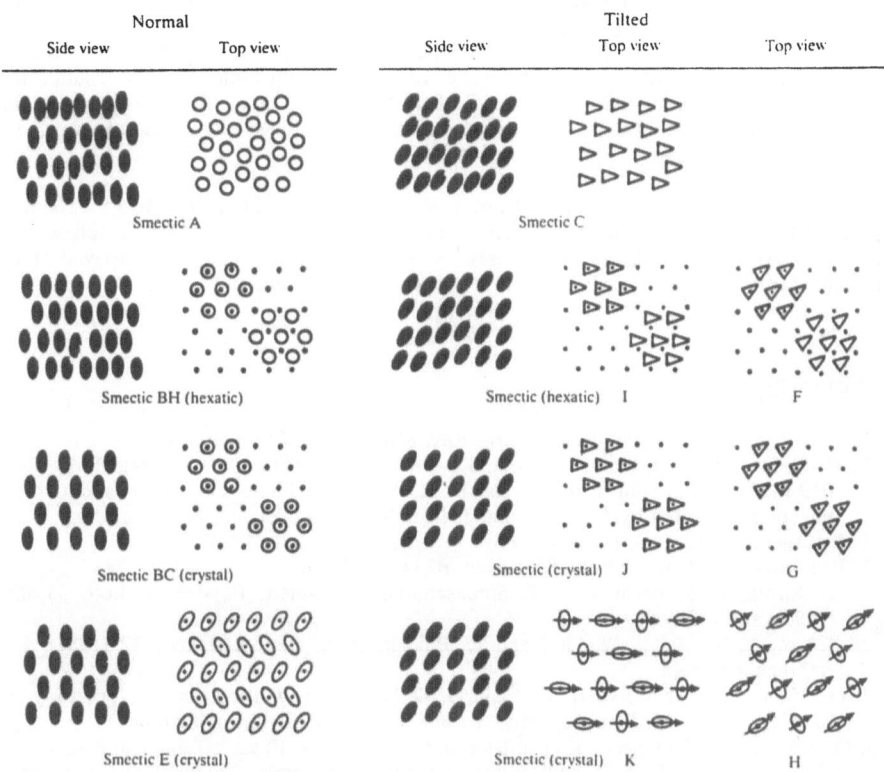

Figure A.1.   Schematic illustration of the order in the principal smectic liquid crystal phases. Figure is form Ref. 10, adapted from Ref. 9.

APPENDIX. SMECTIC PHASES

Smectic liquid crystals are distinguished by having an intermediate degree of positional order in addition to molecular orientational and, in some cases, bond orientational order.

Smectics have historically been identified by the textures they exhibit under a polarizing microscope and by miscibility studies with known phases.[9]

The simplest smectic phase is the smectic A(S_A) phase. This phase has traditionally been described as a system that is a solid in the direction along the director and a fluid normal to the director, or equivalently, as stacked two-dimensional fluids; it is more properly described as a one-dimensional density wave in a three-dimensional fluid with the density wave along the nematic director. The S_C phase is similar except the density wave vector makes a finite angle with the director. In both S_A and S_C phases there is complete translational symmetry perpendicular to the density wave vector.

The remaining smectic phases all possess more order than the S_A and S_C phases. One of two phases, formerly labelled the smectic B phase, is now believed to be an example of a stacked hexatic phase (S_{BH}). In each smectic layer, the director is parallel to the smectic density wave, as in the S_A phase, and there is short-range positional order in the smectic plane; however, there is long-range bond orientational order in the smectic plane. The S_I and S_F phases are similar to the S_B (hexatic) phase, but, as in the S_C phase, the director makes a finite angle with the density wave vector. As illustrated in the figure, the distinction between the S_I and S_F phase is that the projection of the director onto the smectic planes points towards a near neighbor in the S_I phase and between two near neighbors in the S_F phase.

Continuing to increase the amount of order, we leave true liquid crystal phases. The S_{BC} (crystal) phase is actually a three-dimensional crystal. Similarly, the S_J and S_G phases are crystalline versions of the S_I and S_F phases. Further phases (S_E, S_H, S_K) are arrived at by including additional types of ordering such a herring bone packing and so on.

REFERENCES

1. For a recent review, see K. J. Stranburg, Rev. Mod. Phys. 60:161 (1988).
2. D. R. Nelson and B. I. Halperin, Phys. Rev. B19:2457 (1979); A. P. Young, Phys. Rev. B19:1855 (1979). For a review see D. R. Nelson in Phase Transitions and Critical Phenomena, edited by C. Domb and J. L. Lebowitz (Academic, NY, 1983), p.1.
3. R. J. Birgeneau and J. D. Litster, J. Phys. (Paris), Lett. 39:399 (1978).
4. R. Bruinsma and D. R. Nelson, Phys. Rev. B23:402 (1981).
5. E. B. Sirota, P. S. Pershan, L. B. Sorenson, and J. Collett, Phys. Rev. Lett. 55:2039 (1985); and references therein.
6. M. Cheng, J. T. Ho, S. W. Hui, and R. Pindak, Phys. Rev. Lett. 59:1112 (1987); and references therein.
7. J. D. Brock, D. Y. Noh, B. R. McClain, J. D. Litster, R. J. Birgeneau, A. Aharony, P. M. Horn, and J. C. Liang, Z. Phys. (submitted); and references therein.
8. J. D. Brock, R. J. Birgeneau, J. D. Litster and A. Aharony, Physics Today (in press).
9. G. W. Gray and J. W. Goodby, Smectic Liquid Crystals: Textures and Structures (Leonard Hill, Glasgow and London, UK, 1984).
10. A. Aharony, R. J. Birgeneau, J. D. Brock, and J. D. Litster, Phys. Rev. Lett. 57:1012 (1986).
11. J. D. Brock, A. Aharony, R. J. Birgeneau, K. W. Evans-Lutterodt, J. D. Litster, P. M. Horn, G. B. Stephenson, and A. R. Tajbakhsh, Phys. Rev. Lett. 57:98 (1986).
12. A. Aharony and M. Kardar, Phys. Rev. Lett. 61:2855 (1988).
13. D. R. Nelson and B. I. Halperin, Phys. Rev. B21:5312 (1980).
14. e.g., A. Aharony, in Phase Transitions and Critical Phenomena, edited by C. Domb and M.S. Green (Academic, New York, 1976), 6:357.
15. H. Rohrer and Ch. Gerber, Phys. Rev. Lett. 38:909 (1977).
16. A. Aharony, K. A. Müller and W. Berlinger, Phys. Rev. Lett. 38:33 (1977).
17. A. D. Bruce and A. Aharony, Phys. Rev. B11:478 (1975).
18. C.Y. Young, et al., Phys. Rev. Lett. 45:1193 (1980); R. Pindak et al., Phys. Rev. Lett. 45:1193 (1980); C. Rosenblatt et al., Phys. Rev. Lett. 46:140 (1980).

19. D. E. Moncton and R. Pindak, Phys. Rev. Lett. 43:701 (1979); D. E. Moncton et al., Phys. Rev. Lett. 49:1865 (1982).
20. R. Pindak et al., Phys. Rev. Lett. 46:1135 (1981).
21. M. Cheng et al., Phys. Rev. Lett. 61:550 (1988).
22. C. W. Garland, J. D. Litster and K. J. Stine (unpublished).
23. D. J. Scalapino, Y. Imry and P. Pincus, Phys. Rev. B11:2042 (1975). See also J. D. Axe, in Ordering in Strongly Fluctuating Condensed Matter Systems, edited by T. Riste (Plenum Press, New York, 1980), p.399.
24. T. Thio et al., Phys. Rev. B38:905 (1988).
25. J. M. Kosterlitz and D. J. Thouless, J. Phys. C6:1181 (1973); T, M, Kosterlitz, J. Phys. C7:1046 (1974).
26. M. Paczuski and M. Kardar, Phys. Rev. Lett. 60:861 (1988).

MOLECULAR MOTIONS OF HYDROCARBON CHAINS (DECYLAMMONIUM, n-NONADECANE) IN THEIR DYNAMICALLY DISORDERED PHASES : AN INCOHERENT NEUTRON SCATTERING STUDY

F. Guillaume[1], C. Sourisseau[1], and A. J. Dianoux[2]

[1] Laboratoire de Spectroscopie Moléculaire et Cristalline
CNRS URA 124, 351 cours de la Libération, 33405 Talence, France
[2] Institut Laue-Langevin, 156X, 38042 Grenoble Cedex, France

I - Introduction

It is well known that many compounds containing hydrocarbon chains such as liquid crystals or lipid membranes undergo solid state phase transitions prior to melting. In these intermediate phases, between the completely ordered crystal and the isotropic liquid, large amplitude motions of the hydrocarbon chains are effective and give rise to a strong dynamical disorder.

We have undertaken a detailed study on the dynamics of hydrocarbon chains in crystalline model compounds, namely decylammonium manganese tetrachloride and n-nonadecane by means of Incoherent Neutron Scattering (INS). INS technique is a very powerful tool because the molecular motions can be readily analysed in the incoherent quasi-elastic region of the spectra[1] (small energy transfers between the incoming neutrons and the protons). For bonded motions, the "quasi elastic" spectra display an elastic peak and a quasi-elastic broadening centered at the zero energy transfer value. The relative intensities of these bands vary as a function of the Q momentum transfer and the incoherent scattering law for a single scatterer can be written as follows:

$$S(Q,\omega) = A_0(Q).\delta(\omega) + \sum_i A_i(Q) L_i(\omega)$$ [1]

where $A_0(Q)$, called Elastic Incoherent Structure Factor (EISF), can be simply evaluated from the experimental intensities of the elastic $I_{el}(Q)$ and quasi-elastic $I_{qe}(Q)$ contributions ;

$$A_0(Q) = \frac{I_{el}(Q)}{I_{el}(Q) + I_{qe}(Q)}$$ [2]

The second term of equation [1], where $L_i(\omega)$ is a Lorentzian function of amplitude $A_i(Q)$, is responsible for the quasi-elastic broadenings of the spectra, related to the kinetics of the molecular motions. The analysis of an experimental spectra thus consists in simulating the profiles as a function of both the energy and momentum transfer variations by means of equation [1].

Because of space limitations of this paper, the main results will be discussed and more details can be found in forthcoming publications[2,3].

II - Decylammonium tetrachloride compounds $(C_{10}H_{21}NH_3)_2MCl_4$ ($C_{10}M$ in short, with M^{2+} = Mn, Cd,...)

These two dimensional compounds are built of corner sharing MCl_6 octahedra parallel to the (a,b) planes. In between these planes, there are two decylammonium chains linked to the MCl$_6$ octahedra through NH...Cl hydrogen bonds and to each other through van der Waals interactions between the -CH$_3$ end groups : therefore, double layers of organic chains are intercalated between the inorganic sheets. The C$_{10}$Mn compound undergoes a solid state phase transition at 308 K. In the low temperature ordered phase (LT), the hydrocarbon tails are tilted (40°) with respect to the normal of the layers, because of a single gauche form (g) localized in between the the second and the third carbon atom. In the high temperature phase (HT), strong modifications are observed in both the molecular ordering and the structure. The chain axes become, in average, perpendicular to the layers and it has been suggested that kink defects (gt$_{2n+1}$g'), diffusing along the chain axis, are responsible for the conformational disorder. Further works using vibrational spectroscopy on the C$_{10}$Cd derivative have shown that kink forms are effective and well localized within the chains. Furthermore, studies of chain motions in phospholipids have shown that there is a gradient of disorder on going from the hydrophylic core to the -CH$_3$ chain ends and it has been suggested, from computer simulation studies, that cooperative torsions of the -(CH$_2$)- groups along the chain axis were taking place : both models of cooperative torsions and of kink formation should be taken into account. In this context, the results of our calculations for an isolated [C$_{10}$H$_{21}$NH$_3$]$^+$ cation indicated that kink defects are more stable near the methyl end of the chains and that the most probable form was the t$_4$gtg't one. Then, we have elaborated a dynamical model taking account for the gradient of disorder on going from the -NH$_3$$^+$ polar heads to the -CH$_3$ ends, so that the volumes visited by the -(CH$_2$)- units follow a distribution law varying with respect to their position in the chain. Within the framework of a pure kink model, the distribution function should display a step-like behaviour while its profile would be continuous in the framework of cooperative torsions.

Under these conditions, the volume visited by each -(CH$_2$)^i- unit is characterized by a sphere, its radius R$_i$ beeing the parameter characteristic of the disorder :

$$R_i = \frac{R_0}{1 + exp(-X_i)} \qquad with \ X_i = \alpha(i - \beta) \qquad\qquad [3]$$

Considering now the scattering law, we have made use of the diffusion model of a particle in a sphere[5] of radius R$_i$.

Using a wide range of instrumental resolutions, we were able to detect motions over the τ < 700 ps time range. It came out that the onset of all the molecular processes was in the 1-30 picosecond time scale. Fitting our model to the experimental profiles, we were able to extract the distribution functions, as shown in figure 1, at various temperatures. It must be pointed out that, just above the phase transition temperature, the shape of curve (c) confirms a change in the conformation of the cation associated with the structural modification : we conclude that kink defects occur and correspond to the most probable t$_4$gtg't form. As the temperature rises, the profiles (curves a and b) are modified and they correspond to a model in which cooperative torsions are dominant. Finally, a diffusion coefficient of roughly 6.10^{-6}.cm^2.s^{-1} is estimated.

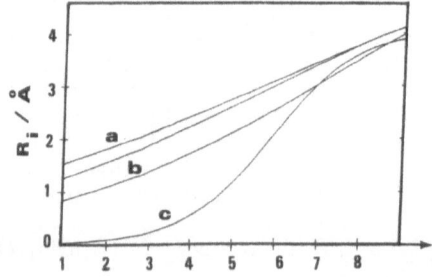

Fig.1. Experimental distribution functions as evaluated from the quasi-elastic profiles of C$_{10}$Mn at 370 K (a), 330 K (b) and 315 K (c).

III- n-nonadecane compound

Nonadecane undergoes a phase transition at 295.2 K from an ordered solid state to a R_I "rotator" phase. In the low temperature crystalline state, $C_{19}H_{40}$ crystallises in an orthorhombic layered structure composed of molecules parallel to each other and in their "all trans" conformation. One can distinguish, in a horizontal section of the layers, two sub-lattices A and B according to two different orientations of the chains, about 85° apart. The a/b ratio of the lattice parameters increases markedly in the R_I phase and tends to the "hexagonal" value $\sqrt{3}$, only reached in the R_{II} phase of longer chain compounds[6].

The infrared and Raman spectra[7] of such systems have shown that in the R_I phase of n-nonadecane, the carbon chains are almost in their "all trans" extended form and display gauche defects localized only at the chain ends.

Many investigations on n-nonadecane using various techniques including ^{2}H-NMR and INS, led to very contradictory results[4]. However, rotational (about the main chain axis) as well as translational (in the chain axis direction) motions occurring on similar time scales were clearly evidenced for longer chain compounds[8]. We have thus intended to clarify the situation by performing new experiments on semi-oriented samples in order to discriminate between the rotational and translational dynamics of n-nonadecane in the R_I phase.

Assuming that translational and rotational processes are not correlated, the total scattering law will take the following form:

$$S(Q, \omega) = A_0^{rot}(Q) . A_0^{trans}(Q) . \delta(\omega) + \sum_{i,j} A_i^{rot}(Q) . A_j^{trans}(Q) L_{i,j}(\omega) \qquad [4]$$

where $A_i^{rot}(Q)$ and $A_j^{trans}(Q)$ are the structure factors corresponding to rotational and translational motions, respectively.

First of all, in agreement with all the previous experiments on n-nonadecane, we did not observe any quasi-elastic broadening in the ordered crystalline phase but large amplitude motions were evidenced in the R_I rotator phase. The comparison between the experimental EISF and various dynamical models already proposed in the literature shows that a model involving rotational diffusion in an effective 2-fold potential (around the chain axis) and a restricted translational diffusion (along the chain axis) should be adopted. The equilibrium distribution function of the positions of the scatterers moving on a circle of radius r will depends on the potential barrier height γ and on the angle between the two preferential sites ; for the translation, we have to consider a particle diffusing between two impermeable walls[9] separated by a length L.

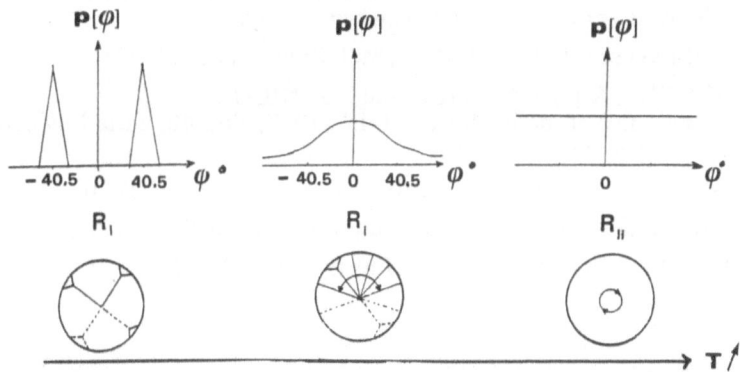

p(φ) is the distribution function of the positions of the chains.

Fig.2. Statement of the results for rotational motions in n-alkanes
(R_I and R_{II} phases).

We have thus carried out the INS study using these models. According to equation [4], very good fits of the spectra were obtained assuming, in agreement with literature data, a rigid chain in its central part and only end-gauche defects. From the whole sets of the best fit parameters, the following conclusions can be drawn :

- For the rotational motions, the angle in between the two prefered orientations is found to be about 81°, a value in agreement with the orientations of the chains in the crystalline phase. Just above the phase transition, the chains display a jump-like behaviour among the two prefered sites and reach, upon increasing the temperature, an intermediate regime corresponding to large oscillations about a mean position. However, preliminary experiments performed in the R_{II} phase of longer chain compounds have shown that this model also applies nicely and that the rotational motions reach the uniform distribution limit (figure 2).

- For translational motions, the mean value of the translational diffusion coefficient is 6.10^{-6} cm^2 s^{-1} at 300 K and it is of the same order of magnitude as those already proposed for n-tricosane or n-tritriacontane. Finally, it is noteworthy that the diffusion length ($L = 2.7$-2.8 Å) corresponds roughly to two -CH$_2$- units, a result which supports the formation of end gauche defects within the interlamellar space, the chains remaining in an almost extended form in their central part.

IV - Concluding remarks

The approach we have developped in this study could be applied to other systems in which the hydrocarbon chains are known to be very flexible (micellar systems, for instance). Further works are needed in particular to developp new theoretical models and new experiments on crystalline models of bilayers would be of first importance.

REFERENCES

1- M.BEE in "Applicaton of quasi-elastic neutron scattering to solid state chemistry, biology and material science", (Adams and Hilger, 1988).

2- F.GUILLAUME, G.CODDENS, A.J.DIANOUX, W.PETRY, M.REY-LAFON and C.SOURISSEAU, Mol. Phys., 1989, (in the press).

3- F.GUILLAUME, J.DOUCET, C.SOURISSEAU and A.J.DIANOUX, J.Chem.Phys., 1989, (in the press).

4- F.GUILLAUME, J.DOUCET, C.SOURISSEAU and A.J.DIANOUX, in "Polymer motions in dense systems", 1987, Vol.29, edited by D.Richter and T.Springer (Springer Proceedings in Physics).

5- F.VOLINO and A.J.DIANOUX, Mol. Phys., 1980, 41, 271.

6- J.DOUCET, A.F.CRAIEVICH and I.DENICOLO in "Dynamics of molecular crystals", 1987, Vol. 46, edited by J.Lascombe (Elsevier Science Publishers).

7- M.MAISSARA and J.DEVAURE, J.Raman.Spect., 1987, 18, 184.

8- B.EWEN and D.RICHTER, J.Chem.Phys., 1978, 69(7), 2954.

9- P.L.HALL and D.K.ROSS, Mol.Phys., 1981, 42, 673.

ANCHORING TRANSITIONS AT CRYSTAL-NEMATIC INTERFACES

P. Pieranski and B. Jérôme

Laboratoire de Physique des Solides
Bât. 510, Faculté des Sciences
91405 Orsay, France

I. ANCHORING OF NEMATICS

"*On obtient presque toujours des orientations définies quand on place une goutte de liquide anisotrope sur les clivages des cristaux ou dans les fissures de clivage*".

In these terms C. Mauguin[1] reported in 1913 the first observation of the orienting action of freshly cleaved surfaces of the muscovite mica on a nematic liquid crystal p-azoxyanisole :

$$CH_3 - O -\!\langle O \rangle\!- N = N -\!\langle O \rangle\!- O - CH_3$$
$$\diagdown_{O}\diagup$$

The discovery of Mauguin was confirmed through an extensive study by Grandjean[2] who tested the orienting action of ten different minerals on five liquid crystalline materials.

Today, the orienting action of any interface (nematic/another phase) on the common direction $\vec{n}$ of molecules in the bulk of the nematic phase, which (in the absence of external fields) would otherwise be arbitrary, is called shortly <u>the anchoring</u>. The subject was revived several times from practical[3] and more fundamental points of view[4,5]. Our purpose here is only to focus on two interrelated typics :

1° symmetries of substrates and types of anchoring,
2° anchoring transitions at nematic/crystal interfaces.

The notions implied in these two topics do not seem to be very obscure,

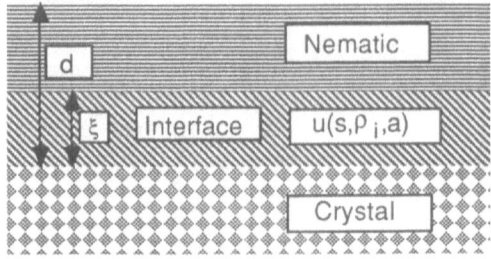

Fig. 1. Nematic layer of thickness d deposited on the surface of an anisotropic substrate. The interface is subjected to heat and particle exchange with the gas above the nematic layer.

nevertheless they deserve a short introduction in terms of the Gibbs-type approach of the surface thermodynamics[6].

We consider an interface between a nematic liquid crystal and a solid substrate (Fig. 1).The interface of thickness ξ has a structure with a composition different from both bulk phases (crystal/nematic) on its two sides and is characterized thermodynamically by the excess intensive variables such as the concentrations ρ_i of chemical species, the entropy s, the internal energy u, etc...[6]. In the present case, where an anisotropic liquid meets an anisotropic solid at the interface, the internal energy u is not only a function of s and ρ_i but also depends on the direction of the nematic with respect to the anisotropic solid :

$$u = u\left(s, \rho_i, \vec{a}\right) \qquad\qquad\qquad \text{I-1}$$

For fixed excess entropy s and densities ρ_i, the anchoring direction $\vec{a}$ minimizes u. More precisely, the interface is subjected to heat and particle exchange with bulk phases so that the grand canonical potential

$$\omega = u\left(s, \rho_i, \vec{a}\right) - \sum_i \mu_i\rho_i - Ts \qquad\qquad\qquad \text{I-2}$$

should be minimized instead of u.With the area A of the interface, the chemical potentials μ_i and the temperature T supposed to be constant, one has:

$$\left(\frac{\partial \omega}{\partial \vec{a}}\right)_{T,\mu_i} = 0 \qquad\qquad\qquad\qquad\qquad\qquad\text{I-3}$$

The two above topics can now be conveniently specified by asking two questions :

1° For fixed T and μ_i, what is the shape of the potential $\omega(\vec{a})$; how many minima $\vec{a}_\alpha^{min}$ has it ?

2° For varying T and μ_i, does the anchoring directions $\vec{a}_\alpha^{min}$ vary and, if so, do anchoring transitions exist ?

II. TYPES OF ANCHORINGS

II.1. Spread droplet patterns[7,8]

The simplest test detecting the number and respective orientations of anchorings produced by some substrate consists in depositing a drop of a nematic on the tested surface. As pointed out in reference [8], in this test the whole space of wetting parameters is explored and all available anchoring directions are revealed by the texture of the droplet observed in a polarizing microscope (Fig. 2).

The upper surface of the nematic droplet being free, the azimuthal angle φ of the director (with respect to some axis in the plane of the substrate) is imposed by the anchoring $\vec{a}_\alpha$ at the nematic/crystal interface. Using this method several types of anchorings (monostable and multistable) have been found on four types of anisotropic substrates :

a – Freshly cleaved crystal surfaces [1,2,7,9,10]: Layered minerals such as phyllosilicates or gypsum ($CaSO_4,2H_2O$) have only one plane of perfect cleavage : (010) in the gypsum and (001) in micas[11]. In such a case, surfaces obtained by cleavage are composed of macroscopically flat (free of steps) terraces. In the absence of surface reconstruction, patterns of atoms on these terraces must be invariant with respect to those of the symmetry axes or planes which are orthogonal to the surface. These symmetry elements form a group G_s which is a subgroup of G_p - the point group isogonal with the space group of the crystal.

As an example of multistable anchorings produced by such cleaved surfaces we show in Fig. 3 a photograph of a nematic droplet (E9 from BDH) spread on a cleaved surface of the phlogopite mica[10]. As explained in reference [10] the surface of the phlogopite

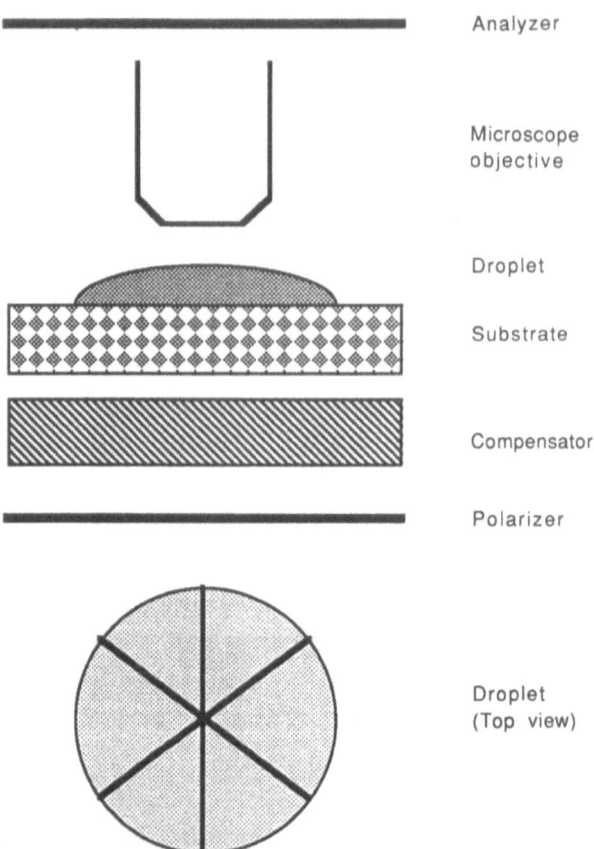

Fig. 2.   Spreading droplet test for multistable anchorings

mica has an approximate symmetry C_{6v}. It is made of oxygen atoms disposed in an almost perfect Kagome lattice; half of the hexagonal cavities of the lattice are filled at random by potassium ions K^+.

The pattern in Fig. 3 consists of six radial sectors. The anchorings $\vec{a}_1,...,\vec{a}_6$ are planar (parallel to the surface), bisect angles of the sectors and make angles $(\vec{a}_\alpha, \vec{a}_{\alpha+1})$ of 60°. Such a pattern is compatible with the C_{6v} symmetry expected for this type of three-octahedral micas [10].

b - <u>Surfaces of crystals obtained by cutting and polishing</u>[12]. Surfaces realized by this method are much less perfect on the atomic scale then those obtained by cleavage. Their average direction can be only approximately parallel to crystal planes, say (hkl). For the disalignment angle $\bar{\alpha} \approx 10^{-2}$ rad, the surface is made of terraces of an average width $w = h_{hkl}/\bar{\alpha}$, where h_{hkl} is the height of steps between adjacent terraces. For $h \approx 1 Å$

Fig. 3.   Pattern in a droplet of the nematic E9 spread on a surface of the phlogopite mica.

one gets $w \approx 10^2$Å. Besides this average disalignement the surface obtained by polishing must also have some microscopic relief corresponding to a local disalignement of amplitude $\delta\alpha$ which is another source of terraces and steps. The thickness ξ of the interface, being defined as the distance necessary for the nematic to recover its perfect order (perturbed by interactions with steps and terraces), one can estimate its order of magnitude as w. The anchoring produced by such a thick interface is a result of an average action of individual terraces. The averaging procedure is simple when all terraces have identical patterns. But this is not the case when crystals have non-symmorphic symmetries (screw axes or glide planes) such as C_{2h}^6 (gypsum and muscovite mica) or O_h^7 (the silicon). For example, on the (100) surface of a polished silicon wafer one expects to have two kinds of terraces, each of them having a structure with only the twofold symmetry C_{2v} but related one to another by operations of the fourfold screw axis 4_1. The process of averaging, performed by the thick interface, makes that the (100) silicon surface behaves as if it had the fourfold C_{4v} symmetry[12]. This remarkable fact is illustrated in Fig.4 which shows a spread droplet pattern obtained with the nematic E8 (from BDH) on the surface of a (100) oriented silicon wafer. The pattern consists of four identical sectors separated by thin walls (white lines). The anchorings $\vec{a}_1,...,\vec{a}_4$ within the sectors bisect the right angles formed by the walls.

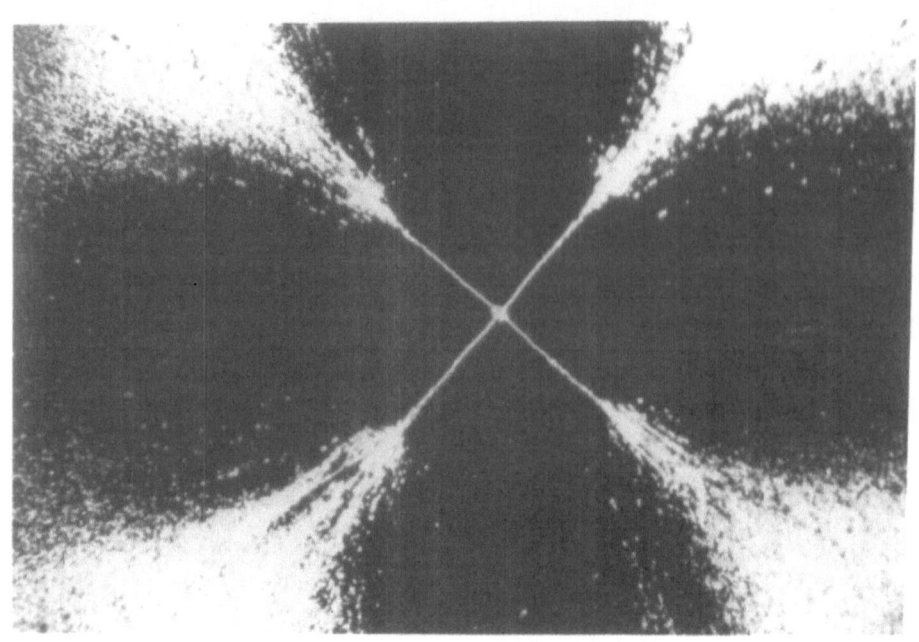

Fig. 4   Pattern in a droplet of the nematic E8 spread on the surface of a (100) silicon wafer.

c - <u>SiO films evaporated under oblique incidence</u>[13,14]. Such films have anisotropic porous columnar structures with the average symmetry C_s. Monostable and bistable anchorings have been obtained on such substrates.

d- <u>Polymer films rubbed in two perpendicular directions</u>[15]

II.2. Types of anchorings

The four above examples show clearly that the types of anchorings occuring on solid substrates are determined by the symmetry G_s of the substrate. *For a given symmetry G_s, if $\vec{a}_1$ is one of the anchorings produced by the substrate at same temperature T and chemical potentials μ_i, then all other anchorings $\vec{a}_\alpha$ are related to $\vec{a}_1$ by symmetry operations of G_s.* In terms of the potential $\omega(T, \mu_i, \vec{a})$ it means that ω must be invariant under all operations g_s of G_s :

$$\omega(T, \mu_i, g\vec{a}) = \omega(T, \mu_i, \vec{a}) \qquad\qquad \text{II-1}$$

Due to the symmetry of the nematic phase, ω must also be invariant with respect to the inversion $\vec{a} \rightarrow -\vec{a}$:

$$\omega(T,\mu_i,-\vec{a}) = \omega(T,\mu_i,\vec{a}) \qquad\qquad \text{II-2}$$

For given T and μ_i, $\omega(\vec{a})$ is a function of the polar and azimuthal angles (θ,φ) defining the direction of $\vec{a}$ with respect to the normal $\vec{z}$ to the surface (θ) and to same axis $\vec{x}$ in the surface (φ). Being a function of (θ,φ) ω can be expanded into a series of spherical harmonics :

$$\omega(\theta,\varphi) = \underset{1,m}{\Sigma}\ Q_{1,m}\ (\mu_i,T)Y_1^m(\theta,\varphi) \qquad\qquad \text{II-3}$$

containing only terms with even l. Number, depth and positions of minima of $\omega(\theta,\varphi)$ depend on values of the amplitudes $Q_{1,m}(\mu_i,T)$.

III. ANCHORING TRANSITIONS

III.1. Types of anchoring transitions

Being a function of the intensive variables μ_i and T the coefficients $Q_{1,m}(\mu_i,T)$ in the above expansion can vary and, consequently, positions and number of minima of ω are expected to change as a function of μ_i and T. In general, to each point (μ_i,T) in the space of intensive variables corresponds some set of anchorings $\vec{a}_\alpha(\mu_i,T)$ which can be represented as points on a unit sphere. A trajectory in the space (μ_i,T) is then modeled by an evolution of $\vec{a}_\alpha$ (μ_i,T) on the unit sphere. Besides smooth variations of the anchorings $\vec{a}_\alpha$ as a function of (μ_i,T) one can expect two types of singularities to occur on trajectories $\vec{a}_\alpha(\mu_i,T)$:

a- <u>First-order anchoring transitions</u> : The anchoring jumps from A to A' (Fig. 5) when the path in the $(\mu_i,$ T) space crosses a first-order anchoring transition boundary. In order to simplify graphical representation (Fig. 5a) it has been supposed that only one of chemical potentials can vary, while all other are kept fixed. The first-order anchoring transition is then represented by a line L_1. Such a line can terminate by a critical point or can join other transition boundaries at multicritical points.

b- <u>Second-order anchoring transitions</u> : The trajectory $\vec{a}(\mu,T)$ can bifurcate at

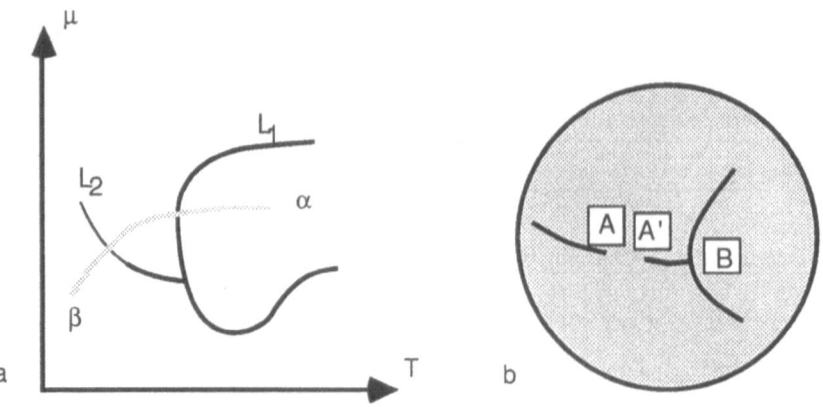

Fig. 5. Definitions of anchoring transitions : a) space of intensive variables (μ,T) ; b) evolution of an anchoring $\vec{a}(\mu,T)$ on the unit sphere, corresponding to the path $\alpha\beta$ in the plane (μ,T).

point B when the path in the (μ,T) space crosses a second-order transition line L_2

III.2. Experimental[10]

In order to control the intensive variables (μ_i, T), the composition and the temperature of the gas, surrounding the substrate with the nematic layer on it, must be controlled. The chemical potentials μ_i in the gas phase, assumed to be a mixture of perfect gases, are :

$$\mu_i = kT \log p_i + \chi_i(T) \qquad\qquad III\text{-}1$$

and can be controlled via two parameters -the temperature and the partial pressures

$$p_i = \frac{N_i}{N_{tot}} p \qquad\qquad III\text{-}2$$

that is, by varying concentrations N_i/N_{tot} for a constant total pressure p. In Fig. 6 we show an experimental set-up which has been used to study the anchoring transition induced by adsorption of water. The composition of the gas mixture $N_2 + H_2O$ is regulated by mixing of two, one dry and the second water saturated, streams of nitrogen i_0 and i_s. The water vapor pressure $p_{H_2O} = \tilde{p}.p_s$ of the resulting mixture is :

$$p_{H_2O} = \frac{i_s}{i_s + i_0} \cdot p_s = \tilde{p} \cdot p_s \qquad\qquad III\text{-}3$$

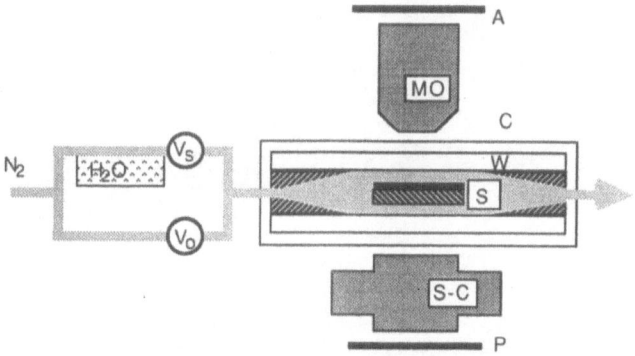

Fig. 6.   Experimental set-up : S-sample, C-cell equiped with glass windows,
          S-C-Soleil compensator, A and P-polarizers, V_0 and V_s-valves,
          M.O.-microscope objective.

III.3. Examples of anchoring transitions

   a- Gypsum/E9 interface : The azimuthal angle φ of the planar $(\theta = \frac{\pi}{2})$ anchoring
varies as a function of the water vapor pressure (Fig. 7). For $\tilde{p}_c \approx 0.75$ a first-order
anchoring transition involves a discontinuity $\Delta\varphi = 82°$ of the azimuthal angle φ (defined

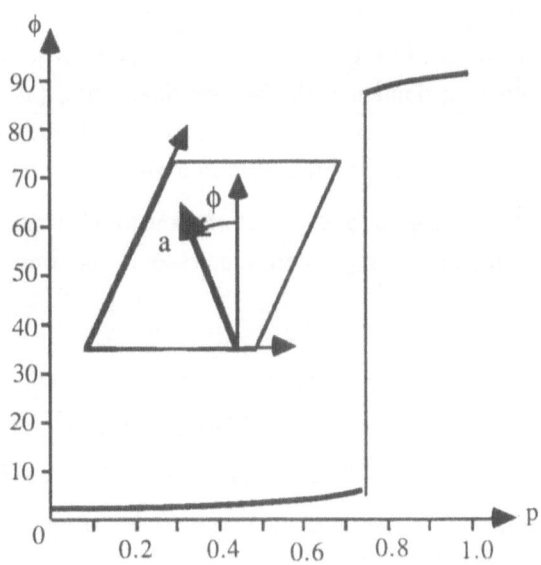

Fig. 7.   Variation of the anchoring direction φ at the E9/gypsum interface.

Fig. 8.   Anchoring transition at the E8/gypsum interface. Nucleation from a point defect.

in the insert of Fig. 7 as an angle between the anchoring $\vec{a}$ and a normal to the fibrous cleavage direction of the gypsum plate).

This transition, illustrated in Figs. 8 and 9 can take place in several different ways. In Fig. 8 we show a growing domain with the new orientation ($\varphi = 87°$) which has nucleated from a dust particle in a matrix with the parental orientation ($\varphi = 5°$). In the example of Fig. 9, the new orientation preexisted in a π-surface wall (visible in Fig. 9a as a loop) pinned at the surface for $\tilde{p} < \tilde{p}_c$. At the anchoring transition ($\tilde{p} = \tilde{p}_c$) the π-well splits into two subwalls which go apart and create in this way a domain with the new orientation.

b- <u>Muscovite mica/E9 interface</u> : This case is much more complex than the above one because not one but three anchoring transitions have been shown to be induced by water vapors in the range $0 < \tilde{p} < 1$ [10]. The first two transitions are illustrated in Figs. 10 and 11. The first transition takes place for $\tilde{p} \approx 0.35$ and involves a discontinuity of 90° between the parental $\vec{a}_1$ and the new orientation $\vec{a}_2$. The photograph in Fig. 10 shows a growing elliptical domain which has nucleated from a point defect.

The second transition is more complicated : it involves a $\Delta\varphi = 60°$ or $\Delta\varphi = -60°$

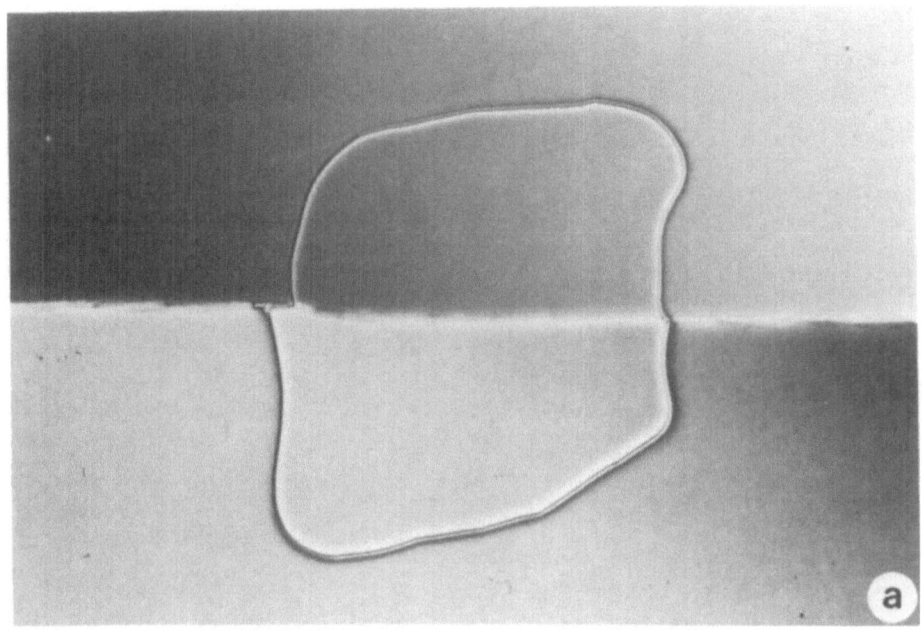

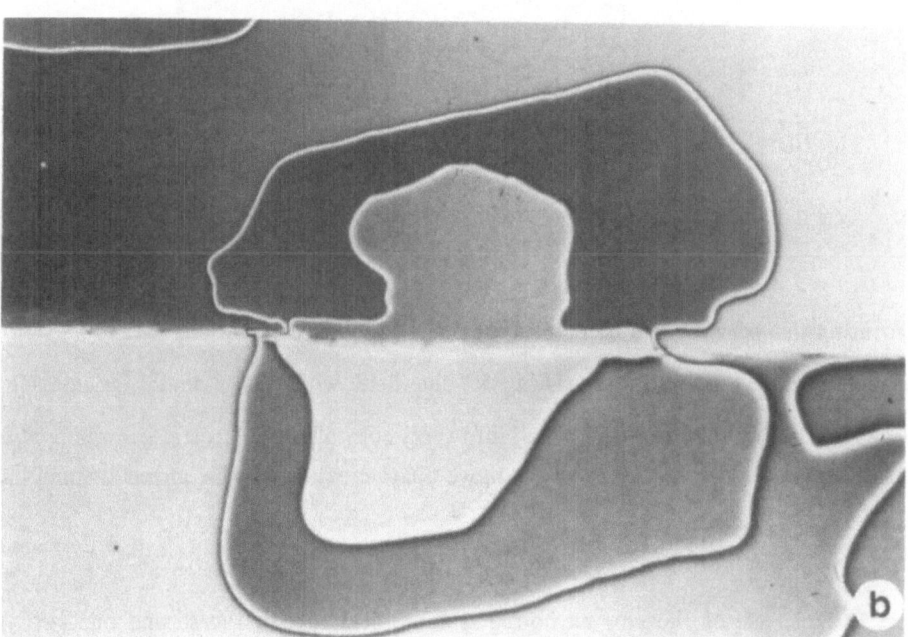

Fig. 9.   Anchoring transition at the E8/gypsum interface. a) a π-surface wall preexists for $\tilde{p} < \tilde{p}_c$. b) the π-wall splits into two subwalls for $\tilde{p} > \tilde{p}_c$.

Fig. 10.   First anchoring transition at the E9/muscovite mica interface.

discontinuities between the parental $\vec{a}_2$ and the <u>two</u> new orientations $\vec{a}_3$ and $\vec{a}_3{}^*$. In the case of a surface π-wall, preexisting in the field with the parental orientation $\vec{a}_2$ (Fig. 11a), the wall splits, at $\tilde{p}_2 = 0.8$, into three subwalls. The central subwall is almost at rest while the two latered subwalls move apart creating by this means domains with orientations $\vec{a}_3$ and $\vec{a}_3{}^*$ (Fig. 11b).

In fact, one of the new anchoring is more stable than the second one ($\omega(\vec{a}_3) < \omega(\vec{a}_3{}^*)$). As a consequence, on a longer time scale the central subwall can be seen to move slowly ; the area of domains with the anchoring $\vec{a}_3{}^*$ decreases for the benefit of domains with the more stable anchoring $\vec{a}_3$. The third anchoring transition is inverse with respect to the second one.

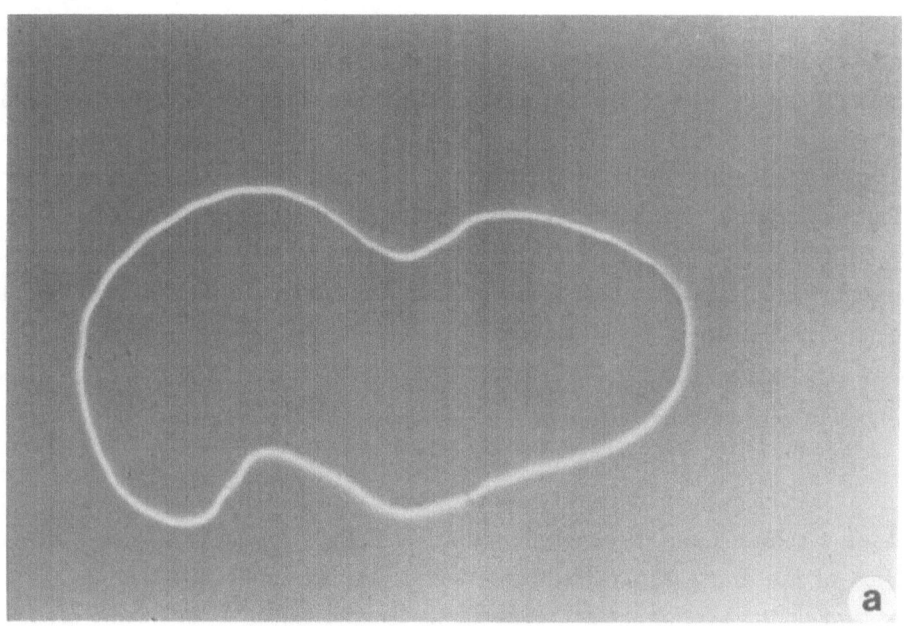

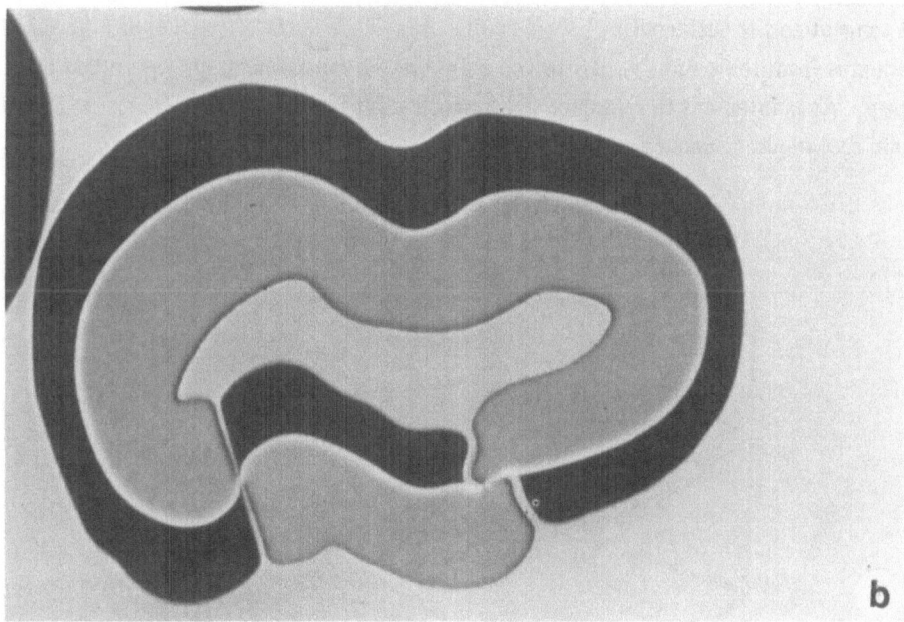

Fig. 11.   Second anchoring transition at the E9/muscovite mica interface.

a) $\tilde{p}_1 < \tilde{p} < \tilde{p}_2$; a surface π-wall preexists in a field with the $\vec{a}_2$

anchoring. b) $\tilde{p} > \tilde{p}_2$; the π-wall splits into <u>three</u> subwalls.

III.4. Models of anchoring transitions[10]

The examples of multistable anchorings and of anchoring transition reported above raise numerous fundamental questions concerning the structure of the interface between the nematic layer and the crystalline support. The existence of the anchoring transitions induced by variations of the water vapor pressure indicates clearly that adsorption of water molecules assists (or induces) variations of the anchoring direction. It must be so for following reasons . For a constant temperature T, only the chemical potential μ_{H_2O} of water varies as a function of the partial pressure p_{H_2O} (variations of the partial pressure of the nitrogen do not induce anchoring transitions), so that the only variable coupled explicitly to μ_{H_2O} is the excess surface concentration of water ρ_{H_2O} (see eq.I-2). Therefore if any variation $\vec{\delta a}$ follows a change $\delta\mu_{H_2O}$ it must be accompanied by some change $\delta\rho_{H_2O}$.

To be more explicit let us consider the geometrical construction shown in Fig. 12. It shows a surface $f(\rho,\varphi)$ representing variation of the free energy $f = u - Ts$ as a function of the excess surface density ρ and of the anchoring direction φ. For the chemical potential μ set by the gas reservoir, the equilibrium values of ρ and φ are found by osculating the surface $f(\rho,\varphi)$ with a plane P parallel to the φ axis, having the slope μ and tangent to the surface $f(\rho,\varphi)$. It is evident that for a surface $f(\rho,\varphi)$ shown in Fig. 12, one jumps from the point (ρ_1, φ_1) to the point (ρ_2, φ_2) when the osculating plane has the slope μ_c. As pointed out in reference [10] for another type of the surface $f(\rho,\varphi)$ one can also obtain second-order anchoring transitions.

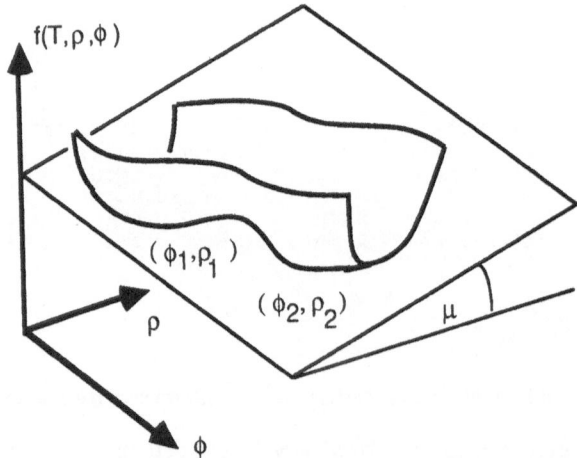

Fig. 12.   Geometrical interpretation of anchoring transitions

IV. DISCUSSION

From a microscopic point of view, the reanchoring action of water is still not understood. Other substances such as alcohols or polyalcohols have also been shown to exert reanchoring action. They have one common characteristic with water : they can form hydrogen bonds. If a molecule of water is hydrogen bonded to both to the substrate and to the nematic molecules, one can imagine that the orientation of the nematic molecule can be different from that which would occur if the nematic molecule interacted directly with the substrate.

The adsorption of molecules at the crystal surface can be more or less strong ; a chemical bond between chemical species i and the substrate can occur. In such a case the anchoring transition looses its reversibility[10].

Beautiful pictures of L.C. molecules adsorbed at graphite surfaces have recently be obtained by STM technique [16]. One can expect that this technique will be usefull for understanding mechanisms of the adsorption induced anchoring transitions.

References
1. C. Mauguin, Bull. Soc. Fr. Cryst. 34 (1911), 71
2. F. Grandjean, Bull. Soc. Fr. Min. Cryst. 39 (1916), 164
3. J. Cognard, Mol. Cryst. Liq. Cryst. Suppl. 1 (1982), 1
4. H. Yokoyama, Mol. Crys. Liq. Cryst. 165 (1988), 265
5. L.M. Blinov, E.I. Kats and A.A. Sonin, Sov. Phys. Usp. 30 (1987) 604
6. J.G. Dash "Films on Solid Surfaces", Academic Press (1975)
7. T.V. Korkishko, V.G. Chigrinov, R.V. Galinlin, A.A. Sonin and N.A. Tikhomirova, Sov. Phys. Cryst. 32 (1987), 263
8. B. Jérôme, P. Pieranski, J. de Phys. 46 (1988), 1601
9. P. Pieranski and B. Jérôme, to be published in Phys. Rev. A.
10. P. Pieranski, B. Jérôme and M.Gabay Proc. of 2nd Int. Conf. on Optics of L.C., Torino, 1988, to be published in Mol. Cryst. Liq. Cryst.
11. G.Friedel "Leçons de Cristallographie", Librairie Scientifique Albert Blanchard, Paris (1964)
12. J. Bechhoefer, A. Bosseboeuf, B. Jérôme and P. Pieranski, to be published
13. B. Jérôme, P. Pieranski and M. Boix, Europhys. Lett. 5 (1988), 693
14. M. Monkade, M. Boix and G. Durand, Europhys. Lett. 5 (1988), 697
15. B. Jérôme and P. Pieranski, Proc. of the 12th Int. L.C. Conf. Freiburg, 1988 to be published in Liq.Cryst.
16. J.S. Foster and J.E. Frommer, Nature 333 (1988), 542.

SPECTRUM OF THE PROPAGATIVE MODES NEAR THE

SMECTIC-A TO HEXATIC-B OR CRYSTAL-B PHASE TRANSITION

Harald Pleiner

Fachbereich Physik
Universität Essen
D 4300 Essen 1, West Germany

INTRODUCTION

Near second order or weakly first order phase transitions the sound spectrum of liquids, liquid crystals and crystals usually shows very pronounced anomalies. The order-parameter modulus S, which describes the degree of ordering in the more ordered phase (and is zero in the less ordered phase), becomes soft near the phase transition. That means, the restoring force acting on δS, the deviation of the order-parameter modulus from its equilibrium value, gets weak (and the dynamics slow) close to the phase transition. Thus, instead of being a fast microscopic variable well inside a phase, δS is a slow and macroscopic variable near the phase transition.[1] Its correlation function even diverges at the transition point, $T = T_c$, in the limit, frequency $\omega \to 0$. Thus, these huge fluctuations give rise to an additional mechanism of dissipation. This is seen as a cusp-like increase of sound absorption with the peak near T_c and a peak height $\sim \omega^{-2}$. In addition, being a macroscopic variable now, δS couples to other scalar variables, like density variations, $\delta \rho$, and entropy variations, $\delta \sigma$, statically.[2] Obviously, this also changes the sound velocity, which indeed exhibits very often a dip near T_c.

Usually these anomalies in the sound spectrum (cusp in the absorption and dip in the velocity) are isotropic (independent of the direction of the sound wave vector), since the couplings discussed above are between scalar quantities. So the experimental results of Gallani et al.[3] came rather unexpected, since they observed a strong anisotropy in the anomalous effects near the smectic-A (Sm-A) to hexatic-B (Hex-B) phase transition. There were pronounced anomalous effects, if the wave vector $\mathbf{k}$ was perpendicular to the layer normal $\hat{\mathbf{n}}^0$ ($\Theta = 90^o$), but no or only very weak anomalous effects for $\mathbf{k}$ parallel to $\hat{\mathbf{n}}^0$ ($\Theta = 0^o$). The Sm-A to Hex-B phase transition is characterized by the onset of bond-orientational order (BOO),[4] which is absent in Sm-A phases. In Hex-B the BOO means that the lines between the centers of gravity of the molecules are ordered hexagonally.[5] There is no in-plane positional order in both, Sm-A and Hex-B phases. In addition, the overall (uniaxial) symmetry due to the layer structure is the same in both phases. Thus, the relevant order-parameter for this phase transition is $S \exp(6i\phi)$,[6] where S describes the degree of BOO and ϕ its structure (i.e. the angle of one of the sixfold degenerate preferred directions of the bonds with respect to an arbitrary reference direction). The additional hydrodynamic variable in Hex-B is $\delta \phi$, while δS has to be considered only near (but on both sides of) the weakly first order phase transition. As discussed above, the usual scalar couplings of δS with density or entropy variations cannot account for the observed anisotropies in the sound anomalies.

ULTRASOUND SPECTRUM NEAR T_c

Let us start with the dynamic equation[7] for δS

$$\dot{S} + \beta_1 \vec{\nabla}_\perp \vec{v}_\perp + \beta_2 \nabla_\parallel v_\parallel = -\eta \Lambda , \tag{1}$$

which shows the reversible dynamical coupling of δS to in-plane compressional flow and to elongational flow across the layers, characterized by the phenomenological (reactive) transport parameters β_1 and β_2, respectively.† The dissipative coefficient η describes the relaxation of the non-conserved variable δS, where Λ is the thermodynamic conjugate of δS, defined as the partial derivative of the free energy with respect to δS. It is given explicitly by[7]

$$\Lambda = \chi^{-1}(\delta S + \gamma_3 \nabla_\parallel u + \gamma_4 \, \delta\rho + \gamma_5 \, \delta\sigma). \tag{2}$$

The susceptibility χ diverges at T_c and this is the reason why δS is a slow variable near T_c, since the characteristic relaxation time is $\tau = \chi/\eta$. The cross susceptibilities $\gamma_{4,5}$ describe the already mentioned static couplings to density and entropy density variations, while γ_3 denotes an additional coupling to layer thickness variations, $\nabla_\parallel u$ (layer compression or dilation). Both, the dynamic and static cross couplings have counterparts in the appropriate currents or thermodynamic conjugates, which are listed as follows

$$\begin{aligned}
\sigma_{ij} &= \ldots + [\beta_1(\delta_{ij} - n_i^0 n_j^0) + \beta_2 \, n_i^0 n_j^0] \Lambda , \\
\Phi &= \ldots + \chi^{-1}\gamma_3 \, \delta S , \\
\delta\mu &= \ldots + \chi^{-1}\gamma_4 \, \delta S , \\
\delta T &= \ldots + \chi^{-1}\gamma_5 \, \delta S .
\end{aligned} \tag{3}$$

Here σ_{ij}, Φ, $\delta\mu$, and δT are the stress tensor, the thermodynamic conjugate to $\nabla_\parallel u$, the variations of the chemical potential, and of the temperature, respectively. The ellipses indicate the usual hydrodynamic terms in Sm-A (Ref. 10) or Hex-B (Ref. 11). The structure of the hydrodynamic equations is the same in both phases, except for the variable $\delta\phi$, which, however, does not contribute to the sound mode spectrum. Thus, the sound mode spectrum also has the same structure in both phases, although the hydrodynamic parameters may have different values (and different critical exponents) at the two sides of the phase transition. The dispersion relation for first sound of frequency ω and wave vector $\mathbf{k}$ is found to read[7]

$$\frac{\omega^2}{k^2} = c_A^2 + \frac{i\omega\tau}{1 + i\omega\tau} \frac{1}{\rho_0\chi} \left[d_1^2 + 2d_1 d_2 - \frac{d_2^2}{i\omega\tau} \right], \tag{4}$$

with $d_1 = \beta_1 \sin^2\Theta + \beta_2 \cos^2\Theta$ and $d_2 = \rho_0\gamma_4 + \sigma_0\gamma_5 - \gamma_3 \cos^2\Theta$, where ρ_0 and σ_0 are the equilibrium values of the mass density and entropy density, respectively. In eq.(4) all dissipative processes have been discarded except for the relaxation of the order-parameter modulus[1] (cf. Eq.(1)). The sound velocity far away from T_c, c_A^2, is only very weakly anisotropic.[10,11]

From the structure of Eq.(4) it is obvious that the experimental results can be explained, if $\beta_1 \gg \beta_2$. Then the anomalies, whose angular dependence is described by the bracket in Eq.(4), are much more pronounced for $\Theta = 90^\circ$ than for $\Theta = 0^\circ$.

---

† Similar reactive parameters are given in Refs. 8 and 9.

The inequality $\beta_1 \gg \beta_2$, however, which is here postulated in order to explain the experiments, also has an intuitive physical meaning: Near T_c an in-plane compressional flow ($\vec{\nabla}_\perp \vec{v}_\perp$) can cause (in-plane) BOO much more efficiently than an elongational flow across the layers ($\nabla_\parallel v_\parallel$). By this description the observed anisotropy of sound anomalies is traced back to the anisotropy of the coupling between in-plane and across-plane flow to BOO. The situation is somehow reminiscent of pre-transitional effects near the isotropic to nematic phase transition, where shear flow can cause nematic ordering.[2] Of course, the picture given here has to be corroborated by additional experiments (currently under way) testing the full Θ- and $\omega\tau$-dependence of Eq.(4).

SPECTRUM OF SECOND SOUND AND TRANSVERSE SOUND

In all smectic systems without in-plane positional order there is, besides ordinary sound, a second propagating mode, called second sound.[2] It consists of elastic deformations of the (one-dimensional) lattice at constant density (incompressibility). It does not exist (as a propagating mode) neither for $\Theta = 90°$ nor for $\Theta = 0°$, since firstly there is no in-plane lattice and since secondly lattice deformations exactly parallel to the layer normal are incompatible with incompressibility. Thus, second sound velocity (in Sm-A and Hex-B) has the angular dependence $c_2(\Theta) = 4c_{max} \sin\Theta \cos\Theta$. From Eqs.(1) and (2) one can read off that the order parameter modulus couples to an incompressible flow through $\beta_1 - \beta_2$ and γ_3 only, and that the anomalous effects have the same angular dependence as the regular ones. Explicitly the dispersion relation reads[7] (in the same approximation as Eq.(4))

$$\frac{\omega^2}{c_2^2(\Theta)k^2} = 1 + \frac{i\omega\tau}{1 + i\omega\tau} \frac{1}{16 \, c_{max}^2 \rho_0 \chi} \left[(\beta_1 - \beta_2)^2 + 2\gamma_3(\beta_1 - \beta_2) - \frac{\gamma_3^2}{i\omega\tau} \right] . \quad (5)$$

Here the predictions to be tested in future experiments are apparent: The anomalies in the second sound spectrum are as pronounced as they are in the first sound spectrum for $\Theta = 90°$ and they have the same angular dependence as the regular second sound spectrum. If, contrary to our expectations, β_2 would be of the same magnitude as β_1, then the anomalous effects in second sound would be as weak as they are in first sound for $\Theta = 0°$. Again, the structure of Eq.(5) is the same for Sm-A and Hex-B , since the additional Goldstone mode in Hex-B does not contribute to second sound in the approximation used.

The situation is much different, if a phase transition from Sm-A to the crystalline B phase (Xtal-B) is considered. A Xtal-B phase does have in-plane positional order and is, thus, a really three-dimensional (albeit rather soft) crystal with still some layered structure.‡ The phase transition from Sm-A to Xtal-B is described by a two-dimensional (in-plane) density wave as order-parameter. The amplitude of this density wave, S', describing the degree of ordering is a scalar quantity with properties quite similar to S of the Sm-A to Hex-B phase transition. Especially the dynamic equation (1) also applies to S', with Λ' only slightly different from Eq.(2) in the Xtal-B phase

$$\Lambda' = \chi^{-1}(\delta S' + \gamma_{3\parallel} \nabla_\parallel u_\parallel + \gamma_{3\perp} \vec{\nabla}_\perp \vec{u}_\perp + \gamma_4 \, \delta\rho + \gamma_5 \, \delta\sigma). \quad (6)$$

Here use is made of the uniaxial symmetry of Xtal-B and $u_\parallel$ (corresponding to u in Sm-A) and $\vec{u}_\perp$ are the across-layer and in-plane components of the three-dimensional

---

‡ In the older literature dealing with the smectic B-phase no discrimination is made between the phases Hex-B and Xtal-B . Thus, various statements given there may apply to one of them, but not to the other.

displacement vector $\vec{u}$, respectively. Its thermodynamic conjugate is now a tensor and the second equation of (3) is replaced by

$$\Phi_{ij} = \ldots + \chi^{-1} \left[\gamma_{3\perp}(\delta_{ij} - n_i^0 n_j^0) + \gamma_{3\parallel} n_i^0 n_j^0 \right] \delta S' \ . \tag{7}$$

The additional Goldstone variables in Xtal-B are $\vec{u}_\perp$ and they do couple to the sound modes. Thus, there is generally a change in the sound mode structure crossing the Sm-A to Xtal-B phase transition. Nevertheless Eq.(4) for the anomalous effects in the first sound spectrum can also be used for the Sm-A to Xtal-B transition, if below T_c the regular sound velocity, c_A^2, of Sm-A is replaced by the appropriate sound velocity of Xtal-B , whose angular dependence is slightly more complicated than c_A^2. There is thus no big difference in the structure of the anomalies of first sound between the Sm-A to Hex-B and the Sm-A to Xtal-B phase transition. Also the anisotropy of the anomalies can be expected to be similar for the two different phase transitions, since $\delta S'$ describes an ordering taking place within the layers and, thus, one can again expect $\beta_1 \gg \beta_2$.

There is however a manifest difference in the anomalies of second sound. Since second sound of Sm-A is transferred into a transverse sound mode (transverse lattice vibrations) in Xtal-B , Eq.(5) has to be replaced in the Xtal-B phase by

$$\frac{\omega^2}{k^2} = c_{2\parallel}^2 \sin^2\Theta + c_{2\perp}^2 \cos^2\Theta + \sin^2\Theta \cos^2\Theta \frac{i\omega\tau}{1 + i\omega\tau} \frac{1}{\rho_0 \chi} \times$$
$$\times \left[(\beta_1 - \beta_2)^2 + 2(\gamma_{3\parallel} - \gamma_{3\perp})(\beta_1 - \beta_2) - \frac{(\gamma_{3\parallel} - \gamma_{3\perp})^2}{i\omega\tau} \right] \ . \tag{8}$$

The angular dependence of the anomalies is now completely different from the angular dependence of the regular parts of the transverse sound spectrum and this is a qualitative different behaviour when compared to the Sm-A to Hex-B transition. This should help in discriminating Sm-A to Hex-B from Sm-A to Xtal-B phase transitions, experimentally.

ACKNOWLEDGEMENTS

Fruitful discussions with H.R. Brand and financial support by the Deutsche Forschungsgemeinschaft are gratefully acknowledged.

REFERENCES

1 I.M. Khalatnikov, *An Introduction to the Theory of Superfluidity* (Benjamin, New York, 1965).
2 P.G.de Gennes, *The Physics of Liquid Crystals* (Clarendon, Oxford, 1967).
3 J.L. Gallani, P. Martinoty, D. Guillon, and G. Poeti, Phys.Rev. **A37**, 3638 (1988).
4 B.I. Halperin and D.R. Nelson, Phys.Rev.Lett. **41**, 121 (1978).
5 R.J. Birgeneau and J.D. Litster, J.Phys.Lett.(Paris) **39**, L399 (1978).
6 D.R. Nelson and B.I. Halperin, Phys.Rev. **A21**, 5312 (1980).
7 H. Pleiner and H.R. Brand, Phys.Rev. **A39**, 1563 (1988).
8 M. Liu, Phys.Rev. **A19**, 2090 (1979).
9 H.R. Brand, Phys.Rev. **A33**, 643 (1986).
10 P.C. Martin, O. Parodi, and P.S. Pershan, Phys.Rev. **A6**, (1972).
11 H. Pleiner and H.R. Brand, Phys.Rev. **A29**, 911 (1984) and
    H. Pleiner, Mol.Cryst. Liq.Cryst. **114**, 103 (1984).

OPTICAL ACTIVITY IN THE ISOTROPIC AND
BLUE PHASES OF A CHIRAL LIQUID CRYSTAL

F. Vanweert and W. Van Dael

Laboratorium voor Molekuulfysika
Katholieke Universiteit Leuven
Celestijnenlaan 200 D
B-3030 Leuven, Belgium

INTRODUCTION

Some liquid crystals exhibit mesophases between the isotropic phase and the chiral nematic (or cholesteric) phase. There may be up to three different spatial structures in a temperature interval as small as 1 K. These phases are optically not birefringent and they reflect preferentially short wavelengths : that is why they are called blue phases[1].

On approaching the phase transition from the isotropic side, pretransitional effects occur producing an incipient chiral ordering. It remains, however, short-range so that the bulk properties keep completely their isotropic character. According to the Landau-de Gennes theory[2] and confirmed by many experiments, this local anisotropy increases sharply on approaching the stability limit of the phase : macroscopic properties like the field induced birefringence or the intensity of the scattered light are correspondingly strongly enhanced. Also the optical activity reflects the pretransitional chiral order[3].

Filev[4] however drew attention to the fact that other fluctuation modes, which are not producing optical rotation in a direct way, may nevertheless produce a contribution to the activity if they couple with the chiral mode.

In this paper we will provide experimental evidence of this mode-coupling phenomenon in the isotropic phase of cholesteryl oleyl carbonate (COC) : an analysis will be given of the temperature and wavelength dependence of the data.

EXPERIMENTAL

The experimental set up is similar to that described earlier[5]. Measurements of the rotation of the polarization vector were carried out in COC for $35\,°C < T < 41\,°C$ and for a wavelength of 589 nm. The samples were obtained from Van Schuppen (The Netherlands) and used without further purification. They were contained in a glass cell with 1 cm path length and regulated in temperature by a three stage oven arrangement. The temperature stability of the copper block in contact with the cell could be kept within 1 mK for a period of several hours. In a measuring cycle temperature is stepwise

91

increased by computer controlled increments : the next reading is only taken after a predetermined lapse of time and provided that no further time dependency is observed.

PRETRANSITIONAL OPTICAL ROTATION

A linearly polarized wave, travelling through a fluid rotates its polarisation plane an angle $\phi = \phi_0 + \phi_s$. ϕ_0 represents the molecular contribution while ϕ_s describes the effect of molecular orientation.

Cooling down from the disordered isotropic phase to the ordered blue phase an enhancement of the molecular optical rotation occurs as a result of short-range ordering. The chirality of the cholesteric liquid crystal gives rise to complex pretransitional phenomena.

Filev[4] drew attention to the fact that the classical mean-field approximation is not sufficient in order to describe the complete pretransitional structural optical rotation. He introduced the effect of the coupling of several fluctuation modes on the optical rotation. Each mode has its own correlation length and transition temperature. This treatment leads to the expression of ϕ_s :

$$\phi_s = \phi_{s_1} + \phi_{s_2} \tag{1}$$

with

$$\phi_{s_1} = \frac{AT}{(T-T_1^*)^{1/2}} \quad : \quad \text{the contribution of the } m = 1$$
$$\text{(canonical spiral) mode.}$$

$$\phi_{s_2} = \frac{cfAT}{(T-T_2^*)^{1/2}} \quad : \quad \text{the contribution of the } m = 2$$
$$\text{(planar spiral) mode.}$$

The $m = 2$ mode has a contribution of opposite sign and a transition temperature T_2^* slightly higher than T_1^*. In COC the higher order contribution of the $m = 2$ mode has a nonnegligible amplitude.

Cooling down in the isotropic phase, the optical rotation due to the $m = 2$ mode becomes more and more important and even dominates the $m = 1$ mode, resulting in a minimum in the (ϕ, T) curve (fig. 1). The line represents a best fit of the temperature dependence of the optical rotation according to eq. (1). In fig. 2 the different contributions ϕ_0, ϕ_{s_1}, ϕ_{s_2} are separated.

In the fitting procedure f, actually a function of T, is considered as a constant. The most probable parameters of the fit are listed below.

$$\phi_0 = -2.46 \text{ deg cm}^{-1}$$
$$A = -0.024 \text{ K}^{-1/2} \text{ deg cm}^{-1}$$
$$cf = 0.78$$
$$T_1^* = 36.39 \,^\circ\text{C}$$
$$T_2^* = 36.44 \,^\circ\text{C}$$

We find $T_2^* - T_1^*$ to be much smaller for COC than other cholesteric esters[6,7]. f is of the same order of magnitude as predicted by Filev (for wavelength pith ratio = 2.3) if the prefactor c is approximately 2, which coincides with Battle[7].

OPTICAL ROTATION IN THE B.P.'S

Decreasing the temperature to $36.47\,^\circ\text{C}$[5] we reach the B.P. region. The optical rotation

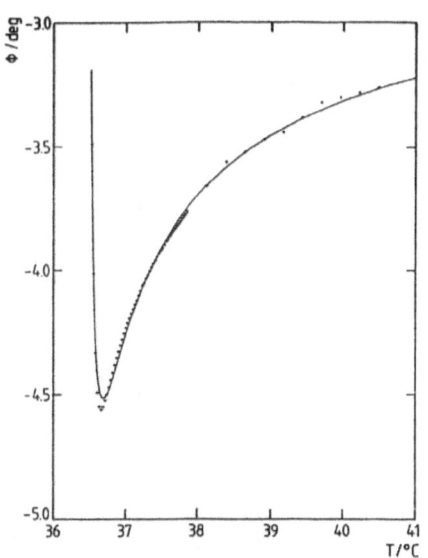

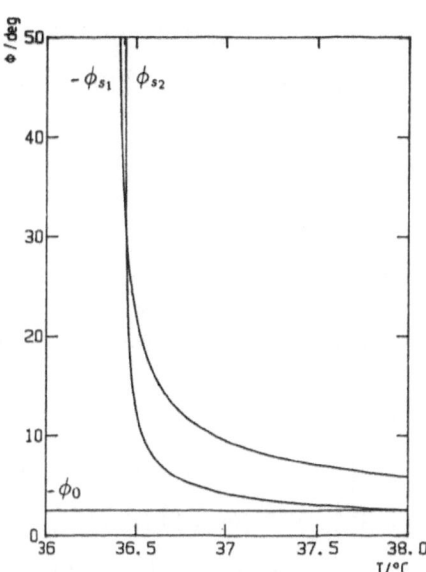

Fig. 1 . Temperature dependence of the pretransitional optical rotation.

Fig. 2 . Temperature dependence of $-\phi_0$, $-\phi_{s_1}$ and ϕ_{s_2}.

and the transmitted intensity in the vicinity of the blue phases are represented in fig. 3. There is a further enhancement of the optical rotation in the same direction as the one of the $m = 2$ mode.

The transition temperatures between the phases were deduced from the combined evidence from the optical rotation and from the transmitted light intensity. We obtained for those temperatures the results in table 1.

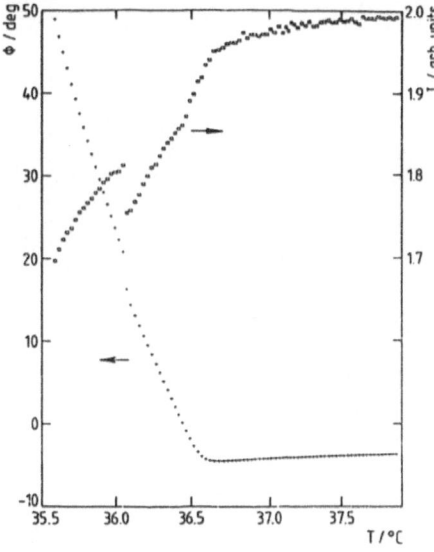

Fig. 3 . Optical rotation of the B.P.'s.

93

Table 1 . Blue phase transition temperatures (°C).

Cholesteric	BPI	BPII	BPIII	Isotropic
35.50	36.05	36.29	36.47	

CONCLUSION

We were able to demonstrate from the temperature dependence of the optical rotation of COC that coupling between the $m = 1$ and $m = 2$ mode is present although both modes diverge at nearly the same temperature.

REFERENCES

1. H. Stegemeyer, Th. Blümel, K. Hiltrop and H.F. Onnseit, Liq. Cryst. 1:3 (1986).

2. P.G. de Gennes, Physics Lett. A 30:454 (1969)

3. J. Cheng and R.B. Meyer, Phys. Rev. A 9:2744 (1974).

4. V.M. Filev, Pis'ma Zh. eksp. theor. Fiz. 37:589 (1983) (JETP Lett. 37:703 (1983)).

5. F. Vanweert, W. Demol, and W. Van Dael, Proc. 12th Intern. Liq. Crystal Conference, Freiburg 1988, to appear in Liquid Crystals 1989.

6. M.B. Atkinson, and P.J. Collings, Mol. Cryst. Liq. Cryst. 136:141 (1986).

7. P.R. Battle, J.D. Miller, and P.J. Collings, Phys. Rev. A 36:369 (1987).

LYOTROPIC LIQUID CRYSTALS,

STRUCTURES AND PHASE TRANSITIONS

Jean Charvolin

Laboratoire de Physique des Solides
bât. 510, Université Paris-Sud
91405 Orsay, France

INTRODUCTION

The most representative lyotropic liquid crystals are those
formed by amphiphilic molecules, such as soaps, detergents and
lipids, in presence of water. These molecules are able to build
structures exhibiting a long range crystalline order, although they
are disordered in a liquid-like manner at the local level. Their
polymorphism is characterized by similar sequences of structures,
whatever the details of the chemical structures of the molecules.
These two characteristics, which will be discussed in the first part
of the lecture, clearly show that the individual molecules can not be
the building blocks of the structures and that the elements of
structures are indeed the interfaces built by the molecules or,
better, the symmetric films built by two facing interfaces. These
structures can therefore be described as crystals of fluid films. In
order to bring out a general basis for the understanding of this new
class of crystals, we developed a model for studying periodic
configurations of symmetric films, which will be presented in the
second part of the lecture. Its basic hypothesis is the existence of
a geometrical frustration resulting from the conflict between forces
normal to the interfaces and forces parallel to the interfaces. We
shall determine its possible solutions, as well as their sequence
when its magnitude varies, following a geometrical approach similar
to those developed for other cases of frustration in condensed matter
physics. The solutions optimizing the frustration appear to be in a
satisfying agreement with the observed structures, as well as their
sequence. This leads to consider crystals of films as structures of
disclinations. In the third part of the lecture, we shall discuss
some recent observations made in the vicinity of phase transitions,
particularly between structures of films with different topologies,
in the light of the above result.

STRUCTURES

Amphiphilic molecules at liquid/liquid interfaces

It is a well known fact that polar and organic liquids are not
miscible. For instance, a dispersion of oil droplets in water
obtained by mechanical agitation is not stable. It has too large an

interfacial area. The system evolves, through the coalescence of its droplets, towards a stable state where the two liquids are separated by one flat interface of smallest area. However, if particular molecules, such as soaps, detergents, lipids and their mixtures with others such as alcohols, are added to the system, even at low concentrations, dispersions with quite large interfacial area may be stabilized. A much discussed example of such a stabilization today is that of microemulsions. We shall not deal with them in this lecture, but with lyotropic liquid crystals, which can be considered as the limit situation when the oil content decreases to zero.They are the structures formed by the molecules in presence of water only.

The action of these molecules is easy to understand from their chemical structures. From the examples shown in Fig. 1 it can be seen that they are built in two parts, one or two paraffinic chains attached to a polar group, which have different affinities for solvents.

a) $CH_3-(CH_2)_{11}-N(CH_3)_3\,Cl$

b) $\begin{cases} CH_3-(CH_2)_3-\overset{\overset{\displaystyle CH_2-CH_3}{|}}{CH}-CH_2-CO_2-\overset{\overset{}{|}}{CH}-SO_3\,Na \\ CH_3-(CH_2)_3-\overset{}{CH}-CH_2-CO_2-\overset{\overset{}{|}}{CH_2} \\ \underset{\displaystyle CH_2CH_3}{|} \end{cases}$

d) $CH_3-(CH_2)_9-SO_3\,Na$

e) $CH_3-(CH_2)_8-\overset{\overset{\displaystyle H}{|}}{\underset{\underset{\displaystyle H}{|}}{C}}H_2OH$

f) $CH_3-(CH_2)_{15}-$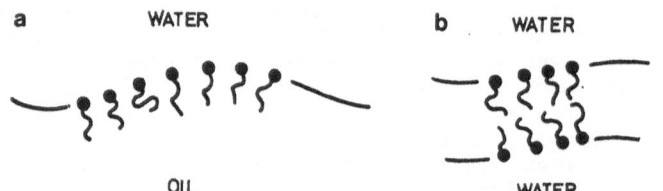

c) $\begin{cases} CH_3-(CH_2)_{15}-CO_2-CH_3 \\ CH_3-(CH_2)_{15}-CO_2-\overset{\overset{}{|}}{CH} \\ \underset{\displaystyle CH_2-PO_4^--CH_2-N^+-(CH_3)_3}{|} \end{cases}$

Fig. 1. Some examples of amphiphilic molecules :a) dodecyl trimethyl ammonium chloride (DTACl), b) aerosol OT, c) dipalmitoyl phosphatidylcholine (DPPC), d) sodium decyl sulfate (SdS), e) decanol, f) cetyl pyridinium chloride (CPCl).

The paraffinic chains have a good affinity for organic solvents, such as oil, but not for polar solvents, such as water. It is the opposite for the polar group. Such molecules are said to be amphiphilic. When put in presence of the solvents they can therefore build 2-D interfaces of the type of those shown in Fig. 2.

a   WATER   b   WATER

OIL   WATER

Fig. 2. Interfacial films built by amphiphilic molecules : a) in presence of oil and water, b) in presence of water only.

The shapes and properties of these interfaces depend upon the structure of the molecules, the temperature and the concentrations, as shown by phase diagram studies. (The molecules shown above have a

rather low molecular mass, a few hundreds, but copolymers with much
larger mass, a few thousands, can exhibit amphiphilic behaviors also
with different solvents).

Phase diagrams and structures

Phase diagrams[1] and structures[2,3] have been extensively studied
for years, mainly by optical observations of textures and small angle
X-ray scattering of polycristalline, or powder, samples. We present
and discuss here a few typical examples only, limiting ourselves to
the high temperature region of the phase diagrams where, as we shall
demonstrate in the following paragraph, the behavior of the
amphiphilic molecules at the interfaces is liquid-like.

The first of these examples is that provided by the binary
mixture dodecyl trimethylammonium chloride / water[4], whose phase
diagram is shown in Fig. 3:

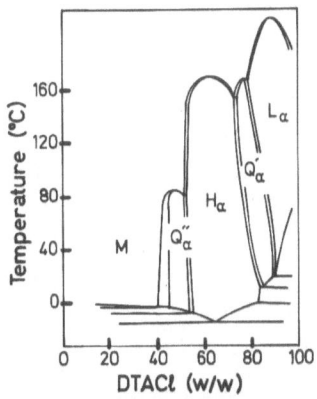

Fig. 3. Phase diagram of the system DTACl/water.

It displays the phases of most general occurence: lamellar (L),
"bicontinuous" cubic (Q'), hexagonal (H), "micellar" cubic (Q") and
the micellar solution. The classical description of the structures of
the phases is given in Fig. 4. The lamellar structure is a periodic
stacking along one dimension of bilayers of amphiphiles and layers of
water. The "bicontinuous" cubic structure is a periodic entanglement
along three dimensions of two labyrinths of amphiphiles separated by
one film of water. The hexagonal structure is a periodic organization
along two dimensions of cylinders of amphiphiles separated by a
honeycomb-like film of water. The structure of the "micellar" cubic
phase is not firmly established at the moment, it has been proposed
to be built of either a labyrinth enclosing spherical micelles[5] or
anisotropic micelles[6], all disposed on cubic lattices. We shall
propose a third possibility in the second part of the lecture. In the
micellar solution finite micelles are dispersed in water without any
long range order. In certain cases giant or worm-like micelles are
possible. The general trend in such phase diagrams is that, when the
water content increases beyond that of the lamellar phase, water
becomes the continuous medium and the interfacial curvature becomes
more and more concave on the paraffinic side.

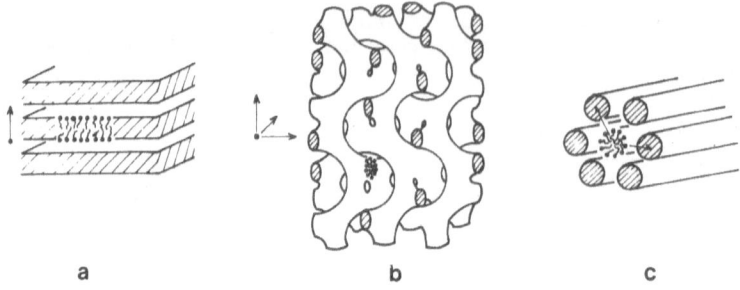

a   b   c

Fig. 4. Schematic representations of the lamellar a), Ia3d
bicontinuous cubic b), and hexagonal c) structures.

The second phase diagram is that of the aerosol OT/water mixture[7]:

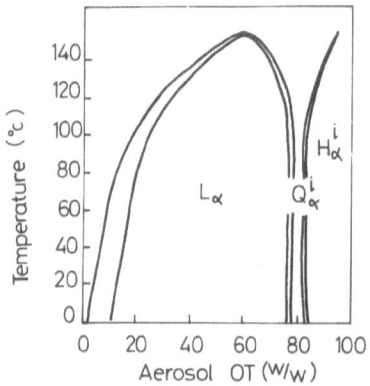

Fig. 5. Phase diagram of the system aerosol OT/water.

As can be seen in Fig. 5, it might be said to be a "mirror" image of
the previous one. The lamellar phase exists for high water content
and the other phases appear for decreasing water contents. Moreover,
if the structures are similar as far as the symmetries are concerned,
their aqueous and paraffinic media are interchanged. For instance,
the labyrinths of the "bicontinuous" cubic phase and the cylinders of
the hexagonal phase are channels of water separated by a film of
amphiphiles. This change is most obviously due to the structure of
the molecule which, because of its two chains, forces the concavity
of the interfaces in the other sense. Lipids with two chains present
similar phase diagrams, with a large biphasic domain between the
lamellar phase and pure water from which vesicles can be obtained.

The addition of a third component, a second amphiphilic species,
may induce interesting behaviors. Good examples are provided by
soap/alcohol/water phase diagrams[1]. Detailed parts of two of these
diagrams are shown in Fig. 6. In sodium decylsulfate/decanol/water
the variation of the relative concentrations of soap and alcohol
enriches the polymorphism by making new phases, rectangular and
nematic, appear[8].

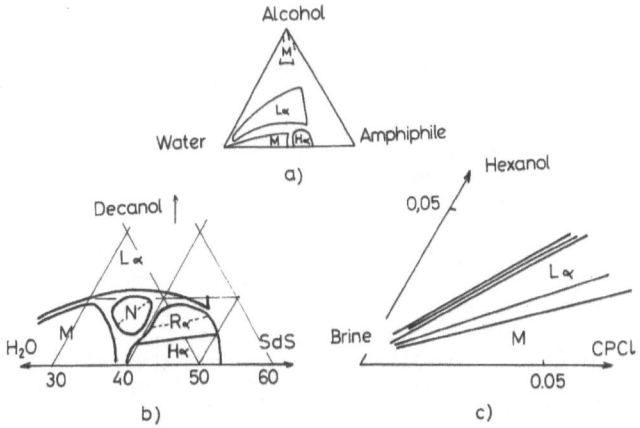

Fig. 6. Schematic ternary phase diagram with an alcohol a),
two particular regions : that of the nematic phases in
the SdS/decanol/water system b), and that of the swollen
lamellar phase in the CPCl/hexanol/water system c).

The structures of the phases are shown in Fig. 7:

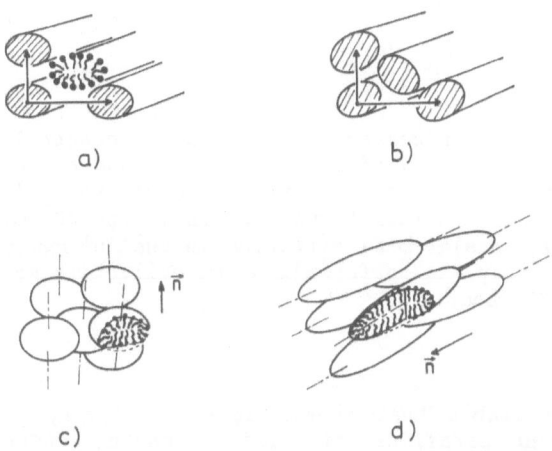

Fig. 7. Schematic representations of the structures found
in diagram 6b : cmm rectangular a), pgg rectangular
b), Nd nematic c), and Nc nematic d).

They are characterized by inhomogeneous interfacial curvatures
associated with inhomogeneous distributions of the two molecules
within the aggregate. For a certain soap/alcohol ratio it has been
shown that the domain of the lamellar phase in these phase diagrams
may extend up to very high water contents. This is quite noticeable
in the cetylpyridinium chloride or bromide/hexanol/brine case[9]. Such
"swollen" lamellar phase are obviously good candidates for studying
forces between interfaces[9,10].

Molecular behavior

The extraordinary polymorphism of the systems described above introduces the question of its relation with the molecular behavior, as it is well known that, in classical molecular crystals, the long range crystalline order is the result of the propagation of the short range order of the molecules.

This question was partly answered by X-ray scattering studies at large angles which showed that the paraffinic medium is disordered[2]. However it was necessary to develop other experiments, using techniques sensitive to local fluctuations, to describe this disorder in a more accurate manner. Among them NMR proved to be remarkably simple and efficient[11]. Relaxation studies showed first that the disorder is dynamical, with similar characteristic times in all structures[12,13,14]: the chains are deformed by isomeric rotations around their C-C bonds, with correlation times of about 10^{-9}-10^{-11} sec, and diffuse along the interfaces, with translational diffusion coefficient of about 10^{-6}-10^{-7} cm^2/sec, as shown in Fig. 8.

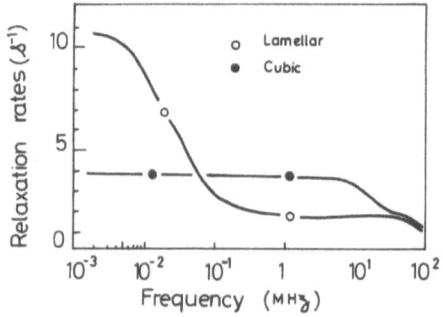

Fig. 8. NMR relaxation curves, related to the spectral densities, obtained in the lamellar and cubic phases of the system potassium laurate/water, the relaxation due to rapid deformations appears in the 10^2 range, that due to translational diffusion in the 10^1 range and that due to very slow deformations or collective motions in the 10^{-2} range.

It is quite noticeable that these figures are very close to those known for melted paraffins of similar chain lengths. Then the development of NMR studies of deuterated molecules[15] and of C^{13} NMR[13] permitted to show that the amplitude of the deformations are about the same in all structures, ordered liquid crystals[11,16] as well as in disordered micellar solutions[13]. Thus there does not appear any definite correlation between one structure and the molecular behavior in it. The chains are about equally disordered in all structures, in spite of important changes of interfacial curvature and mean area per molecule at the interface.

It is therefore not possible to analyze liquid crystalline structures in the same terms than those used for classical molecular crystals. These structures are to be described as ordered entanglements of two disordered liquids separated by an interface, and their element of structure is not the individual molecule but the interface built by the molecules or, to respect the basic symmetry, the film formed by two facing interfaces. This point of view is equally supported by the already emphasized fact that similar structures appear along a similar sequence in phase diagrams, whatever the details of the chemical structures of the molecules. For instance, in the phase diagram of Fig. 3, the lamellar phase is followed by a "bicontinous" cubic phase then by a hexagonal one and, finally, by a micellar phase when the water content increases. Similar sequences, or part of it , can be found in other diagrams too, when their dominant parameter varies. These similitudes can not but hold to the common amphiphilic nature of the molecules and to their ability to build interfaces organized in symmetric films. We therefore analyze lyotropic liquid crystals as crystals of fluid films, the film being defined by its middle surface, as shown in Fig. 9.

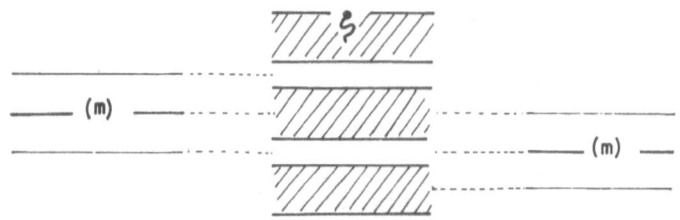

Fig. 9. Definition of the film and its middle surface in a
lamellar structure, it can be either a film of
amphiphile or of water.

PERIODIC SYSTEMS OF FRUSTRATED FLUID FILMS

The question we want to address now is that of the understanding of the structural polymorphism of films in interaction. Can it be described on a simple basis common to all systems?

To explain our point of view on this problem let us consider the classical methods which were developed to analyze organizations of atoms and molecules. It is well known that it is not possible to deduce their structures from their interactions potentials directly, in a simple manner. Two directions were taken. The first consists in calculating the free energies of interacting particles assembled on lattices of given symmetries. The comparison of the results with the observed symmetries is a test of the validity of particular potentials. In this case the catalogue of lattices is provided by crystallography, which can be defined as the investigation of possible geometrical configurations to pack together particles of certain symmetries. The second direction is that of the simulation of the system of particles in a computer which build a structure. This is a very heavy method, particularly in the case of complex 3-D structures. We are now facing the same problem in the case of our systems, but at a higher level of complexity because we are dealing with 2-D objects, films, and not 0-D objects, atoms or molecules.

Up to now, the direction of simulation on a computer could not be followed, because of the high numbers of molecular degrees of freedom and molecules per unit cells. The first direction could not be strictly followed either, in the absence of any crystallographic information about film organizations. Nevertheless this information could be replaced by considering the structures as they are given by the observations, at least for the simplest of them, lamellar,

hexagonal and micellar. Free energy calculations were developed on
this basis with a precise account of the forces acting in the system,
between molecules, ions, interfaces, particularly those of
electrostatic origin[17]. A quite reasonable agreement was observed
between the calculated locations of lamellar, hexagonal and micellar
structures in some phase diagrams and the experimental data,
particularly for ternary diagrams of the type of that shown in Fig.
6. This gave a strong weight to electrostatic forces. However, recent
observations of rectangular or nematic phases within domains of
certain of these phase diagrams thought before to be hexagonal or
micellar[8], and of intense hexagonal or rectangular fluctuations
within their lamellar phases thought before to have infinite
lamellae[18], should temper this first conclusion. In anyway, the
general success of this approach pushed us to think of a real
crystallographical basis to support it. As for atoms or molecules,
the way films can be packed are obviously driven by their "shape", or
their interfacial curvatures, and their distances; but, in the case
of films, strong topological constraints appear which impose states
of frustration.

Interplay of forces and frustration

The forces acting in a system of films of amphiphiles are many of
different origins[19,20]. They are not all well known at the moment.
However their detailed knowledge is not necessary at this stage. We
just need to use the fact that the role of their components normal to
the interfaces is to maintain constant distances between them, if the
interfaces are supposed homogeneous, and that the role of their
components parallel to the interfaces is to determine the interfacial
curvatures, as they do not vary necessarily in concordance at
different levels within the layers limiting the film. Owing to the
symmetry of the film with respect to its middle surface it is clear
that the fact that two facing interfaces may have symmetric
curvatures is not always compatible with constant distances between
the interfaces, if a lamellar-type stacking is kept, as shown in Fig.10.

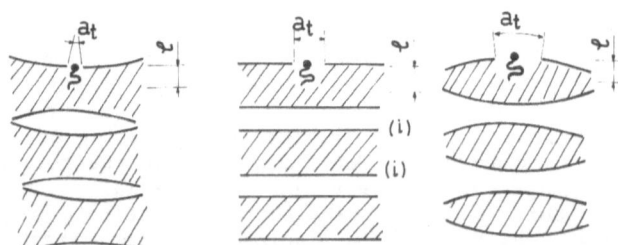

Fig. 10. Representation of a periodic system of films with
flat interfaces (center), constant interfacial distances
and zero curvatures are compatible, the same with
curved interfaces (right and left), constant
interfacial distances and non-zero curvatures are no
longer compatible and the system becomes frustrated.

The left and right drawings of this figure present situations of
conflict between forces normal to the interfaces and forces parallel
to them. Such a situation is a typical case of frustration. The
forces are therefore obliged to compromise and new structures should

be built. We now look for the possible geometrical configurations of the films optimizing this frustration.

Search for solutions

Method. A frustration is directly related to the structure of the space containing the system under study. As shown in Fig. 10, where a system embedded in the usual flat 3-D Euclidean space R_3 is represented, the situation with flat interfaces is the only situation compatible with constant interfacial distances in this space. However, if the system is embedded in an adequately curved space, the situations with curved interfaces may become compatible with constant interfacial distances. This is formally similar to problems of bi-dimensional tilings with regular polygons and tri-dimensional packing of regular polyhedra[21]. Indeed, ideal lamellar-like structures without frustration, conciliating the two antagonistic constraints, can be built in curved spaces. Such ideal structures have of course no reality but are useful, as starting points, to generate possible configurations in the flat Euclidean 3-D space. For this it is necessary to map the curved space onto the flat space. A classical example of such a process with 2-D spaces is that of the mapping of a sphere onto a plane. A sphere can not be deformed into a plane without being torn, except if matter is added during the process. If a structure is present on the sphere the introduction of matter must respect the symmetry of this structure. This is generally known as the Volterra process in condensed matter physics, and corresponds to the creation of discontinuities, or defects, in the structure. Such defects can be defects of translation, or dislocations, and rotation, or disclinations[22]. As the direct displacements on a sphere are rotations the defects needed to map it onto a flat space are disclinations only. In a similar way, the mapping of the 3-D curved space, containing the lamellar-like ideal structure, onto the 3-D flat space, where the real structure should be found, is obtained by introducing a network of disclination lines. This network must respect the symmetries of the structure without frustration in the curved space and it density of disclination is determined by the curvature of the curved space. The possible configurations obtained that way can therefore be seen as structures of disclinations. The application of this method to frustrated films can be found in a series of recent articles[23-27]; we do not detail it here but just give its results.

Results. The first result is that there are three disclination processes possible giving access to three topological classes[23]. One corresponds to structures where a unique film separates our space in two identical subspaces or cells, this is the topology of the "bicontinuous" cubic phases. The second corresponds to structures where a connected film defines an infinite number of infinitely long cells, this is the topology of the hexagonal and rectangular phases. The third topology corresponds to structures where a connected film defines an infinite number of finite cells, this is that of the nematic, "micellar" cubic and micellar phases.

The second result concerns the fact that any interfacial curvature can not be associated with any interfacial distance when an ideal structure is built in a curved space, they are indeed related by a specific relation in each topological class[26]. This set of relations is represented in Fig. 11, it is such that, when the frustration varies monotonically with the parameters of the phase diagram, the topologies must appear along a sequence similar to that observed in phase diagrams. For instance, when the water content is increased in a system with an amphiphile with one chain, a "bicontinuous" cubic and/or hexagonal phases and, finally, micellar phases are met after the lamellar phase.

The last result concerns the fact that the possible symmetries in each topological class compare well with those repertoried for ordered structures. The agreement is particularly striking in the case of cubic phases, "bicontinuous"[24] and micellar"[25], as shown with two particular examples in Fig. 12 and 13. In the first case, that of the "bicontinuous" cubic phases, we demonstrate that the surfaces supporting the film are indeed 3-D infinite periodic minimal surfaces (or IPMS), well known to mathematicians for a century[26], and we show that the idea contained in the disclination process provides a way to develop their crystallography in the 2-D hyperbolic plane along a relatively simple formalism[27]. It is interesting to quote here that surfaces of zero potential in ionic crystals have been shown to be IIPMS also[29]. Thus, in our systems, IPMS could relax mechanical constraints as well as electrical ones when they are present.

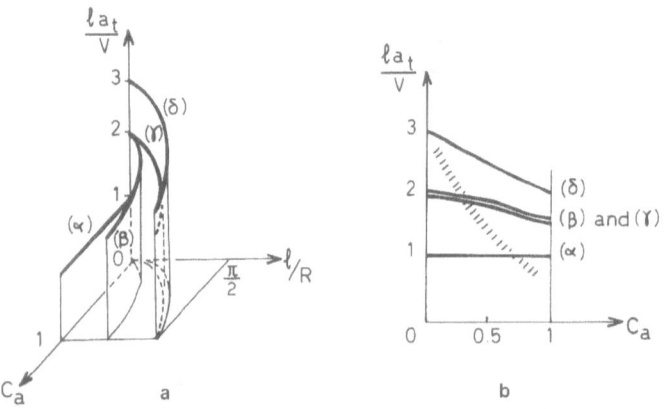

Fig. 11. Relations between curvature (la/v) and distances
(c) imposed by spatial constraints in curved spaces of
radii R a): α corresponds to lamellar, β to bicontinuous,
γ to the cylindrical, and δ to the micellar topologies;
in b) these geometrical relations are confronted with
a physico-chemical one.

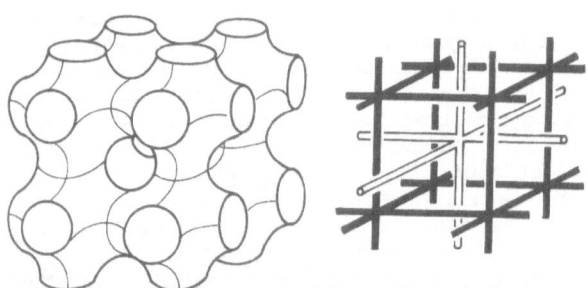

Fig. 12. Part of Schwarz' P surface (IPMS) supporting the
film separating the two labyrinths of the bicontinuous
cubic structure Im3m.

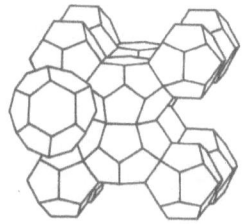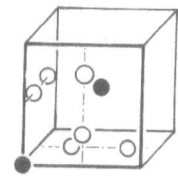

Fig. 13. Proposition for the micellar cubic structure Pm3n,
the surface supporting the film is built by the faces of
slightly distorted dodeca- and tetrakaidecahedra.

Discussion

   We think important at this stage to define the real nature of our
approach again. We analyze the conflict between antagonistic physical
forces in the Euclidean 3-D space as a frustration. This frustration
is expressed in terms of curvatures and distances, it is relaxed by
transfering the structure into a curved space, a Volterra process is
used to map this space onto the Euclidean space and this creates
disclination structures in this space. The terms and the operations
used in this method are purely geometrical so that the structures of

disclinations obtained that way can not be but the possible
geometrical configurations imposed by the spatial constraints. They
can not be considered as the real structures immediatly as energetic
terms, stretching and curvature elasticities of the film, and
entropic terms, distributions of disclinations and fluctuations, are
not included. Those terms, and their interplay in the free energy,
should determine the configurations which can be the real structures
and, in this view, the possible role of elasticity terms in the
stability of IPMS was considered recently[30]. However the striking
agreement existing between the results of our geometrical approach
and the observed structures and their sequence suggests that the
minima of the free energy are closely related to the geometrical
solutions.

PHASE TRANSITIONS

   Except for a very limited number of examples, which are not
relevant to this discussion, most of the phase transitions in
amphiphile/water systems are first order, the monophasic domains of
the phase diagrams being separated by large polyphasic domains where
several phases coexist according to Gibbs' phase rule. First order
phase transitions are well understood and it is generally considered
that there is no much interest in their study. However, in the
particular cases discussed here, the transformation of one structure
into another may eventually imply dramatic changes in the structure
of the film if the two structures have different topologies. This is
a situation which was never investigated before, as this problem does
not exist in classical atomic or molecular crystals built with point
objects.We describe here recent experiments giving some insight into
the processes occuring at the lamellar/cubic/hexagonal phase

transitions of the $C_{12}EO_6/H_2O$ system[31,32]. The phase diagram of this system[33,34] is shown in Fig. 14.

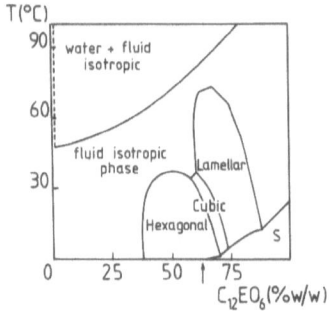

Fig. 14. Phase diagram of the system $C_{12}EO_6$/water.

Information about the lamellar/ isotropic phase transition, which we do not discuss here, can be found in[35]. With respect to the preceding structural studies, which could be made with polycristalline samples, these new studies required the rather delicate preparation of monocrystalline samples.

Epitaxial relationships

Epitaxial relationships concern the relative orientations, most often parallel, of reticular planes of two structures when one grows within another or at its contact, they are well known phenomena in atomic and molecular crystals[36]. Their existence in crystals of films was suggested by very close values of particular reticular distances in powder patterns of neighboring structures of lipids[37], more recently they have been seen in electron micrographs of structures of lipids and amphiphilic copolymers[38,39].

The X-ray scattering studies of monocrystals of $C_{12}EO_6/H_2O$ have precised the orientations and distances relationships[31]. It appears that the [001] direction of the hexagonal structure grow along the <111> directions of the Ia3d cubic structure, whose {211} planes grow parallel to the (001) planes of the lamellar structure. These relationships are summarized in Fig. 15.

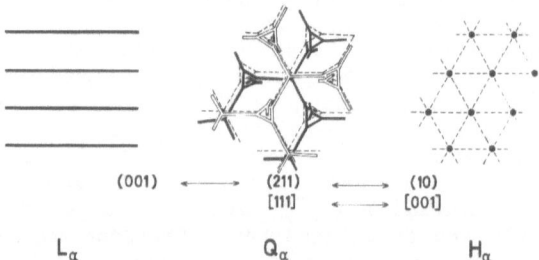

Fig. 15. Epitaxial relationships in the system $C_{12}EO_6$/water.

These results were not expected when this study was started. We expected rather that the points of equal Gaussian curvature of the film would be kept from one structure to the other. If this had been the case the cylinders of the hexagonal structure, which are parallel to its [001] direction, would have grown parallel to the rods of the cubic structure, which are parallel to the <110> directions of the latter, and the junctions of three rods in the cubic structure, which are parallel to the {111} planes and correspond the points of zero curvature of the film separating them, would grow parallel to the lamellar planes. The relationships observed concern indeed the planes of highest density of matter, as usual in classical crystals. It should not be forgotten that this fact concerns a Ia3d cubic phase of a particular system and may be not general.

Heterophase fluctuations

These are the fluctuations observed in a structure, when a phase transition is approached, and which can be considered as precursors of the transformation. In general their local symmetry is closely related to that of the stucture to come. For instance, in the case of the solid/lquid transition for which this term was proposed first[40], they are local perturbations of the crystalline order when the transition to the liquid is approached. Similar phenomena have indeed been observed in the case of the lamellar/cubic transformation in $C_{12}EO_6/H_2O$. They manifest themselves, on the X-ray pattern of the monocrystalline lamellar phase, under the form of diffuse scatterings away from the Bragg's spots of the structure, whose intensity increases when the temperature decreases towards the cubic phase[32]. An example of such diffuse scattering is given in Fig. 16.

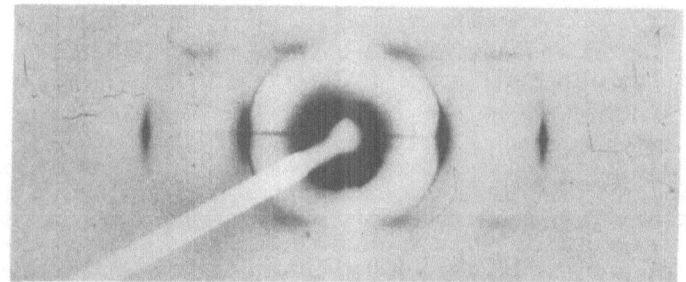

Fig. 16. Diffuse scatterings in the X-ray pattern of a
monocrystal of the lamellar phase of the system
$C_{12}EO_6$/water.

At the beginning of our investigations, we expected these fluctuations to be precursors of the very dramatic topological change of lamellae into labyrinths, so that they should have had a local symmetry related to that of the cubic structure. We could not but find out that this is not the case, their symmetry is indeed related to that of the hexagonal structure. We shall spend sometime discussing this fact as it contains in itself the manifestation of another well known phenomenon of solid state physics.

Going more into the details, the experimental facts obtained from direct optical observations and X-ray scattering experiments are the following. When the temperature is decreased, from the core of the domain of the lamellar phase towards that of the cubic phase, diffuse scattering appear away from the Bragg's spots of the lamellar structure. They were analyzed as corresponding to local fragmentations of the lamellae in cylinders, with a local ordering of the hexagonal type, as shown in Fig. 17.

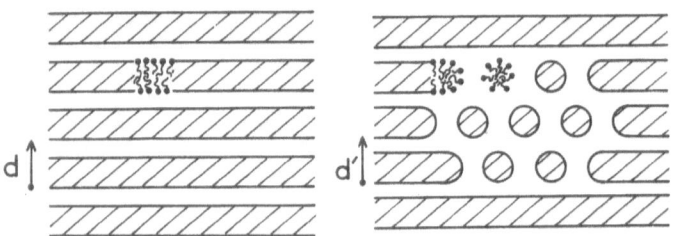

Fig. 17. A fluctuation with local hexagonal symmetry in a lamellar phase.

If some precaution is taken, particularly a slow cooling, it is possible to supercool this situation at temperatures which should correspond to the domain of existence of the cubic phase. This can be maintained for a time long enough to permit the observation of the supercooled lamellar phase and its fluctuations. While the temperature in the cubic phase goes on decreasing slowly, without going out of the domain of this phase, the diffuse scatterings become less and less diffuse and transform into Bragg's spots which, associated with those of the lamellar structure, build a pattern of a hexagonal structure. Thus the system has been forced to build a hexagonal structure at a temperature where a cubic one is more stable. Finally, if a time long enough is waited for, a sudden transformation of the hexagonal structure into the cubic one occurs, the system jumps into its state of lower energy.

Thus, it looks easier for the lamellar structure of $C_{12}EO_6/H_2O$ to transform into a hexagonal structure rather than into a cubic one, although the latter has a lower free energy in this region of the phase diagram. This is a typical example of metastability whose understanding requires the introduction of the phenomena of nucleation and growth. When a structure nucleates and grows within another, the process is not determined by the respective energies of the two structures only, the energy of the interfacial region, or boundary, separating them must be also taken in account. It plays a dominant role at the early stage of the process, when the domains of the new structure are so small that the energy cost localized in their surface for building the boundary with the host structure is comparable with the energy gain in volume. When the domains are large enough the interfacial energy cost may become smaller than the gain in volume, as their surface grows as the square of their size while their volume grows as its cube. In the case of the system under study it is obvious, from Fig. 12 and 17, that the boundary between a lamellar structure and a hexagonal fluctuation can be easily built, with limited distortions of the host structure, whereas the boundary between a lamellar structure and a cubic fluctuation implies important distorsions, as it can be shown that there is no flat line with a constant normal orientation on the surface supporting the film in the cubic structure. Thus, the energy cost of the lamellar/cubic boundary is certainly higher than that of the lamellar/hexagonal one and it may overcome the gain of energy in volume when the domains of fluctuation are small. It is only when the sizes of the domains become large that the volume energy of the cubic structure can win over the interfacial one.

Topological changes

The above discussion explains the observation of a metastable hexagonal structure within the region of existence of the cubic structure, when cooling from the lamellar one, but does not say much about the nature of the processes permitting the topological changes. It suggests that, at least in the case of the $C_{12}EO_6/H_2O$ system studied here, the transformation of a lamellar structure into a cubic one may eventually proceed through a first easy step which is that of the formation of the cylinders of a hexagonal structure. But this is not enough, and a second more difficult step is in anyway needed, as the topologies of the hexagonal and cubic structures are different. The cylinders of the first must become connected to build the two labyrinths of the second, as can be seen in Fig. 4, or the honeycomb-like organization of the film in the first must transform into the complex minimal surface of the second, as can be seen in Fig. 12. This is certainly a complex process as some cylinders are to be broken into pieces which are to be connected to other pieces and cylinders. We do not have much information to describe it at the moment, except one which was obtained when studying the other transformation of the system, that of the stable low temperature hexagonal structure into the cubic one when the temperature increases. Diffuse scatterings around the Bragg's spots of the hexagonal structure show the presence of structural modulations along the cylinders, they are most likely the precursors of the transformation[32].

CONCLUSION

The structures formed by amphiphilic molecules in presence of water have been considered for a long time as tricky particular cases of solutions and, therefore, have been approached mainly following the thermodynamics of solutions and aggregation. This type of approach proved to be extremely useful to understand diluted micellar solutions, when the aggregates of amphiphiles may be assumed to have simple shapes with a simple topology of the spheroïd type, and when they are so diluted and disordered that the notion of a film with constant thickness separating them is of no significance[41]. The situation is different in the case of the concentrated liquid crystalline structures of complex topologies, where the notion of aggregate is not very operative as there is no simple argument to propose any particular shape for the aggregates. Fortunately enough, the proximity of the aggregates gives the notion of film a sense and, this new element of structure being defined and the systems considered as crystals of films, a new approach could be tried, following a program very much inspired of that developed in solid state physics, i.e. dealing with crystallography, defects, epitaxial relations, phase transitions. The classical developments of solid state physics can be applied to this problem of solution because we are concerned with ordered structurations of space, or that a great importance is given to geometry. However, it must not be forgotten that these crystals of films hold part of their originality from the fact that the film is a fluid film. There lies the source for the extraordinary polymorphism. The fluidity is the only state permitting topology changes through the variation of the number of the first neighbors of one molecule and, when two amphiphiles of different characters are present, it permits their relative concentrations to fluctuate, extending the polymorphism by the possibility of inhomogeneous interfacial curvatures, as shown in Fig. 7.

REFERENCES

1. P. Ekwall, in "Adv. Liq. Cryst."1:1, edited by G.H. Brown, Academic Press, New York (1975).
2. V. Luzzati,in "Biological Membranes" 1:71, edited by D. Chapman, Academic Press, New York (1968).
3. A. Skoulios, Ann. Phys.3:421 (1978).
4. R.R. Balmbra, J.S. Clunie and J.F. Goodman, Nature 222:1159 (1969).
5. P. Mariani, V. Luzzati, and H. Delacroix,J. Mol.Biol.204:165 (1988).
6. K. Fontell, K. Fox and E. Hansson, Mol. Cryst. Liq. Cryst.1:9 (1985).
7. J. Rogers and P. A. Winsor, J. Coll. Interface Sci.30:247 (1969).
8. Y. Hendrikx and J. Charvolin, J. de Physique 42:1427 (1981).
9. G. Porte, J. Marignan, P. Bassereau and R. May, J. de Physique 49:511 (1988); J. Marignan, J. Appell, P. Bassereau, P. Delord, F. Larché and G. Porte, Mol. Cryst. Liq. Cryst. 152:153 (1987).
10. C. Safinya, D. Roux, Smith G.S.,Sinha S.K., Dimon P., Clark N.A. and Bellocq A.M., Phys. Rev. Lett. 57:2513 (1986).
11. J. Charvolin and Y. Hendrikx, in "Nuclear Magnetic Resonance of Liquid Crystals", p. 449, edited by J.W. Emsley, Reidel Publishing Company, Dodrecht (1985).
12. J. Charvolin and P. Rigny, J. Chem. Phys. 58:3999 (1973).
13. H. Walderhaug, O. Soderman and P. Stilbs, J. Phys. Chem. 88:1655 (1984).
14. W. Kühner, E. Rommel, F. Noack and P. Meier, Z. Naturforsch. 42a:127 (1987).
15. J. Charvolin, P. Manneville and B. Deloche, Chem. Phys. Lett. 23:345 (1973).
16. E. Sternin, B. Fine, M. Bloom, C. Tilcock, K. Wong and P. Cullis, Biophys.J. 54: 689 (1988).
17. B. Jönsson and H. Wennerström, J. Phys. Chem. 91:338 (1987) and references therein.
18. Y. Hendrikx, J. Charvolin, P. Kekicheff and M. Roth, Liq. Cryst. 2:677 (1987).
19. J.N. Israelachvili, "Intermolecular and surface forces", Academic Press, New York (1985).
20. W. Helfrich, Z. Naturforsch.33a:305 (1978).
21. J.F. Sadoc and R. Mosseri, Pour la Science 87:10 (1985).
22. W. F. Harris, Sc. Am. 237:130 (1977).
23. J.F. Sadoc and J. Charvolin, J. de Physique 47:683 (1986).
24. J. Charvolin and J.F. Sadoc, J. de Physique 48:1559 (1987).
25. J. Charvolin and J.F. Sadoc, J. de Physique 49:521 (1988).
26. J. Charvolin and J.F. Sadoc, J. Phys. Chem. 92:5787 (1988).
27. J.F. Sadoc and J. Charvolin, Acta Cryst.A 45:10 (1989).
28. A.H. Schoen, NASA Tech. Note D-5541 (1970).
    L.E. Scriven,Nature 263;123 (1976).
    S.T.Hyde and S. Andersson, Z. Krystall. 168:221 (1984).
    A. L. MacKay, Nature 314:604 (1985).
    E.L. Thomas, D.M. Anderson, C.S. Henke, and D. Hoffman, Nature 334:598 (1988).
29. H.G. von Schnering and R. Nesper, Angew. Chem. 26:1059 (1987).
30. W. Helfrich, J. de Physique 48:291 (1987).
    D.M. Anderson,S.M. Gruner and S. Leibler, PNAS 85:5364 (1988).

M. Kleman, _Liq. Cryst_.3:1355 (1988).

J. Charvolin and J.-F. Sadoc, in preparation.

31. Y. Rançon, and J. Charvolin, _J. Phys. Chem_. 92:2646 (1988).
32. Y. Rançon, and J. Charvolin, _J. Phys. Chem_. 92:6339 (1988).
33. D.J. Mitchell, G. Tiddy, L. Waring, T. Bostock, and M.P. McDonald, _J. Chem. Soc., Far. Trans_. 79:975 (1983).
34. Y. Rançon, and J. Charvolin, _J. de Physique_ 47:683 (1987).
35. M. Allain, _Europhys. Lett_. 2:597 (1986).
36. I. Markov and S. Stoyanov, _Cont. Phys._ 28:267 (1987).
37. K. Larsson, K. Fontell, and N. Krog, _Chem. Phys. Lipids_ 27:321 (1980).
38. S.M. Gruner, K.J. Rotschild, W.J. DeGrip, and N.A. Clark, _J. de Physique_ 46:193 (1985).
39. E.L. Thomas, D.M. Anderson, C.S. Henke, and D. Hoffman, _Nature_ 334:598 (1988).
40. J. Frenkel, "Kinetic theory of liquids", Dover, New York (1955).
41. C. Tanford, "The hydrophobic effect", Wiley, New York (1980).
    J.N. Israelachvili, S. Marcelja, and R. Horn, _Quat. Rev. Biophys_. 13:121 (1980).
    W.M. Gelbart, A. Ben Shaul, and A. Masters, in "Physics of amphiphiles: micelles, vesicles and microemulsions", edited by V. Degiorgio and M. Corti, North Holland, Amsterdam (1985).

X-RAY REFLECTIVITY AND DIFFRACTION STUDIES OF LIQUID

SURFACES AND SURFACTANT MONOLAYERS

Jens Als-Nielsen and Kristian Kjær

Physics Department
Risø National Laboratory
DK-4000 Roskilde, Denmark

1. INTRODUCTION

The structure of matter on a molecular length scale can be revealed by diffraction studies using radiation with wavelength in the Ångstrøm region. If the radiation couples weakly to the scattering objects, as is the case for neutrons and X-rays (but not for electrons), the interpretation of the diffraction pattern in terms of the underlying structure becomes particularly simple and reliable due to the validity of the Born approximation, which in this context is also called kinematical diffraction. On the other hand, if the coupling is weak, the beam has to be sufficiently intense and/or the sample must be of sufficient size to obtain an accurate diffraction pattern within a reasonable time. In a surface structure the number of diffracting atoms, confined within a nanometer thick surface layer, is comparatively small.

For this reason neutron diffraction has not yet been developed to be a significant tool in surface science although neutrons are sensitive to magnetism and to hydrogen locations by the use of a controlled isotopic ratio of protons to deuterons.

Very intense X-ray beams are now available in synchrotron radiation laboratories, and surface X-ray diffraction methods have been developed concomitantly. In this paper we shall confine ourselves to describing methods for studying surfaces of **liquids**. These can be simple liquids such as water or methanol for which the surface diffuseness due to thermal fluctuations has been determined[1], they can be liquid crystals with smectic layering near the free surface[2], or they can be heterogeneous systems such as a monolayer of amphiphilic molecules on a water surface[3].

It is useful to distinguish between two geometries of the diffraction experiment as shown in Fig. 1.1. **Specular reflection**, shown in the top part has a wavevector transfer **Q** perpendicular to the liquid surface and thus measures the average density variation across the surface. In the **diffraction** geometry, shown in the bottom part, surface sensitivity is obtained by means of grazing incidence of the incoming beam. In-plane structure of the surface is probed by scanning the angle 2θ or the corresponding horizontal wavevector transfer Q_{hor}. For fixed 2θ or Q_{hor} the intensity variation with Q_z as observed, e.g., with a vertical position sensitive detector reflects the structure perpendicular to the surface just as does specular reflection. Nevertheless the two methods do not necessarily provide identical information. If, for example, we consider a monolayer film on water with coexisting two-dimensional

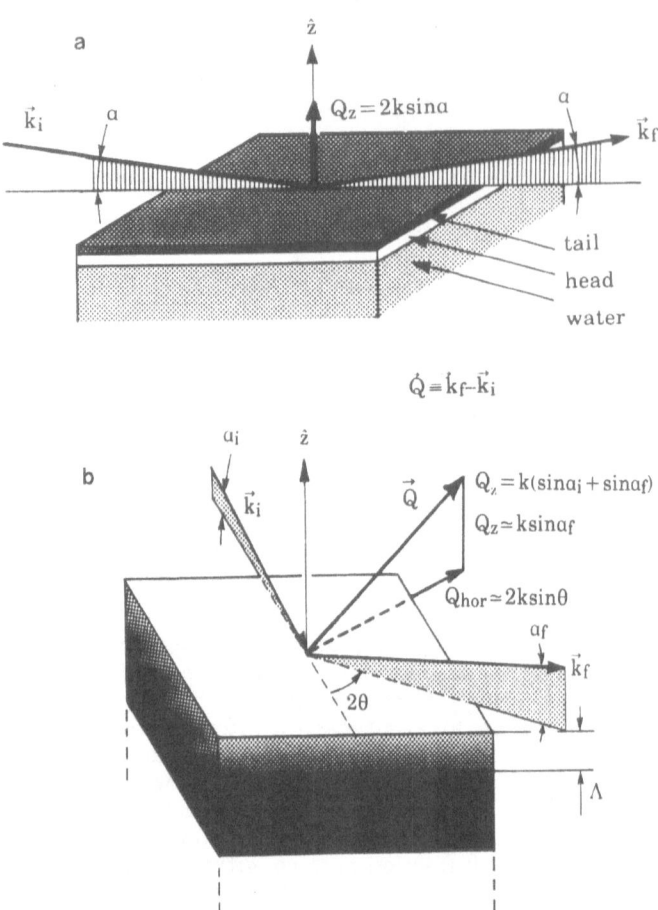

a

$\hat{z}$

$\vec{k}_i$ a $Q_z = 2k\sin a$ a

                                                                        $\vec{k}_f$

                                                                tail

                                                                head

                                                                water

$\dot{Q} = \dot{k}_f - \vec{k}_i$

b

a_i $\hat{z}$

   $\vec{k}_i$ $\vec{Q}$ $Q_z = k(\sin a_i + \sin a_f)$

                                $Q_z \simeq k\sin a_f$

                        $Q_{hor} \simeq 2k\sin\theta$

                                                a_f

                                                $\vec{k}_f$

        2θ

                                                Λ

Fig. 1.1. Top: Specular reflection geometry probes the density profile across
the interface.
Bottom: By grazing incidence the X-ray penetration depth Λ can be
limited to a few nanometers. The evanescent wave is diffracted by
the in-plane structure of the Langmuir film.

crystalline and liquid phases, the Q_z variation of intensity at 2θ corresponding
to Bragg reflection from the crystalline phase reflects the thickness of the
crystallites only, whereas the specular reflection pattern measures some sort of
average thickness of the liquid and crystalline film.

The paper is organized as follows. First we recall and discuss Snell's and
Fresnel's laws for X-ray optics. We then derive the general relation of the
density profile across the surface to specular reflectivity (Fig. 1.1a) and to the
Q_z-variation in grazing incidence diffraction (Fig. 1.1b). Specular reflectivity
is illustrated by two examples. The first is reflection from a bare water surface
and the determination of the diffuseness of the air-water interface due to
thermally excited capillary waves. In the second example we consider a
monomolecular film of an amphiphilic molecule, arachidic acid, floating on
water, as the area per molecule is varied by a moveable barrier in a Langmuir
trough[4].

The reflectivity data, analyzed in terms of a smeared box model of the molecular density, suggests that the hydrocarbon tails are close-packed but tilt uniformly as more area becomes available per molecule.

The grazing incidence diffraction (GID) technique is used to examine the molecular structure of the arachidic acid film in more detail. By mapping the intensity variation with both 2θ and α_f (Fig. 1.1b) we find:

(i)   In the most compressed phase the molecules are upright and form a hexagonal lattice.

(ii) As the pressure is released the molecules tilt towards their nearest neighbours and the hexagonal lattice becomes uniaxially distorted in the direction of tilt.

(iii) The tilt angle and the density profile deduced from the GID data are consistent with the interpretation of the specular reflection data.

2. X-RAY OPTICS

For X-rays or neutrons of wave vector k the refractive index, n, of a medium can be simply related to the scattering properties of the medium, see, e.g., appendix of Ref. 3. For X-rays the relevant parameters are the Thompson scattering length of a single electron, r_o, and the electron density, ρ_{el}:

$$n = 1 - \delta, \qquad \delta = 2\pi\rho_{el}r_o k^{-2}$$

Here, for simplicity we have neglected absorption effects and effects occurring when the photon energy is close to a resonance between the electron shells of the atom. The geometry of refraction at a sharp interface is as depicted in Fig. 2.1 with the angles α and α' being related by Snell's law: $\cos(\alpha)/\cos(\alpha') = n$. Since n is only very slightly less than unity (and can be written as $n = 1 - \alpha_c^2/2$)

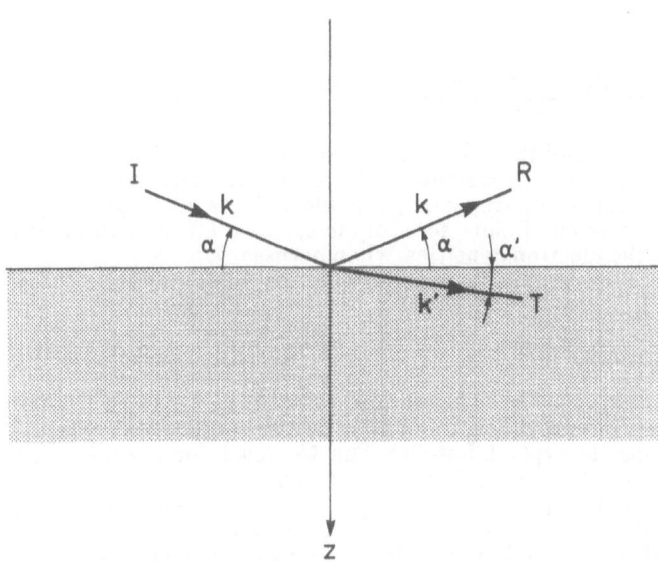

Fig. 2.1.   The geometry of glancing incidence reflection and transmission at a discontinuous interface.

all glancing angles are small and expansion of Snell's law yields

$$a^2 = a'^2 + a_c^2$$

The reflectivity R_F is the square of the ratio of the reflected and incident wave amplitudes. For small angles the Fresnel law for a sharp interface becomes particularly simple. The corresponding Fresnel reflectivity R_F is

$$R_F = |(a-a')/(a+a')|^2 \xrightarrow[a \gg a_c]{} (a_c/2a)^4 . \tag{2.1}$$

Similarly one finds the transmitted or refracted wave intensity T_F, normalised to the incident intensity,

$$T_F = |2a/(a+a')|^2 . \tag{2.2}$$

Snell's and Fresnel's laws are derived by satisfying the boundary conditions for the X-ray wave fields at the interface.

Next, to include absorption effects in the refractive index, n, consider a plane wave, $\exp(-ikz)$, at normal incidence on a semi-infinite medium with linear absorption coefficient μ. On entering the medium the wavevector is changed to nk. We would like to write the plane wave in the medium as $\exp(-inkz)$, but we require an exponentially decaying **amplitude** $\exp(-\mu z/2)$. Formally, this is obtained by letting n be a complex number with an imaginary part $-i\beta$, with the relation to μ given by equating $-\beta kz$ with $-\mu z/2$ or $\beta = \mu/(2k)$. Including absorption effects thus leads to a complex index of refraction

with
$$n = 1 - \delta - i\beta \tag{2.3}$$

and
$$a_c^2 = 2\delta = 4\pi r_0 k^{-2} \rho_{e\ell}(1 + f'/Z) \tag{2.4}$$

$$\beta = \mu/(2k) = 2\pi r_0 k^{-2} \rho_{e\ell}(f''/Z) \tag{2.5}$$

In Eqs. (2.4) and (2.5), f' and f'' are the real and imaginary parts of the anomalous dispersion correction to the atomic scattering factor[5], which can be important when the X-ray energy is close to an absorption edge. Z is the number of electrons. Below, for simplicity, such effects will be assumed to be included in the electron densities, when necessary.

Including absorption effects, Snell's law for small angles becomes

$$a^2 = a'^2 + a_c^2 + i2\beta \tag{2.6}$$

whereas the expressions (2.1) and (2.2) remain valid using a' from Eq. 2.6.

The z-dependence of the transmitted wave amplitude is $\exp(-ik'a'z)$ which is proportional to $\exp(-k'\mathrm{Im}(a')z)$. The 1/e depth for the intensity, Λ, is thus given by $\Lambda^{-1} = 2k'\mathrm{Im}(a')$. However, for X-rays the deviation of n from unity is very small and the difference in **length** between k and k' can be neglected. The results for R_F, T_F and Λ depend on several parameters: The incident angle a, density and absorption in the medium, as well as the wavevector. In order to get an overview of this multi-parameter problem it is convenient to use suitable units and to estimate orders of magnitudes. The natural unit for angles is the critical angle a_c. However, in connection with diffraction and reflection phenomena from non-homogeneous media, the wavevector transfer $Q_z = 2k\sin(a) \approx 2ka$ is a more useful variable than just the grazing angle a. The natural unit for Q_z is $Q_c = 2ka_c$ which, incidentally, varies only slightly with substance: $Q_c = 0.0217$ Å^{-1} for a light material such as water and $Q_z = 0.0678$ Å^{-1} for a heavy material like mercury. In terms of the dimensionless quantities x and x' for wavevector transfer and b for absorption

$$x \equiv Q_z/Q_c = a/a_c, \quad x' \equiv 2ka'/Q_c = Q_z'/Q_c = a'/a_c \quad \text{(complex)}, \quad b \equiv -(2\mu k/Q_c^2), \quad (2.7)$$

with x' determined from the dimensionless form of Eq. 2.6

$$x^2 = x'^2 + 1 + i(2b) \tag{2.8}$$

and recalling explicitly the formula for Q_c (cf. Eq. 2.4):

$$Q_c = 4(\pi\rho_{e\ell} r_0)^{1/2} \tag{2.9}$$

the final formulas for R_F, T_F and Λ become

$$R_F(x) = |(x - x')/(x + x')|^2, \tag{2.10}$$

$$T_F(x) = |2x/(x + x')|^2, \tag{2.11}$$

$$\Lambda^{-1}(x) = Q_c \,\text{Im}(x'). \tag{2.12}$$

Fig. 2.2 (left part) shows from top to bottom graphs of R_F, ΛQ_c, T_F and the phase of the reflected wave for different absorption parameters b. For $x \geq 1.4$, the dependence on absorption is very small and the four quantities are compared to their asymptotic forms in the right part of Fig. 2.2. It may be useful to discuss separately the two limiting cases $a \gg a_c$ and $a \ll a_c$ as well as the special case of $a = a_c$.

(i) $x >> 1$ or $a >> a_c$:

In this case the solution for x' in Eq. 2.8 yields $\text{Re}(x') \approx x$ and $\text{Im}(x') \approx -b/x$. From Eq. 2.10 we find $R_F(x) \approx 1/(2x)^4$, in phase with the incident wave. The incident wave is almost completely transmitted though the interface, and the penetration depth is a/μ. The reflected intensity falls off as

$$R_F \to (Q_c/2Q_z)^4 \tag{2.13}$$

(ii) $x << 1$ or $a << a_c$:

In this case x' is almost purely imaginary with $\text{Im}(x') \approx 1$, implying a reflectivity $R_F(x)$ close to unity. The reflected wave is out of phase with the incident wave, so the transmitted wave becomes very weak. It propagates nealy **parallel** to the surface with a minimal penetration depth of Q_c^{-1}, independent of a for $a << a_c$. Due to its small penetration depth, the wave field below the interface is called an **evanescent wave**.

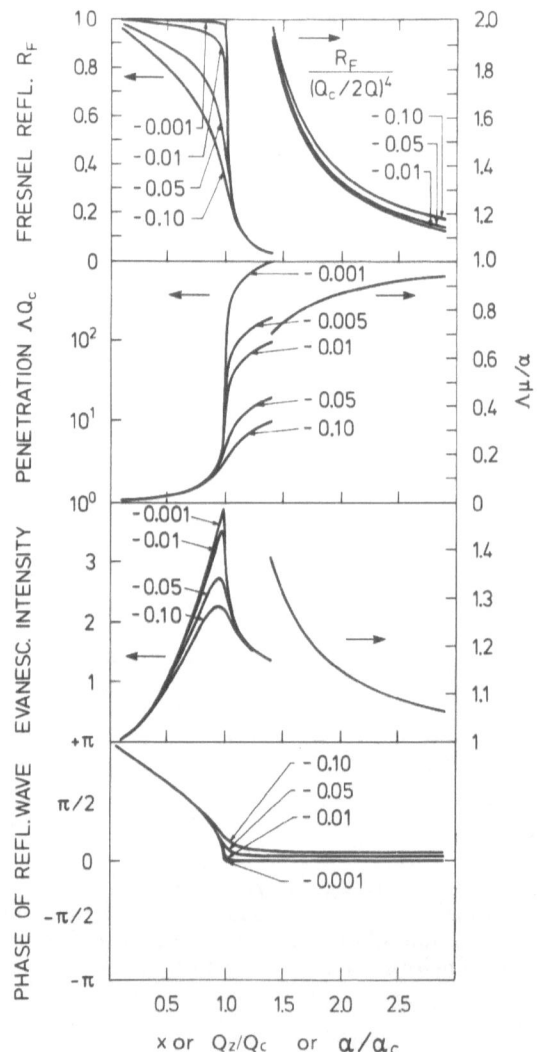

Fig. 2.2. *Dependence on* $x = Q_z/Q_c = \alpha/\alpha_c$ *of reflectivity, penetration depth, evanescent wave intensity and phase difference between reflected and incident wave for various values of the absorption parameter b.*

(iii) $x = 1$ or $\alpha = \alpha_c$:

From Eq. 2.8 we find $x' = |b|^{1/2}(1 + i)$. Since $b \ll 1$, $R_F(x) \approx 1$. The reflected wave is in phase with the incident wave implying that the evanescent wave **amplitude** approaches twice the incident wave amplitude. The penetration depth from Eq. 2.12 becomes $|b|^{-1/2}$ times larger than that for $x \ll 1$.

In summary then, we have considered reflected and transmitted waves for a grazing X-ray beam incident on a homogeneous, planar substance with a sharp interface. The transmitted wave intensity has a finite $1/e$ penetration depth Λ, partly due to ordinary absorption in the medium, but mainly due to the phenomenon of total external reflection. When the grazing angle α is less than the critical angle α_c the transmitted wave propagates exactly parallel to the surface when ordinary absorption is neglected, and almost parallel to the surface for the absorption occurring in practice. This wave is called the **evanescent** wave. For $\alpha > \alpha_c$ the reflected wave has a finite intensity approaching $(\alpha_c/2\alpha)^4$ for $\alpha \gg \alpha_c$.

For quantitative results at given wavevector k and a given material one first calculates the critical wavevector Q_c using Eq. 2.9 and the absorption parameter b using Eq. 2.7. With this information figure 2.2 can be used directly to estimate penetration depth, reflectivity or evanescent intensity. For more accurate work use formulas 2.8 and 2.10-12 recalling that the quantity x' is a complex number.

3. SPECULAR REFLECTIVITY AND INTERFACIAL DENSITY PROFILE

On an atomic length scale the interface between the liquid and the vapour above it is not sharp. In this section we shall see how the specular reflectivity $R(Q_z)$ at wavevector transfer Q_z is changed accordingly from the Fresnel reflectivity $R_F(Q_z)$. The electron density profile is denoted $\rho(z)$ and the density gradient $\rho'(z)$, see Fig. 3.1.

In order to derive $R(Q_z)$ we consider the reflected wave as a superposition of waves reflected from infinitesimal planes at varying depth z, implying the phase factor $\exp[iQ_z \cdot z]$. At first we neglect refraction and absorption effects. The reflectivity of a thin plate[6] with thickness Δz can be derived from the following simple dimensional argument. The reflected wave is the result of Thompson scattering of the incident photon wave by the individual electrons. The reflected **amplitude** ΔA_r must be proportional to the incident amplitude A_i, to the scattering length of a single electron and to the number of electrons per unit area perpendicular to the incident beam, $\rho(z)\Delta z/\sin\alpha$. Since $\Delta A_r/A_i$ is dimensionless and the dependence on quantities with dimensions of length such as scattering length, density and plate thickness is exhausted by their product, the only additional length in the problem, the X-ray wavelength λ must enter linearly. Hence,

$$\frac{\Delta A_r}{A_i} = c \cdot \lambda \cdot r_0 \cdot \rho(z)\Delta z/\sin\alpha$$

$$= c(4/Q_z)|\pi\rho_\infty r_0| \cdot \rho(z)/\rho_\infty \cdot \Delta z$$

$$= c(Q_c^2/4Q_z) \cdot \rho(z)/\rho_\infty \cdot \Delta z \tag{3.1}$$

$$k = 2\pi/\lambda$$

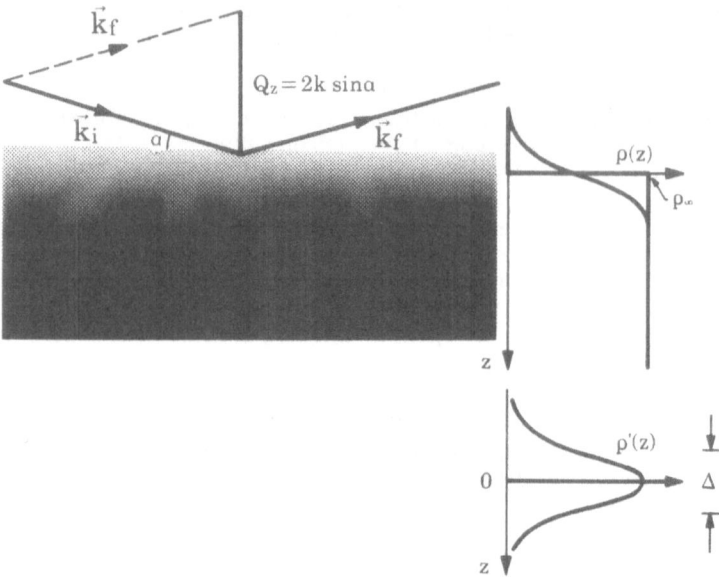

Fig. 3.1. *The density variation across a surface is indicated by shading in the left part and more quantitatively by the function ρ(z) in the right part. The reflectivity versus wavevector transfer Q_z is related to the Fourier transform of the gradient of the density, ρ'(z).*

using Eq. (2.9) for Q_c. Here, c is a dimensionless constant to be determined, and ρ_∞ is the electron density below the interface region.

The resulting reflected wave amplitude A_r is obtained by integration of Eq. (3.1) with the appropriate phase factor $\exp[iQ_z z]$ included. We find

$$\frac{A_r}{A_i} = (c/i)(Q_c/2Q_z)^2 \phi(Q_z) \tag{3.2}$$

with

$$\phi(Q_z) = iQ_z \rho_\infty^{-1} \int \rho(z) \exp[iQ_z z]\, dz \tag{3.3}$$

$$= \rho_\infty^{-1} \int \frac{d\rho(z)}{dz} \cdot \exp[iQ_z \cdot z]\, dz \tag{3.4}$$

being the Fourier transform of the gradient of the density profile across the interface. In the limit of a sharp interface the density gradient approaches a delta-function, $\phi(Q_z)$ approaches unity and the reflectivity is

$$(A_r/A_i)_F^2 = |c|^2 \cdot (Q_c/2Q_z)^4 \cdot 1 \tag{3.5}$$

In this derivation we have neglected refracting effects, which is equivalent to $Q_z \gg Q_c$. We saw in the previous section that in this limit $R_F = (Q_c/2Q_z)^4$ so we conclude that $|c|^2 = 1$ and are lead to the conjecture

$$R(Q_z) \simeq R_F(Q_z) \cdot |\phi(Q_z)|^2 , \tag{3.6}$$

replacing $(Q_c/2Q_z)^4$ by the general form $R_F(Q_z)$.

We discuss the limits of validity of the **master formula** (3.6): In the superposition of reflected waves from thin plates we used the phase factor $\exp[iQ_z z]$ with $Q_z = 2k\sin\alpha$. A more accurate phase than $Q_z z$ would be $Q'_z z = 2k\sin\alpha'\cdot z$ with $\alpha' = (\alpha^2 - \alpha_c^2)^{\frac{1}{2}}$.

Furthermore, multiple scattering effects have been neglected, e.g., a reflected wave from a thin plate at z_2 might be reflected back into the substrate from another thin plate at z_1 closer to the surface. Such multiple scattering effects are not important for $Q_z \gg Q_c$ as the reflectivity for a plate gets very small in this limit, but for Q_z approaching Q_c the validity of Eq. (3.6) might be questioned. In order to elucidate this problem we show in Fig. 3.2 the

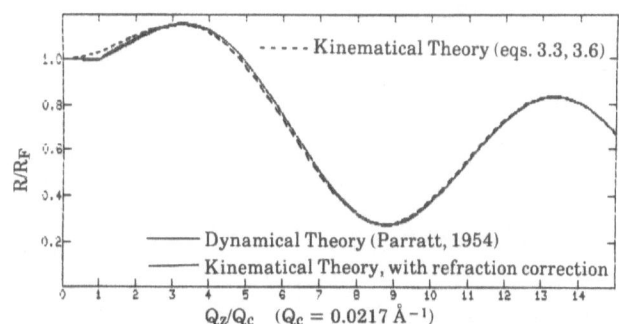

Fig. 3.2. *Normalized reflectivity R/R_F versus normalized wave vector transfer Q_z/Q_c, as calculated:*

 ---- *(i) using the kinematical* **master formula**, *Eq. (3.4);*

 ——*(ii) using Eq. (3.4) with* **refraction correction** *(replacing Q_z by Q_z'); and*

 ——*(iii) using the* **dynamical** *theory by Parratt[7].*

Methods (ii) and (iii) give almost identical results, while (i) differs noticeably for $Q_z < 10\cdot Q_c$, and significantly for $Q_z < 2\cdot Q_c$.

reflectivity of a particular density model corresponding to a typical amphiphilic monolayer on water as calculated using Eqs. (3.4) and (3.6), the so-called kinematic approximation. We now compare this to an exact calculation method devised by Parratt[7] in 1954. The model used in both calculations consists of two stratified layers of different densities. The top layer corresponds to the tail density of the film, the next layer to the head density, cf. Fig. 1.1a. At each of the three interfaces (air-tail, tail-head, head-water) both incoming and outgoing rays can be reflected or transmitted and one imposes the usual boundary conditions for the electric field of the electromagnetic wave at each interface. This results in a set of coupled linear equations which can be solved[7] for the overall reflectivity. We conclude from Fig. 3.2 that the simple kinematic approximation, Eq. (3,4), is adequate for interpreting reflectivity data from amphiphilic monolayers on water for grazing angles exceeding twice the critical angle and that the remaining discrepancy can be repaired by inclusion of the refraction correction (Q_z' instead of Q_z in Eq. (3.4)).

Before closing this section let us go back and discuss in more detail the quantity $\phi(Q_z)$, the Fourier transform of the density gradient, in the case of a monolayer on a substrate. It is convenient to separate the density into two parts

$$\rho(z) = \rho_1(z) + \rho_2(z) \tag{3.7}$$

where $\rho_1(z) = \rho_\infty \cdot H(z)$ derives from the subphase or substrate, $H(z)$ being the step function, and $\rho_2(z)$ is the density due to the molecules in the monolayer. The effect of interfacial diffusiness will be discussed below in sections 5 and 6. Denoting the electron density in the molecule by $\rho_m(x,y,z)$ and the average molecular area by A, $\rho_2(z)$ becomes

$$\rho_2(z) = A^{-1} \int \rho_m(x,y,z) \, dx dy \tag{3.8}$$

It follows from Eqs. (3.3), (3.4) and (3.8) that

$$\phi(Q_z) = 1 + \rho_\infty^{-1}(iQ_z) \int \rho_2(z) \exp[iQ_z z] dz$$

$$= 1 + iQ_z(\rho_\infty A)^{-1} \int \rho_m(x,y,z) \exp[iQ_z z] dx dy dz. \tag{3.9}$$

In terms of the molecular form factor

$$F(\mathbf{Q}) \equiv \int \rho_m(\mathbf{r}) \exp[i\mathbf{Q} \cdot \mathbf{r}] d^3 r \;, \tag{3.10}$$

Eq. (3.9) becomes

$$\phi(Q_z) = 1 + iQ_z(\rho_\infty A)^{-1} F(0,0,Q_z) \;. \tag{3.11}$$

Using this form of $\phi(Q_z)$ in the master formula Eq. (3.6) for the reflectivity it is particularly transparent that the specular reflection along $\mathbf{Q} = (0,0,Q_z)$ is formed by the *interference* of waves scattered from the substrate, the first term in (3.11), and from the molecular film, the term with the molecular form factor $F(0,0,Q_z)$.

In particular, we note in passing that the substrate scattering (measured perhaps with an uncovered surface) should **not** be subtracted from experimental data as a background. The correct experimental background is found by off-setting the detector laterally from the specularly reflected beam.

4. GRAZING INCIDENCE DIFFRACTION AND BRAGG RODS

In this section we discuss the similarities and differences between information obtained by specular reflection (XR) and by grazing incidence diffraction (GID). Experimental examples are discussed in section 7 below.

For simplicity, we assume a 2D–periodic structure (”2D–crystallinity”) in the monolayer film floating on the subphase: The molecules are arranged in identical unit cells which form a regular lattice. Then, in GID, Bragg diffraction occurs when the lateral scattering vector $\mathbf{Q}_{hor}$, c.f. Fig. 1.1b, coincides with a reciprocal lattice vector $\mathbf{G}_{hk}$: The scattering is concentrated in so–called **Bragg Rods** (parallel to the Q_z–axis), defined by **two** Laue conditions or, in vector notation, by the equation $\mathbf{Q}_{hor} \equiv \mathbf{G}_{hk}$. By constrast, XR is characterised by the condition $Q_{hor} \equiv 0$. The substrate gives no Bragg diffraction for $Q_{hor} \neq 0$ so in GID there is no interference between the scattering from the substrate and that from the film. The substrate scattering just contributes to a flat background which can be subtracted from the total intensity to obtain the GID signal. The purpose of using grazing incidence is to minimize the background level by illuminating a depth of only a few nanometers, c.f. section 2. The GID signal, $I_{hk}(Q_z)$, is proportional to the square of the unit cell structure factor (identical to the molecular form factor, Eq. (3.10), for the case of one molecule per unit cell):

GID:
$$I_{hk}(Q_z) \propto A_c^{-2} |F(G_{hk}, Q_z)|^2 ,$$
(4.1)

where A_c is the unit cell area. Compare Eq. (4.1) for GID with the XR result:

XR:
$$I_{00}(Q_z) = R_F(Q_z) \left| 1 + iQ_z (\rho_\infty A)^{-1} F(0,0,Q_z) \right|^2 .$$
(4.2)

Thus, XR and GID measure different parts of the monolayer structure factor, corresponding to different projections of the monolayer structure. XR corresponds to the projection of the monolayer density onto the z–axis and includes also scattering from the sub–phase. GID – with $Q_z \approx 0$, i.e. , grazing exit as well as grazing incidence – measures the structure of the "2D–crystalline" part of the monolayer, as projected onto the x–y plane. Finally, measurement of the GID signal $I_{hk}(Q_z)$ versus Q_z (so–called Bragg Rod scans) gives three–dimensional information about the 2D–crystalline part of the monolayer.

We end this section with a couple of examples which further illustrate the similarities and differences between the XR and GID methods.

(i) Assume, for example that only the aliphatic tails order laterally whereas the polar heads are laterally "disordered". This means that the Debye–Waller factor (implicit in Eq. (3.10)) will be large and anisotropic for the "head" part of the molecule, so as to effectively make the heads invisible for $\approx$horizontal wavevector transfers Q. Thus, effectively, for GID the formfactor to be used is that describing the tails, - the polar heads will just contribute to the background together with the substrate. In XR the lateral disorder of the polar heads is irrelevant; they do indeed contribute to the average density modulation across the surface.

(ii) Another example where XR and GID give complementary rather than identical information is that of a heterogeneous film: Islands of 2D–solid phase coexisting with a non-diffracting 2D–liquid phase. The GID measures the form factor of the molecules in the solid islands whereas XR yields some sort of average density profile across the surface. The wording "some sort of average" is deliberately vague, because one must distinguish between averaging the formfactor before squaring in Eq. (4.2) (coherent averaging) or averaging the reflected intensities which is obviously required if the size of the islands is much larger than the X-ray coherence length given by $1/k_x$ and $1/k_y$ of Fig. 5.2 in the following section.

5. THERMAL ROUGHNESS OF LIQUID–VAPOUR INTERFACE

Consider in Fig. 5.1 a liquid surface confined within the area $L \times L$. A capillary wave with amplitude u_q and wavelength λ or wave vector q has been excited. The excitation energy has two origins: the surface has been enlarged, which requires the surface tension energy E_c, and liquid has been lifted from troughs to crests, which requires the gravitational energy E_g. In the bottom

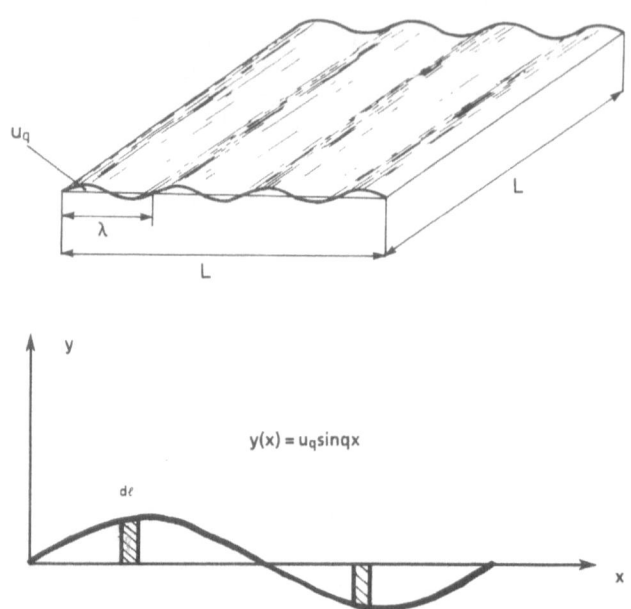

Fig. 5.1. *Top:* *A capillary wave, confined within the area* $L \times L$, *of wavelength* λ *or wavevector* q *and amplitude* u_q.
Bottom: *Side view of one wavelength. The arc-element has a length* $d\ell = [1 + (dy/dx)^2]^{1/2}dx$ *or approximately* $[1 + \frac{1}{2}(u_q q \cos(qx))^2]dx$.

part of Fig. 5.1 we consider one period of the wave. Note in passing that there will be L/λ such periods along the x-axis. The excess surface is $L(u_q q/2)^2 \lambda (L/\lambda)$ and the associated energy $E_c = L^2 \gamma (u_q q/2)^2$ where γ is the surface tension. The gravitational energy for one period is the integral of dE_g over half a period as the liquid in the trough in the second half of the period is lifted into the crest in the first half of the period. The entire gravitational energy is $E_g = L^2 \rho g (u_q/2)^2$. Since ρg has dimension of energy/area times an inverse length squared, it can be written as

$$\rho g \equiv \gamma k_g^2 \tag{5.1}$$

where k_g is a wave vector and therefore

$$E_q = (E_c + E_g) = \tfrac{1}{2} L^2 (u_q^2/2) \gamma (q^2 + k_g^2). \tag{5.2}$$

Equipartition of gives the thermal average value $\langle E_q \rangle = k_B T/2$ or

$$\tfrac{1}{2} \langle u_q^2 \rangle = k_B T L^{-2} [\gamma (q^2 + k_g^2)]^{-1}. \tag{5.3}$$

Summing over all q-modes gives $\langle u^2 \rangle$:

$$\langle u^2 \rangle = \tfrac{1}{2} \sum_q \langle u_q^2 \rangle = k_B T (L/2\pi)^2 L^{-2} \int_0^{k_{max}} [(\gamma (q^2 + k_g^2)]^{-1} 2\pi q \, dq \tag{5.4}$$

or

$$<u^2> = k_B T (2\pi)^{-2}(\pi/\gamma)\log_e\left(\frac{k_{max}^2+k_g^2}{k_g^2}\right) \simeq \frac{k_B T}{2\pi}\frac{1}{\gamma}\log_e\frac{k_{max}}{k_g} \quad . \tag{5.5}$$

In the integral over q-space we have introduced an arbitrary cut-off wavevector k_{max} which is of order π/(molecular radius).

In this derivation we have assumed that the effective surface tension is independent of wavevector. In reality this is not the case. In deriving the excess surface area of mode q we used that the line element $d\ell$ along the curved interface $y(x)$ is $(1 + (dy/dx)^2)^{1/2}$ and we expanded the square root to first order as $(1 + (dy/dx)^2/2)$. Within this approximation the total energy is then a sum of independent, harmonic q-modes. However, with the square root expanded to higher orders one realizes that the modes are **not** harmonic: the energy of a q-mode contains all even powers of q. As long as $(dy/dx)^2 = (u_q q)^2 << 1$ the harmonic approximation is accurate for describing the excitation of one single q-mode out of the ground state, but when it comes to excite this mode out of a general, thermally excited state, the population of the other q-modes matters for calculating E_q. It is convenient to write E_q as proportional to $\gamma(q)q^2$ with an effective surface tension $\gamma(q)$ depending on q. By symmetry there cannot be any term linear in q, and the coefficient to q^2 must be proportional to $k_B T$, as it reflects the thermal population of the other q-modes. To order q^2 the effective surface tension $\gamma(q)$ is therefore of the form:

$$\gamma(q) = \gamma + a k_B T q^2, \tag{5.6}$$

a being a dimensionless constant. Meunier[8] finds $a = 3/(8\pi)$. Rewriting Eq. (5.6) as

$$\gamma(q) = \gamma(1 + (q/k_m)^2), \tag{5.7}$$

$$k_m^{-2} = a k_B T/\gamma \ , \ a = 3/(8\pi), \tag{5.8}$$

we find, in analogy with Eq. (5.4)

$$<u^2> = k_B T/(4\pi^2\gamma)\int_0^\infty [(1 + (q/k_m)^2)(q^2 + k_g^2)]^{-1} 2\pi q \, dq$$

$$\simeq k_B T/(4\pi^2\gamma)\int_0^\infty [q^2 + k_g^2 + k_m^{-2} q^4]^{-1} 2\pi q \, dq \simeq k_B T/(2\pi\gamma)\log_e\frac{k_m}{k_g} \quad , \tag{5.9}$$

i.e. the same form as Eq. (5.5) but with the arbitrary cut-off wavevector k_{max} replaced by the mode-mode coupling parameter k_m known from Eq. (5.8). For water the numerical values are 1.63 Å^{-1} and 1.23 Å^{-1}, respectively. The relative difference between $\log_e(k_{max}/k_g)$ and $\log_e(k_m/k_g)$ is thus only 1.6 per cent. We now generalize this result to the case where the base surface is covered by a monolayer[8]. This layer has a certain stiffness against undulations so a fluctuation as given in Fig. 5.1 will require an additional energy of the form $\frac{1}{2}\cdot K\cdot u_q^2 q^4 \cdot L^2$. Expressing K in units of $k_B T$ by the dimensionless number κ

$$K \equiv \kappa \cdot k_B T \tag{5.10}$$

125

we see immediately from (5.9) that k_m^{-2} in the integral must be replaced by

$$k_M^{-2} \equiv k_m^{-2} + K/\gamma = (a+\kappa)k_B T/\gamma. \tag{5.11}$$

The lower limit of zero in the two-dimensional integral (5.9) is an idealization which cannot be fulfilled in an actual experiment. Here one must distinguish between intensity which is specularly reflected and intensity scattered out of specular reflection by the surface roughness. This, then is a matter of the lateral wavevector **resolution** widths k_y along the direction of the beam projected onto the surface and k_x perpendicular to this direction. In practice both k_x and k_y are much larger than k_g, so the second term in the integrand in Eq. 5.9 can be neglected. The observed roughness σ is then given by

$$\sigma^2 = k_B T/(4\pi^2\gamma) \int_{k_x}^{\infty} \int_{k_y}^{\infty} [q^2 + k_M^{-2}q^4]^{-1} dq_x dq_y. \tag{5.12}$$

The integration area is indicated in Fig. 5.2 as the shaded area. In a synchrotron X-ray reflectivity experiment the resolution may well be entirely determined by the detector apertures (width w_d, height h_d) relative to the distance D between sample and detector since the incident beam collimation usually is very narrow. In that case the resolution function is box-like with dimensions as indicated in Fig. 5.2. Utilizing that the resolution perpendicular to the beam is much broader than along the beam, cf. Fig. 5.2, the integral 5.12 can again be carried out analytically[1]. The result is a roughness parameter σ which depends logarithmically on the wavevector transfer Q_z because the resolution rectangle varies linearly with Q_z:

$$\sigma^2 \simeq k_B T/(2\pi\gamma)\log_e(k_M/k_y), \qquad k_y = Q_z(h_d/D). \tag{5.13}$$

Note that neither the gravity term k_g nor the width of the detector aperture w_d appears in this final result.

Daillant et al. find in their study of a behenic acid film on water[9], that for surface pressures below 17 mN/m the action of the behenic acid film on the thermal roughness is just to diminish the surface tension from the pure water value of 72 mN/m to (72-17)mN/m = 55 mN/m and indeed they find a rms. roughness σ_{exp} varying as $\gamma^{-1/2}$ as shown in the left part of Fig. 5.3 which is reproduced from Ref. 9. Furthermore, σ_{exp} agrees with σ as calculated above without any adjustable parameters. In their original study of thermal fluctuations on a water surface Braslau et al. found[1] σ_{exp} 10% larger than σ from (5.13). Recent measurements using in situ monitoring of surface tension and ultra pure water[10] indicates that our original results[1] might have been influenced by an impure surface, both for water and carbon-tetra-chloride, and it seems as though the capillary wave model indeed accounts for the entire roughness of the surface of simple liquids like water, methanol and carbon-tetra-chloride.

Most interestingly, Daillant et al.[9] find a discontinuous decrease in $\sqrt{\gamma} \cdot \sigma$ around a surface tension of 53 mN/m. They ascribe this observation to a first order phase transition of the monolayer from a soft layer with a small value of the bending constant to a more rigid layer with a bending constant of around 200 kT. Because the observed effect only varies as the square root of the logarithm of k_M/k_y, cf. Eq. 5.13, it requires either a very rigid layer ($\kappa \geqslant 1$) or a low surface tension γ to significantly reduce the pure capillary wave roughness.

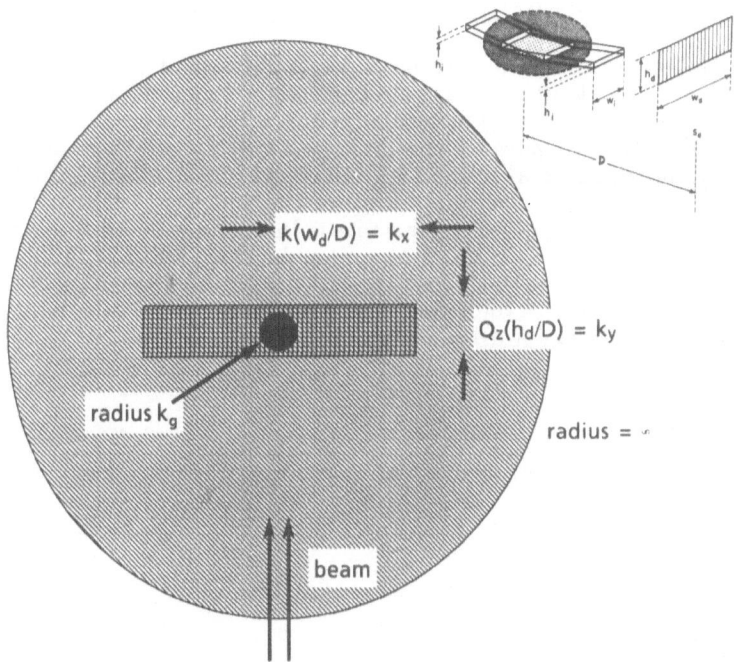

Fig. 5.2. *The capillary r.m.s. roughness σ is obtained by integrating Eq. (5.12) over the whole plane except for the inner rectangle determined by the experimental resolution.*

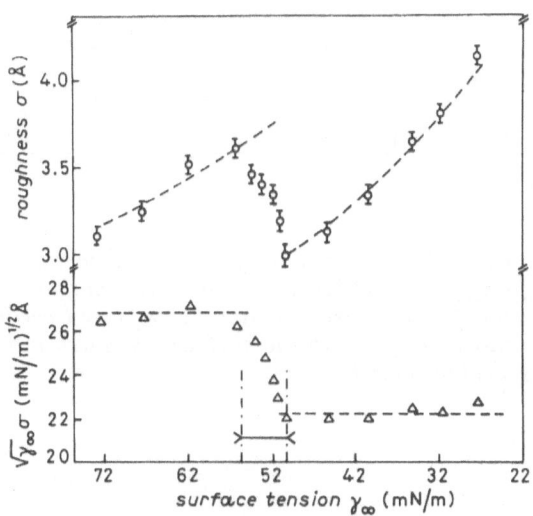

Fig. 5.3. *Roughness parameter σ for a behenic acid monolayer, ref. 9. The jump at γ ~ 53 mN/m corresponds to the monolayer attaining a large rigidity against bending for γ < 53 mN/m.*

6. REFLECTIVITY OF ARACHIDIC ACID FILM

The simplest monolayers may be those composed of fatty acids. The molecules consist of a hydrocarbon chain and a carboxylic acid headgroup. The lateral pressure as a function of molecular area A for one representative of this class, arachidic acid on a pure water subphase is given in the inset of Fig. 6.1. On compressing the monolayer to a molecular area of $A_t = 24$ Å^2, the lateral pressure remains below the detection limit of 1 mN/m. On further increasing the molecular density the pressure increases almost linearly with decreasing A. At a distinct pressure $\Pi_S = 25$ mN/m and molecular area $A_S = 19.8$ Å^2 the pressure/area isotherm becomes very steep. The phases above and below Π_S have previously been called solid and condensed liquid[4], respectively. We will show to what extent X-ray scattering provides a better picture.

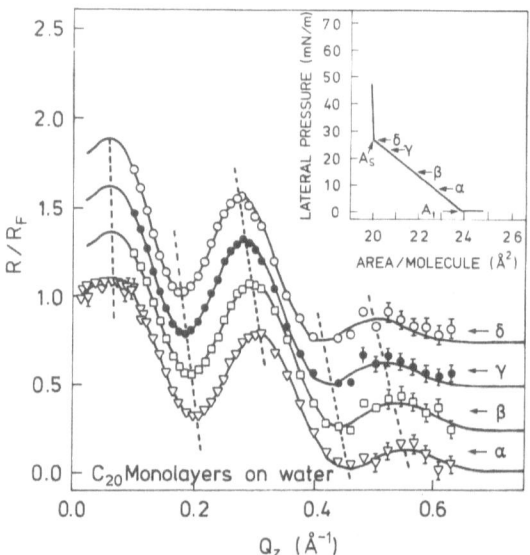

Fig. 6.1. Normalised X-ray reflectivity R/R_F vs. vertical wave vector transfor Q_z, for arachidic acid monolayers on pure water (pH 5.5, T = 20°C). The measurements are displaced vertically by 0.25 units and correspond to the surface pressures indicated on the isotherm of the insert.

Fig. 6.1 gives the reflectivity vs. wave vector transfers Q_z perpendicular to the surface for various surface pressures indicated as by arrows in the isotherm. One clearly observes a shift of the extrema to lower Q_z with increasing surface pressure.

The full lines through the data points represent a simple box model of the monolayer density $\rho(z)$ with some adjustable parameters. In the model, the aliphatic tail region has a constant density ρ_T and a certain thickness l_T whereas the polar head region has density ρ_H and thickness l_H, see figure 6.2. The sharp box edges are smeared by one common Gaussian function so the

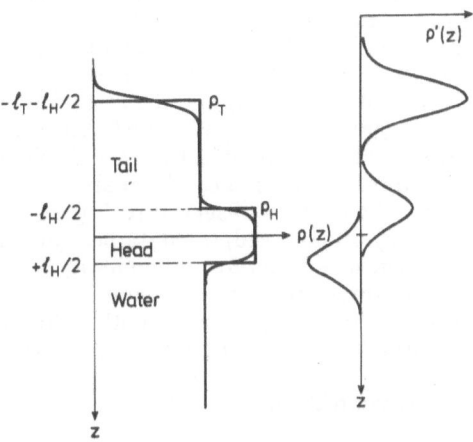

Fig. 6.2. Two-box density profile model of a Langmuir film. The boxes, describing the tail part and the polar head part respectively, are smeared by a Gaussian, as indicated by the full line. The density gradient in this model is two positive Gaussians followed by one negative Gaussian.

model contains five parameters: Two thicknesses, two densities and a smearing parameter. This model has the virtue that the Fourier transform of the density gradient can readily be written down. Let the origin be in the middle of the head group region which extends from $-\ell_H/2$ to $+\ell_H/2$ with a density of ρ_H. The tail region extends from $-\ell_T-\ell_H/2$ to $-\ell_H/2$ with a density of ρ_T. The density **gradient** is thus a set of Gaussians, all with the same width parameter σ, and located at the edges of the boxes at $z = -\ell_T-\ell_H/2$, $-\ell_H/2$ and $+\ell_H/2$. The height of each Gaussian is the **difference** between consecutive box densities. The Fourier transform of a Gaussian is a Gaussian itself, so altogether we find for the Fourier-transform, $\phi(Q) \equiv (1/\rho_\infty)\int \rho'(z)\exp(iQz)dz$, of the density gradient $\rho'(z)$ to be:

$$\phi(Q) = (1/\rho_\infty) \cdot \exp(-Q^2\sigma^2/2)$$
$$\{\rho_T\exp[-iQ(\ell_T+\ell_H/2)]+(\rho_H-\rho_T)\exp[-iQ\ell_H/2]-(\rho_H-1)\exp[iQ\ell_H/2]\}. \qquad (6.1)$$

In order to obtain a direct qualitative understanding of the features in the normalized reflectivity, $R(Q)/R_F(Q) = |\phi(Q)|^2$, we shall make one further approximation in taking the tail density to be almost that of water, i.e. $\rho_T \simeq \rho_\infty$. In that case we get $\phi(Q) \simeq \phi_1(Q)$ with

$$\phi_1(Q)\exp(Q^2\sigma^2/2) = \exp[-iQ(\ell_T + \ell_H/2)] - 2i(\rho_H/\rho_\infty - 1)\sin[Q\ell_H/2].$$

(6.2)

In the complex plane it is easy to visualize the two terms on the right hand side of Eq. (6.2) versus Q. The first term starts out at (1,0) for Q = 0 and then moves clockwise around the unit circle as Q increases . The second term is bound to the imaginary axis. It starts out at (0,0), then increases almost linearly with Q along the negative axis and then continues along the imaginary axis as a sine wave versus Q. Suppose for the sake of argument that $\ell_H = \ell_T$ and $\rho_H/\rho_\infty = 1.5$. For $Q(\ell_T + \ell_H/2) = \pi/2$ there is maximal constructive interference with the first term being at (-i,0) and the second term being at $(-i/2,0)$. At a 3 times larger Q there is complete destructive interference as the first term is at $(+i,0)$ and the second term is at (-i,0). At a 5 times larger Q we have again constructive interference, the first term being again at (-i,0) and the second term being at (-i/2,0) etc. When, in reality, ℓ_H is considerably smaller than ℓ_T this second constructive interference will obviously be larger than the first, as observed in Fig. 6.1. If the minimum in the reflectivity data is very pronounced, the data become very sensitive to the values of parameters, because a deep minimum simply reflects a very delicate balance in the destructive interference phenomenon. From the second term in Eq. (6.2) one also deduces that the height and depth of the reflectivity extrema depend sensitively on $(\rho_H - \rho_\infty)$. The head group density is therefore determined very accurately.

Best fit results are shown in Table 1.

Table 1. Fitted parameters of the model densities $\rho(z)$ of Fig. 6.2, corresponding to the reflectivity data of Fig. 6.1. The densities are normalized to the density of water, $\rho_\infty = \rho_W = 0.334\ e/Å^3$.

	π	A	N_T	ℓ_T	ρ_T/ρ_W	ℓ_H	ρ_H/ρ_W	N_H	σ
label	mN/m	Å²	-	Å	-	Å	-	-	Å
δ	25.0	19.8	157	≡24.2	0.983	3.07	1.59	32	3.38
γ	21.6	20.5	≡157	23.4	0.979	3.48	1.48	35	3.22
β	15.9	21.7	≡157	22.2	0.977	3.88	1.38	39	2.99
α	11.0	22.7	≡157	32.2	0.977	4.43	1.31	44	2.93

What are reasonable dimensions of the boxes of constant density? First, consider the hydrocarbon tail. According to Ref. 11 the average distance between two CH_2 groups, projected onto the molecular axis is 1.265 Å. Each CH_2 group contains 8 electrons and is in Fig. 6.3 represented by a Gaussian distribution of width $\sigma = 1$ Å (which is certainly smaller than the final fitted smearing parameter). The neighboring CH_2 group contributes a similar Gaussian but displaced 1.265 Å along the z-axis. The last hydrocarbon group of the aliphatic tail is a CH_3 group with 9 electrons. If this is also represented by a Gaussian of the same width the height must be 9/8 of the height of the CH_2 Gaussians. As is apparent from Fig. 6.3, the molecular density is well approximated by the two-box model, provided only that the terminal CH_3 group is represented by a segment (9/8)·1.265 Å long, to give the correct number of electrons in the tail. The tail length in the all-trans configuration is thus

$$\ell_T^0 = [\, n + (9/8)\,]\cdot\ 1.265\ \text{Å} \tag{6.3}$$

The z-positions of the atoms of the COOH head group are more difficult to assign. A plausible choice was made in Fig. 6.3.

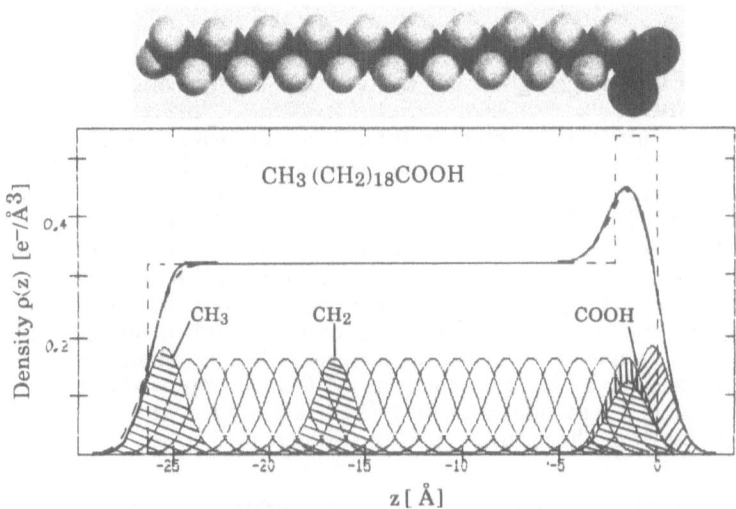

Fig. 6.3. Top: *Space-filling model of the molecule.*
Bottom: *Full lines: Each CH_2 group is represented by a Gaussian succesively displaced by 1.265 Å. For the terminal CH_3 group and the polar head group COOH, see text.*
Dashed lines: Box model and smeared box model of the density. The width parameter $o = 1$ Å was chosen for display purposes. In reality, the fitted $o \approx 3$ Å.

For arachidic acid the number n of CH_2 segments is 18 and one obtains $\ell_T^0 = 24.2$ Å. Thus the number of fitting parameters was reduced to four. With the density ρ_T determined from the parameter fit one then calculates the number N_T of electrons in the hydrocarbon moiety according to

$$N_T = \rho_T A \ell_T \tag{6.4}$$

For the particular case one obtains $N_T = 157\,e^-$/molecule which has to be compared with the number $N_T = 153\,e^-$/molecule derived from the molecular formula $(CH_3-(CH_2)_{18})$ of the tail. This discrepancy may be an artefact introduced by the simplicity of the two box model or indicates that part of the hydrocarbon moiety may be penetrated by water (the discrepancy corresponds to 40% of one water molecule). Now keeping the number N_T constant, independent of molecular area, ρ_T and ℓ_T are related according to Eq. (6.4) and the model parameters of table 1 are deduced using four independent fitting parameters.

The tail region thickness, ℓ_T, decreases monotonically as the area per molecule increases. This, together with the constancy of ρ_T, strongly indicates that the aliphatic tails are predominantly in the all-trans configuration and uniformly tilted. From geometric considerations one can then derive the tilt angle t by comparing the length ℓ_T of the tilted tail with the value ℓ^0_T for vertical orientation:

$$\cos t = \frac{\ell_T}{\ell^0_T} \tag{6.5}$$

One derives that on going from A_t to A_s the tilt angle t continuously decreases from about 30^0 to 0^0.

Considering the parameters in table 1 for the head group, the correlation with molecular models is more ambiguous since the hydrophilic region not only contains the carbonyl group but probably also some water. In analogy with Eq. (6.4) N_H is determined by the product of ℓ_H and ρ_H. The number of electrons in the carbonyl group COO^- is 23, so the data indicate a hydration of one water molecule per carbonyl group at the highest pressure increasing to two water molecules per carbonyl group at the lowest pressure. The decrease of ℓ_H with increasing pressure may indicate a gradual confining of the head group moieties in a plane parallel to the surface. If this ordering were perfect, ℓ_H would correspond to the dimension of a COO^- group in the direction of its symmetry axis. From molecular models one estimates values between 2.5 and 3.0 Å, not far from the value of ℓ_H at the highest pressures.

7. IN-PLANE DIFFRACTION AND BRAGG ROD DATA FROM ARACHIDIC ACID FILM

In this section we discuss grazing incidence diffraction (GID) from a monolayer of arachidic acid on water.

The geometry for GID experiments is shown in Fig. 7.1. The top view shows the footprint of the grazing incidence beam on a water/film surface as well as the specular reflected beam. The grazing incidence angle is typically 0.8 times the critical angle for total reflection so according to Eqs. (2.12), (2.8) and Fig. 2.2 the penetration depth of the evanescent wave (EW) is around $\Lambda = 1.7/Q_c$ or 77 Å. The EW is diffracted by the monolayer, and we select for detection a horizontal scattering angle of 2θ by the Soller collimator and a vertical scattering angle of α_f by the Position Sensitive Detector PSD. The signal is diffrated from the "crossed-beam-area" ABCD of the monolayer and it is clear that the signal rate is proportional to the widths of the two crossed beams. A broad Soller collimator is thus much more efficient than a slit geometry.

Let us first consider the compressed state where the molecular area is A_s and where according to the specular reflectivity data the molecules are upright. The ordered structure forms a hexagonal lattice with a corresponding reciprocal lattice as indicated by the broken lines in Fig. 7.2 (top part). The Bragg scattering selection rule (the horizontal component, Q_{hor}, of the scattering vector must coincide with a reciprocal lattice vector G_{hk}) implies that the scattering seen in a side view (bottom part) is confined to vertical Bragg rods through G_{hk}. Due to the finite length L of the molecule the intensity along a Bragg rod is not constant but peaks at $Q_z = 0$ with a width of

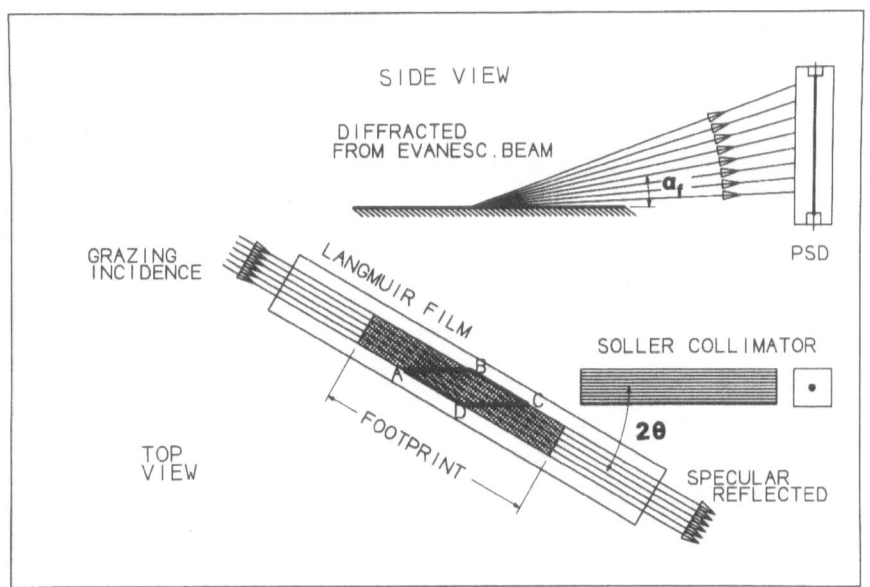

Fig. 7.1 Top view and side view of the GID geometry. The footprint of the grazing incidence beam is indicated by the darker area of the Langmuir film. The position sensitive detector PSD has its axis along the vertical direction. Only the crossed-beam-area ABCD contributes to the scattering.

order π/L. If, more specifically, we model the molecule by a cylinder of length L and diameter a, the molecular form factor along the molecular axis Q_z' is

$$S(u) = \sin(u)/u \tag{7.1}$$

with $u = Q_z'L/2$.

The radial molecular formfactor, $R(|Q_r|)$, at radial Q_r reflects the electron distribution of the CH_2 groups projected onto a plane perpendicular to the molecular axis. Each CH_2 group has an electron distribution similar to that of oxygen and therefore its Fourier transform is approximately the atomic scattering factor of oxygen. The CH_2 groups are connected in a zig-zag line with a ~110° opening angle and a distance of 1.54 Å. The molecule is assumed to rotate freely, so the center of each CH_2 group is evenly distributed on a circle of radius $R = (1.54/2)\cos(110/2) = 0.44$ Å. The final electron distribution is the **convolution** of each CH_2 distribution with the center distribution so the Fourier transform is the **product** of the center distribution, $J_0(Q_rR)$, and the oxygen scattering factor $f_O(Q_r)$. Within the range of interest, 1.5 Å^{-1} < Q_r < 2.0 Å^{-1}, the radial formfactor is well approximated by a Lorentzian in terms of the dimensionless variable $v \equiv Q_r a$

$$R(v) \propto (1+v^2)^{-1} \tag{7.2}$$

with $a = 0.38$ Å.

We now consider the model proposed on the basis of the reflectivity data:

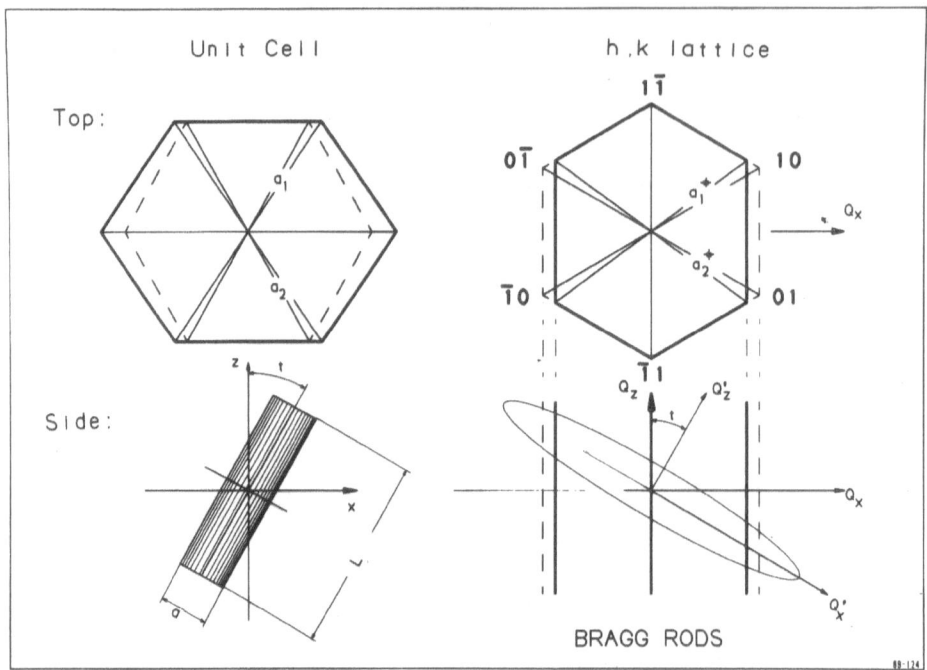

Fig 7.2 *At high pressures the molecular electron density is modelled by an upright cylinder of length L and diameter a, the molecules forming a hexagonal lattice (broken line). At lower pressures the molecules tilt and the unit cell becomes distorted to a centered orthorhombic cell (full line). The lower, right panel indicates the Bragg selection rule (rods) and the formfactor of a tilted molecule (ellipse).*

For pressures lower than π_s the area per molecule , A, becomes larger than A_s and the molecules tilt but **remain closed packed**, i.e., the tilt angle t is given by

$$\cos(t) = A_s/A, \qquad (7.3)$$

Tilt angles t deduced from Eq. (7.3) were previously shown[12] to agree well with the values deduced from Eq. (6.5). With tilted molecules, the Bragg selection rule still holds but the molecular formfactor, indicated by the ellipse in the right bottom part of Fig. 7.2, tilts. In Fig. 7.2 a particular tilt direction (towards nearest neighbour) was chosen but in general we define the Q_x-axis as the tilt direction. In the particular case of tilting towards nearest neighbour we see from Fig. 7.2 that along the $(1,-1)$ and $(-1,1)$ Bragg rods the intensity still peaks at $Q_z=0$, whereas along the $(0,-1)$ and $(-1,0)$ rods the intensity peaks at $Q_z>0$ and that along the $(1,0)$ to $(0,1)$ rod at $Q_z<0$ — the latter not being observable as only scattering away from the water surface can be detected. In the cylinder model the molecular formfactor is still as given in Eq. 7.1 and 7.2 in the **molecular** frame (Q_x',Q_y,Q_z') so in the laboratory frame we insert u and v from

$$v = a \cdot |Q_y^2 + (Q_x\cos t + Q_z\sin t)^2|^{1/2} \qquad (7.4)$$

$$u = (Q_z \cos t - Q_x \sin t)L/2 \tag{7.5}$$

with

$$Q_x = |G_{hk}| \cos \psi_{hk}, \qquad Q_y = |G_{hk}| \sin \psi_{hk}, \tag{7.6}$$

G_{hk} being the reciprocal lattice point considered and ψ_{hk} the angle from G_{hk} to the tilt direction. Since the sample is a two-dimensional powder the observed Bragg rod intensity, $I_{obs}(Q_z)$, contains contributions from several (h,k) reflections. The tilting of closed-packed molecules implies a distortion or strain of the hexoganal lattice to a centered orthorhombic lattice as is also indicated in Fig. 7.2, so that the optimal 2θ position for observing different (h,k) rods may split beyond the experimental resolution for sufficiently large tilt – in the example of Fig. 7.2 into a high angle peak for the $\{1,-1\}$ rod and a low angle peak for the $\{0,1\}$ and $\{1,0\}$ rods. After discussing this tutorial model let us look at the actual data in Fig. 7.3, left column. At the two highest pressures (panels c and d) there is no observable splitting in 2θ but at the two lowest pressures (panels a and b) the optimum 2θ position for the Bragg rod centered around $Q_z = 0$ (open circles) is larger than for the Bragg rod which peaks at $Q_z \approx 0.5 \text{ Å}^{-1}$ (crosses).

The data exhibit a very sharp peak at $Q_z \gtrsim 0$. The width is of order Q_c – the critical scattering vector for total reflection. This is an interference effect analogous to that discussed in section 2: Recall that for $2\theta = 0$ and $a_i = a_f \simeq a_c$, the incident and total reflected beams interfere constructively to produce maximum intensity above and below the interface (cf. Fig. 2.2, third panel). In the present case 2θ is around $19°$ so clearly neither the incident nor the specular reflected beams contribute directly. Around $2\theta \simeq 19°$ the beam(s) must have undergone diffraction by the in-plane ordered structure of the monolayer. However, the diffracted rays are distributed over a range of angles a_f, c.f. Fig. 7.1, and, in particular, the diffracted rays with $a_f = +a_c$ and the rays diffracted into $a_f = -a_c$ and then total-reflected in the interface will interfere constructively. Once this mechanism is appreciated it can of course be accounted for in model calculations. In terms of $x \equiv a_f/a_c$ the interference effect implies a factor $V(x)$

$$V(x) = \begin{cases} 2x & \text{for } 0 < x < 1 \\ 2x/(x + (x^2-1)^{\frac{1}{2}}), & \text{for } x > 1 \end{cases} \tag{7.7}$$

in the diffracted amplitude, cf. Eqs. (2.6) and (2.11) and Ref. 13.

The data are compared with the tutorial model of tilted cylinders in columns 2 and 3 of Fig. 7.3. In model 1 the molecules tilt towards nearest neighbours, in model 2 towards next nearest neighbours. In both models 1 and 2 the intensity is calculated as the sum over the Bragg points $(1,0),(1,0),(1,-1)$ etc. of $(S(u)R(v)V(x))^2$ from Eqs. 7.1-7 using tilt angles t of $7°$, $14°$, $20°$, and $22°$ respectively from bottom to top. The squared structure factors are multiplied by the Gaussian smearing $\exp(-(Q_z\sigma)^2)$ with a smearing parameter of $\sigma = 1 \text{ Å}$ – considerably smaller than the over-all smearing of around 3 Å found in the reflectivity data as discussed further below. As cylinder length L we used the tail length of 24.2 Å from Eq. 6.3.

The tilt angles, t, were determined from Eq. 7.3, inserting for A the unit cell area A_{cell}, calculated from the observed 2θ peak positions. Apart from the smearing parameter and one over-all scale factor for all the model curves of Fig. 7.3, the only free parameter in generating the intensity profile of the Bragg rods is thus the direction of tilt. In comparing the data[14] with the model we note the following points:

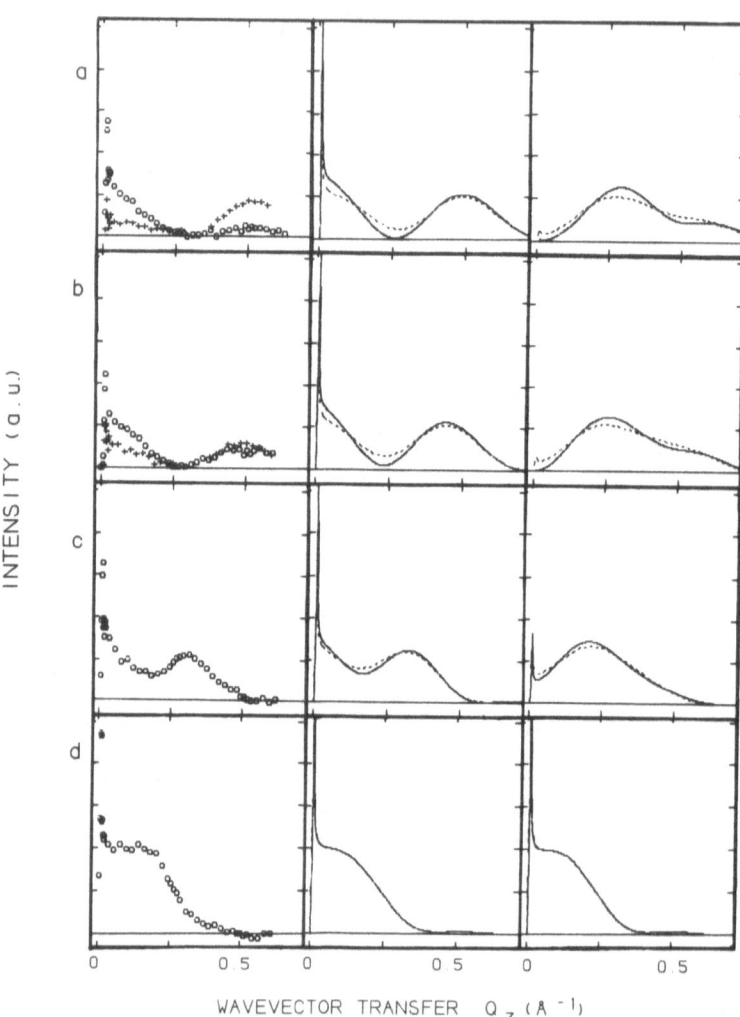

Fig. 7.3. DATA.   Arachidic acid. 2D-powder Bragg rods observed at nominal pressures of 10,16,20,21 mN/m (panels a to d, respectively). In case c and d the optimal 2θ settings for the merging peaks at $Q_z = 0$ and $Q_z > 0$ coincide, but in case a and b they split as indicated by open circles (optimum 2θ for $Q_z = 0$) and crosses (optimum 2θ for $Q_z \simeq 0.5\,\text{Å}^{-1}$).

MODELS.   Both models are the tutorial model of Fig. 7.2. In model 1 the molecules tilt towards nearest neighbours, in model 2 towards next nearest neighbours. All other parameters are essentially determined from the reflectivity data. The dashed lines represent a perturbation of the models: The molecules were tilted in a direction 8° from the symmetry directions.

(i)   Model 1 approximates the data quite well at all four tilt angles, whereas model 2 cannot at all account for the considerable intensity observed around $Q_z = 0$ for large tilt angles. The model demonstrates in a simple way that GID data are quite sensitive in determining the tilt direction as well as its magnitude.

(ii) The smearing parameter σ_{GID} is only 1 Å compared to $\sigma_{XR} = 3$ Å for the XR data. This may be related to the different coherence lengths probed in XR and GID experiments. The coherence length defines the size of the area inside which the squared deviations from ideal flatness are averaged to yield a roughness parameter σ. For XR, this has already been discussed at length in section 5. The relevant quantity for XR is the lateral Q-resolution area. In Fig. 5.2 this area is a rectangle with dimensions of the order 10^{-2} Å on one side and 10^{-3} on the other side. In GID the 2θ peak width is typically $\delta \simeq 0.3° \simeq 5$ mrad, to be compared with an instrumental resolution width of $0.1°$. The **intrinsic** width is therefore of the order $k\delta \approx 2\cdot10^{-2}$ Å^{-1}, so the coherence length is much shorter in the GID experiment than in in the XR experiment. Consequently σ_{GID} may be expected to be smaller than σ_{XR}, as indeed observed experimentally.

8. CONCLUSION

The application of X-ray scattering methods to the study of the liquid-vapour interface has been developed theoretically and illustrated by experimental examples. The surfaces of simple liquids are rough due to thermally excited capillary waves. The interface can be characterized by one parameter, the rms. diffuseness, σ, which can be determined by X-ray reflectivity measurements, as illustrated for the case of water.

A liquid surface with a surfactant monolayer requires more structural parameters for its characterization. By X-ray reflectivity the densities and thicknesses of constituent sub-layers can be deduced. For rod-like arachidic acid molecules, the results could be interpreted in terms of tilted, close-packed molecules, with the tilt angle determined from a **cosine** relation, Eq. (6.5). Grazing Incidence Diffraction gives information about the lateral order in the interface: lattice spacings and correlation length. By measurement of the intensity variation along the Bragg rods, structural information complementary to that from X-ray reflectivity can be obtained. For arachidic acid monolayers, this allows determination of the tilt angle by a **sine** relation, Eq. (7.5) and figure 7.2.

Finally, we note that the methods here presented are applicable as well to **hard** interfaces, e.g., surfaces of crystalline solids or Langmuir-Blodgett films[4] of surfactant molecules on solid supports.

REFERENCES

1.   A. Braslau, P.S. Pershan, G. Swislow, B.M. Ocko, J. Als-Nielsen, Phys. Rev. **A38**, 2457 (1988) and A. Braslau, M. Deutsch, P.S. Pershan, A.H. Weiss, J. Als-Nielsen and J. Bohr, Phys. Rev. Lett. **54**, 114 (1985).
2.   P.S.Pershan, A.Braslau, A.H.Weiss, J.Als-Nielsen, Phys. Rev. **A35**, 4800 (1987) and references therein.
3.   J. Als-Nielsen and H. Möhwald, in Handbook of Synchrotron Radiation, Vol. 4 (North Holland, eds. S. Ebashi, E. Rubinstein and M. Koch).
4.   G. Gaines "Insoluble Monolayers at Liquid-Gas Interfaces", Interscience N.Y. (1966).
5.   The International Tables for X-ray Crystallography, vol. 3, Kynoch Press (1962) and vol. 4 (1974).

6.   B.E. Warren "X-Ray Diffraction", appendix A, Addison-Wesley (1969).
7.   L.G. Parratt, Phys. Rev. **95**, 359 (1954).
8.   J. Meunier, J. Phys. (Paris) **48**, 1819 (1987).
9.   J. Daillant, L. Bosio, J.J. Benattar, J. Meunier, Europhys. Lett. **8**, 453 (1989).
10. P. Pershan, private communication (1989).
11. C. Tanford "The hydrophobic effect: Formation of micelles and biological membranes", Wiley, N.Y. (1973).
12. K. Kjær, J. Als-Nielsen, C.A. Helm, P. Tippmann-Krayer and H. Möhwald, J. Phys. Chem. **93**, 3200 (1989).
13. G. Vineyard, Phys. Rev. **B26**, 4146 (1982).
14. K. Kjær, J. Als-Nielsen, P. Tippmann-Krayer and H. Möhwald, to be published (1989).

MARANGONI EFFECT, INSTABILITIES AND WAVES AT INTERFACES

M. G. Velarde and X.-L. Chu

UNED– Facultad de Ciencias
Apartado 60.141
E–28.071–MADRID (Spain)

1. INTRODUCTION

Capillary-gravity waves have been well studied since Laplace, Kelvin and Stokes , and their properties were shown to be mainly determined by the stress conditions at liquid surfaces[1-3]. Specifically, the major influence of the normal stress to the surface has been emphasized. While these oscillations are rather *transverse* motions due to the deformation of the surface and surface tension action there is yet another type of wave. It refers to mostly *longitudinal* motion along the surface[3-5]. Its existence is not surprising considering that a strong analogy is expected between a monolayer-covered surface and a stretched elastic membrane. The coverage with a surfactant–either by adsorption from solution or by spreading-gives elastic properties to a surface so that it tends to resist the periodic expansion and compression which appears as wave motion. Normally, an interfacial motion has both transverse and longitudinal components, and they are not separable. Only in the case of small amplitude wave motion , transverse and longitudinal waves can be considered as two genuine and independent modes of oscillation.

Longitudinal waves are, to a major extent, related to the tangential stress with a frequency that depends on the viscosity and surface elasticity. They do not exist in non-viscous liquids. Capillary-gravity waves, however, have a frequency that depends on gravity and on surface tension(Laplace-Kelvin overpressure) but not on the viscosity and are admissible in the absence of dissipation. The latter only appears in the damping factor and frequency-deviation in the dispersion relation.

Generally, these waves are damped *albeit* differently by viscosity. However, if a sufficiently strong non-equilibrium distribution of surfactant/temperature is imposed in the system that drives mass/energy transfer across and/or along the interface, the surface tension will not be uniform along the interface thus creating surface stresses leading to motion. This is the *Maragoni effect* that can be considered as the "interfacial engine" to convert chemical or thermal energy into convection[4-5].

2. DISTURBANCE EQUATIONS

We consider a liquid-liquid interface initially at rest and located at z=0 (z is the vertical axis and x is the only horizontal coordinate we shall take in the simplest two-dimensional description). For small amplitude motions the evolution equations for disturbances upon the quiescent state are

$$\nabla \cdot \vec{v}_\varphi = 0 \qquad (\varphi = 1, 2) \tag{2.1}$$

$$\rho_\varphi \frac{\partial \vec{v}_\varphi}{\partial t} = - \nabla p_\varphi + \eta_\varphi \nabla^2 \vec{v}_\varphi \tag{2.2}$$

$$\frac{\partial c_\varphi}{\partial t} - \beta^c_\varphi w_\varphi = D_\varphi \nabla^2 c_\varphi \tag{2.3}$$

together with boundary conditions (b.c.) at z=ζ(x,t)

$$\frac{\partial \zeta}{\partial t} = w_\Sigma = w_\varphi|_\Sigma \quad (\varphi = 1, 2) \tag{2.4}$$

$$\vec{v}_1|_\Sigma = \vec{v}_2|_\Sigma = \vec{v}_\Sigma \tag{2.5}$$

$$\rho_\Sigma \frac{\partial u_\Sigma}{\partial t} - \left(\frac{\partial \sigma}{\partial \Gamma}\right)_0 \frac{\partial \gamma}{\partial x} - (\eta_{dil} + \eta_{sh}) \frac{\partial^2 u_\Sigma}{\partial x^2} - [\eta(\frac{\partial w}{\partial x} + \frac{\partial u}{\partial z})] = 0 \tag{2.6}$$

$$\rho_\Sigma \frac{\partial w_\Sigma}{\partial t} - \sigma_0 \frac{\partial \zeta^2}{\partial x^2} - g[\rho] \zeta + [p] - 2[\eta \frac{\partial w}{\partial z}] = 0 \tag{2.7}$$

$$\frac{\partial \gamma}{\partial t} + \Gamma_0 \nabla_\Sigma \cdot \vec{v}_\Sigma - D_\Sigma \nabla_\Sigma^2 \gamma - [D \frac{\partial c}{\partial z}] = 0 \tag{2.8}$$

$$\gamma = k^l_\varphi (c_\varphi - \beta_\varphi \zeta) \tag{2.9}$$

where v=(u,w) is the disturbance velocity; p the disturbance pressure; c the disturbance surfactant concentration; ζ the deviation of the interface from the motionless level initial position(the subscript Σ denotes the interface); Γ the excess accumulation of surfactant at the interface and γ the disturbance on Γ. ρ accounts for liquid density; $\eta = \rho v$ is the dynamic shear viscosity with v the kinematic viscosity; D the surfactant diffusivity; g the gravitational acceleration; β the surfactant gradient; ρ_Σ the surface exess density and η_{dil} and η_{sh} the surface dilational and shear viscosity, respectively (for simplicity,here, we shall disregard these viscosities). σ is the surface tension and k^l the Langmuir adsorption constant [3]. The subscript φ indicates the liquid: φ=1 the liquid below and φ=2 the liquid above the interface. The subscript "0" accounts for the initial reference state and the bracket [f] for the jump across the interface, [f]=$(f_2-f_1)_\Sigma$. β is positive when supply is from "1" to "2".

Now with suitable units we rescale the original quantities. We can choose the capillary length as the unit of length: $l=(\sigma_0/g|\rho_1-\rho_2|)^{1/2}$, and the following scales for velocity, time, pressure, solute concentration and excess surface concentration, respectively, v_1/l, l^2/v_1, $v_1^2\rho_1/l^2$, $\beta_1 l$ and Γ_0. Then, in dimensionless form we have

$$\nabla \cdot \vec{v}_1 = \nabla \cdot \vec{v}_2 = 0 \tag{2.10}$$

$$\frac{\partial \vec{v}_1}{\partial t} = - \nabla p_1 + \nabla^2 \vec{v}_1 \tag{2.11}$$

$$N_\rho \frac{\partial \vec{v}_2}{\partial t} = -\nabla p_2 + N_\eta \nabla^2 \vec{v}_2 \qquad (2.12)$$

$$\frac{\partial c_1}{\partial t} - w_1 = S^{-1} \nabla^2 c_1 \qquad (2.13)$$

$$\frac{\partial c_2}{\partial t} - w_2 = S^{-1} N_D \nabla^2 c_2 \qquad (2.14)$$

Note that , as no confusion is expected ,we have maintained the same symbols for dimensionless and dimensional variables. Now the b.c are at z=0,

$$w_1 = w_2 = w_\Sigma \qquad (2.15)$$

$$\frac{\partial w_1}{\partial z} = \frac{\partial w_2}{\partial z} = \frac{\partial w_\Sigma}{\partial z} \qquad (2.16)$$

$$\frac{\partial \zeta}{\partial t} = w_\Sigma \qquad (2.17)$$

$$\frac{1}{SC} \nabla_\Sigma^2 \zeta - \frac{Bo}{SC} \zeta - [p] + 2(N_\eta - 1)\frac{\partial w_\Sigma}{\partial z} = \Gamma_t \frac{\partial w_\Sigma}{\partial t} \qquad (2.18)$$

$$\frac{H E}{H_z S} \nabla_\Sigma^2 \gamma - (\nabla_\Sigma^2 - \frac{\partial^2}{\partial z^2}) (N_\eta w_2 - w_1) = (\Gamma_t \frac{\partial}{\partial t} - \eta_\Sigma \nabla_\Sigma^2)\frac{\partial w_\Sigma}{\partial z} \qquad (2.19)$$

$$HS(\frac{\partial \gamma}{\partial t} + \nabla_\Sigma \cdot u_\Sigma - S_\Sigma^{-1} \nabla_\Sigma^2 \gamma) + \frac{\partial c_1}{\partial z} - \frac{\partial c_2}{\partial z} = 0 \qquad (2.20)$$

$$\gamma = \frac{H_z}{H}(c_1 - \zeta)\big|_\Sigma = \frac{H_z}{H} N_z (c_2 - \zeta)\big|_\Sigma \qquad (2.21)$$

We have used the following dimensionless groups: $N_\rho = \rho_2/\rho_1$, density ratio; $N_\eta = \eta_2/\eta_1$, visicosity ratio; $N_D = D_2/D_1$, diffusivity ratio; $N_z = k^l_2\beta_2/ k^l_1\beta_1$, ratio of Langmuir adsorption coefficients; $S = v_1/D_1$, Schmidt number; $S_\Sigma = v_1/D_\Sigma$, surface Schmidt number; $C = \eta_1 D_1/\ell\sigma_0$, capillary number; $Bo = -[\rho]g\ell^2/\sigma_0$, Bond number; $E = -(\partial\sigma/\partial\Gamma)_0 k^l_1\beta_1\ell^2/D_1\eta_1$, elasticity (surfactant Marangoni) number; $H = \Gamma_0/\beta_1\ell^2$, surface excess solute number; $H_z = k^l_1/\ell$, Langmuir adsorption number; $\Gamma_t = \rho_\Sigma/\ell\rho_1$, ratio of surface excess density to liquid density; $\eta_\Sigma = (\eta_{dil} + \eta_{sh})/\ell\eta_1$, ratio of surface viscosity to liquid visicosity. Note that by choosing the capillary length as the unit of length, $Bo = 1$. However , we shall indicate the Bond number dependence in our results.

3. INSTABILITY THRESHOLDS

With Eqs. (2.10)-(2.14) and b.c. (2.15)-(2.21) we have carried out a standard hydrodynamic stability analysis and the following results have been obtained [6-10] :

1) Capillary-gravity (Kelvin-Laplace) *transverse* waves (denoted by superscript T) can be sustained beyond the critical elasticity Marangoni number

$$E_c^T = \frac{4 S N_\eta \sqrt{Bo} \left(1 + \frac{1}{N_z \sqrt{N_D}}\right)}{C (N_\eta - \sqrt{N_D})(N_\rho + 1)} \tag{3.1}$$

with critical wavenumber

$$a_c^T = \sqrt{Bo} \tag{3.2}$$

and critical frequency

$$\omega_c^T = \left[2 Bo^{3/2}/SC(N_\rho + 1)\right]^{1/2} \tag{3.3}$$

provided the following condition

$$D_f/D_t > [\eta_f/\eta_t]^2 \tag{3.4}$$

is fulfilled. Note that "f" and "t" indicate *from* which *to* which liquid phase the transport proceeds.

2) Long wavelength ($\omega \gg a^2$) Lucassen *longitudinal* waves (denoted by superscript L)[3,5,11] can be sustained beyond the critical elasticity Marangoni number

$$E_c^L = -\left\{ \frac{(\sqrt{N_D} + N_\eta)(\sqrt{N_\rho N_\eta} - 1)}{H_z \pi_1 \pi_2} \right\}^2 \frac{S}{\pi_1}(1 + \sqrt{N_\eta N_\rho})(1 + 1/N_z\sqrt{N_D}) \tag{3..5}$$

with a dispersion relation

$$\frac{\omega_c}{a_c} = \frac{(\sqrt{N_D} + N_\eta)(\sqrt{N_\rho N_\eta} - 1)}{\sqrt{S} H_z \pi_1 \pi_2} \tag{3.6}$$

provided the following conditions

$$D_f/D_t > v_f/v_t \qquad \text{and} \qquad v_f/v_t < 1 \tag{3.7}$$

are fulfilled . We have used

$$\pi_1 = \sqrt{N_D/N_v} - 1 \quad \text{and} \quad \pi_2 = (1 + \sqrt{N_\mu N_\rho})/(1 + 1/N_z\sqrt{N_D}) .$$

3) Scriven and Sternling ($a^2 \approx \omega$) *longitudinal* waves [3,5,11] can be sustained beyond a critical elasticity Marangoni number(which is a cumbersome function of all parameters of the problem, N_ρ, N_D, N_η, etc) provided the following necessary conditions are fulfilled

$$D_f/D_t > 1 \qquad \text{and} \qquad v_f/v_t > 1 \tag{3.8}$$

4. NON LINEAR RESULTS

Once we know threshold values, dispersion relations and necessary conditions to be fulfilled in order to excite and eventually sustain oscillations we must show that the nonlinear terms saturate the exponential growth thus stabilizing the interfacial waves. This is a formidable task and,for simplicity,we now restrict consideration to capillary-gravity (Kelvin-Laplace) *transverse* waves at an *air*-liquid interface.In this case we know that in order to have sustained oscillations the elasticity Marangoni number must be negative [7]. When due consideration is taken of the nonlinear terms[1-3]-that we have omitted in Section 2-we have shown that , indeed, past the instability threshold the evolution corresponds to a nonlinear wave propagation.

In the simplest single-Fourier-mode approximation the nonlinear evolution corresponds to a *limit cycle* governed by the following nonlinear *ordinary* differential equation [9]

$$\frac{d^2\zeta}{dt^2} + \Delta\frac{d\zeta}{dt} + \zeta = -\zeta^2 + \alpha\zeta^3 + \left(\frac{d\zeta}{dt}\right)^2 - \beta\zeta\frac{d\zeta}{dt} - \zeta\left(\frac{d\zeta}{dt}\right)^2 - \gamma\zeta^2\frac{d\zeta}{dt} \tag{4.1}$$

where $\Delta = [E\, a_c^3/\sqrt{2}(S\omega)^{3/2} + 4a_c^2]/\omega_c$, $\alpha = 3a_c^3/2CS\omega_c^2$, $\beta = 16a_c^2/\omega_c$ and $\gamma = 6a_c^2/\omega_c$.

When, however, due consideration is given to the full space dependence, the *transverse* Kelvin-Laplace wave motion is described by the following set of nonlinear *partial* differential equations

$$\frac{\partial\zeta}{\partial t} = w - \frac{1}{a_c}\, w\,\frac{\partial^2\zeta}{\partial x^2} \tag{4.2a}$$

and

$$\frac{\partial w}{\partial t} = -\frac{a_c}{SC}\left\{Bo - \frac{\partial^2}{\partial x^2}\right\}\zeta - \frac{3\,a_c}{2\,S\,C}\left(\frac{\partial^2\zeta}{\partial x^2}\right)\left(\frac{\partial\zeta}{\partial x}\right)^2 - 2\,a_c^2\,w + 2\frac{\partial^2 w}{\partial x^2}$$

$$+ 4\,a_c^2\,w\left(\frac{\partial\zeta}{\partial x}\right)^2 + 4\,a_c(1 - 2\,a_c\,\zeta)\frac{\partial}{\partial x}\left\{w\frac{\partial\zeta}{\partial x}\right\} + \frac{E_0\,a_0}{S\,\omega_0\,\sqrt{2\,S\,\omega_0}}\left\{ -a_c\frac{\partial^2\zeta}{\partial x^2}\right.$$

$$\left. + \frac{3}{2}\left[\frac{\partial}{\partial x}\left(\frac{\partial\zeta}{\partial x}\right)^2\right]\frac{\partial}{\partial x} + \left[1 - 2\,a_c\,\zeta + \frac{3}{2}\left(\frac{\partial\zeta}{\partial x}\right)^2\right]\frac{\partial^2}{\partial x^2}\right\}(\frac{\partial\zeta}{\partial t} - \omega_c\,\zeta) \tag{4.2b}$$

which constitute our extension of Boussinesq water wave theory [12] to a dissipative liquid layer subjected to Marangoni stresses [13]. Note that , for the two-component order parameter (ζ,w), Eqs.(4.2 a,b)provide the typical Landau-Ginsburg description of a bifurcated state in a non-uniform system [14].

ACKNOWLEDGMENTS

This research has been sponsored by CICYT (Grant PB 86-651)and by an EEC Grant.

REFERENCES

1. L.D. Landau and E.M.Lifshitz, *Fluid Mechanics*, Pergamon Press, Oxford,1987
2. B.G.Levich, *Physicochemical Hydrodynamics* , Prentice Hall, N.J., 1962, Chap. XI. See also M.G. Velarde (editor), *Physicochemical Hydrodynamics:Interfacial Phenomena* , Plenum Press,New York, 1988
3. C.A. Miller and P. Neogi, *Interfacial Phenomena*, M.Dekker, N.Y., 1985
4. M.G. Velarde and C. Normand, *Sci. American* 243(1980)78
5. H. Linde, in *Convective Transport and Instability Phenomena*, (J. Zierep and H. Oertel, editors), Braun-Verlag, Karlsruhe, 1982, pp.256-296
6. M.G. Velarde and X.-L. Chu, *Phys. Lett.* A131(1988)430
7. X.-L. Chu and M.G. Velarde, *Physicochemical Hydrodyn.*, 10(1988)727
8. X.-L. Chu and M.G. Velarde, *J. Colloid Interface Sci.* 131(1989) to appear
9. X.-L. Chu and M.G. Velarde, *Phys. Lett.* A136(1989)126
10. M.G. Velarde and X.-L. Chu, *Physica Scripta* T25(1989)231. Here we correct for some unfortunate omissions in Sect.4-Eqs. (4.1) and (4.2)-of this reference.
11. J.C. Legros, A. Sanfeld and M.G. Velarde, in *Fluid Sciences and Materials Science in Space* (H.U. Walter, editor), Springer-Verlag, Berlin, 1987,pp.83-140
12. See,*e.g.*,G.B. Whitham, *Linear and Nonlinear Waves*,J.Wiley, N.Y.,1974, Chap.13
13. X.-L. Chu and M.G. Velarde, *Phys. Lett* .A (1989) to appear
14. C. Normand,Y. Pomeau and M.G. Velarde, *Rev. Mod. Phys.* 49(1977)581

THE PHASES AND PHASE TRANSITIONS OF LIPID MONOLAYERS

H. Möhwald

Univ. Mainz, Inst. Phys. Chem
Welder Weg 11
D6500 Mainz, FRG

Summary

The structure and phase transitions of lipid monolayers are described combining recent data from X-ray, surface potential and quantitative analysis of fluorescence microscopic measurements. It is shown that peculiar domain shapes and superstructures observed can be understood from a competition between line energy and long-range electrostatic repulsion. The latter can be calculated from the surface potential difference measured for the coexisting phases. It is shown that the phase coexisting with the fluid phase undergoes an ordering transition at a distinct pressure π_s. This may correspond to ordering and dehydration of the lipid head groups.

I. Introduction

This seminar will consider two interesting aspects of phase transitions in soft matter:
- Structure formation in a two-dimensional (2D) system undergoing a phase transition.
- 2D ordering of molecules with different moieties requiring different phases and thus being comparable to incommensurate or frustrated systems.

The systems that we study are monolayers of phospholipids and fatty acids at air/water interfaces. This interface is physically very interesting since it enables a high degree of freedom to vary many parameters. On the other hand films manipulated at the water surface are also the precursors of Langmuir-Blodgett films, i.e. mono- or multilayers of these lipids on solid support. Phase transitions in these systems, which are also of technical relevance, will be discussed at the end of the second part.

II. Monolayers on water surfaces

II. 1. Pressure/area isotherm

Fig. 1 gives a schematic pressure/area isotherm for surfactant molecules with aliphatic tails. Decreasing the average molecular area the lateral pressure begins to rise to a defined value π_s, above which the isotherm assumes almost horizontal slope. On further compression the lateral pressure again increases steeper and the compressibility becomes very low at a molecular area of about 20Å^2 per aliphatic chain. From a closer inspection of the isotherm one also deduces a distinct point (π_s, A_s), where the slope becomes steepest (see Fig. 1).

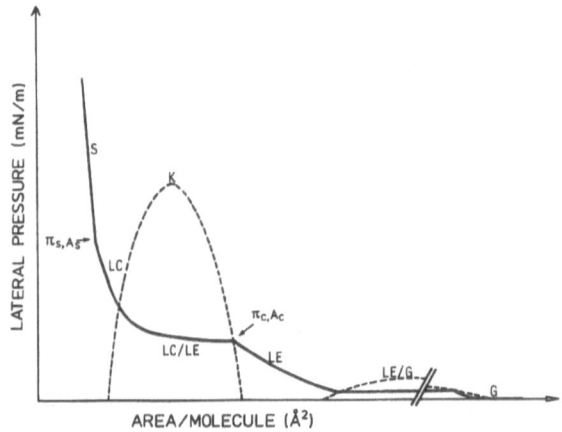

Fig. 1. Sketch of characteristic features of a pressure/area isotherm indicating corresponding phases. The gaseous phase (G) and the critical point K are not considered further.

The pressures π_c and π_s correspond to the onset of phase transitions and there has been much discussion and confusion about the assignment of corresponding phases /1,2/. This was also due to the fact that there was much intuition but little data on the microscopic molecular arrangement available. There exist now many more microscopic studies and in the later part of this lecture I will try to summarize these to contribute to a clarification of the situation. For the first part I will use the nomenclature of Gaines /3/, or Cadenhead et al /1/, calling the condensed phase below π_c liquid expanded (LE). This phase is sometimes called isotropic fluid /2/ since it exhibits a reasonably high diffusion coefficient ($\sim 10^{-7}\text{cm}^2/\text{sec}$) /4/, no positional ordering and also no preferential tail alignment within the plane. Above π_c there is a first order phase transition to a more ordered phase which we will call liquid condensed (LC). This phase is now shown to be different from the one above π_s which we will call solid (S). Justifications and additional complications will be presented later, but for the discussion on structure formation it suffices to keep in mind that on increasing the pressure above π_c the more condensed LC phase is formed and coexists with the LE phase.

The pressure π_c can be varied via environmental conditions in a rather predictable way:
- It decreases with temperature T and for low enough T the LE phase is suppressed /2/. Increasing T there is a critical temperature above which the first order phase transition disappears /2/.
- Increasing the head group charge density via pH or ionic strength is similar to a temperature increase /5/. Conversely binding of divalent ions reduces π_c extension of the tail length increases π_c by an amount on the order of $10mN/m$ per CH_2 group.

II. 2. Domain Structure

The formation of the LE phase can be followed by fluorescence microscopic observation of the monolayer at the air/water interface /4,7,8/. One incorporates a small amount (<1mole%) of a surface active fluorescent dye, which, like most impurities, is less soluble in the LC compared to the LE phase. Hence domains of the LC phase appear bright under the microscope. Fig. 2 gives a series of fluorescence micrographs obtained on increasing the surface pressure from a to f. Whereas for pressures below π_c the surface appears homogeneous (not shown) LC phase domains appear, increasing in

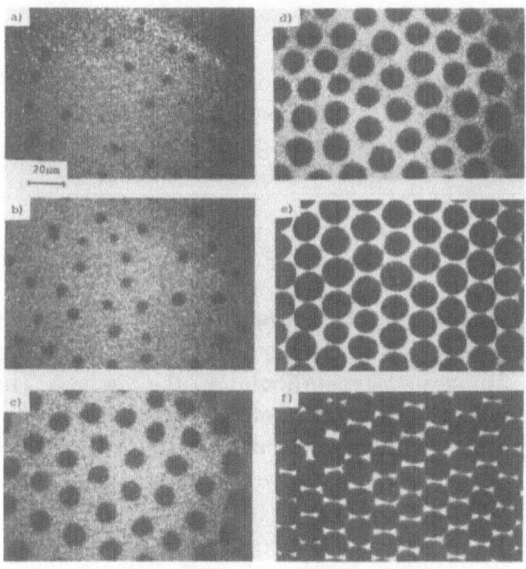

Fig. 2. Fluorescence micrographs of a monolayer of L-α-dimyristoyl-phosphatidic acid (DMPA) on the water surface at surface pressures between π_c and π_s.

size with lateral pressure. These domains, however, exhibit interesting structures and superstructures which will be the main subject of the first part of this lecture. The striking feature of the images in Fig. 2 to be explained, is that all domains are of uniform size and form a superlattice. Also domains can have different shapes and a collection of shapes is given in Fig. 3. There exist even thermally or chemically induced transitions between circular and spiral shapes (Fig. 3b) /9,10/. The shapes in the lower part of Fig. 3 will not be considered further. They are clearly nonequilibrium shapes with line tension being not large enough to minimize the phase boundary. In the lower left the domains initially grew as a fractal that eventually would like to anneal into a circular shape. The domains on the lower right correspond to a two-dimensional dendritic growth, and the axes of fast growth have been shown to be parallel to the diagonals of a nearly orthorhombic lattice /11/.

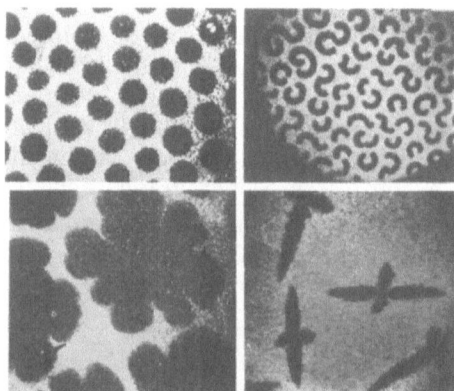

Fig. 3. Fluorescence micrographs of
different lipid monolayers
containing about 1mole% of a dye
probe in the phase coexistence
range of the LE and LC phase.

The peculiar shapes of the images in Fig. 3 can qualitatively be explained by the existence of long-range electrostatic forces with the following arguments (see Fig. 4):
The two coexisting phases differ by their molecular densities. Hence the LC phase exhibits either an excess charge (if the lipid is charged) or an excess density of dipole moments perpendicular to the surface. Thus two domains repel one another, and assuming realistic data one can show that the interaction energy is large enough to allow formation of a Wigner lattice (see below) /12/.

Principally electrostatic forces can also cause a uniform domain size since the electrostatic energy density, which is positive, increases with domain size. The latter counteracts line tension the energy corresponding to the LC/LE phase boundary. The density of the latter decreases with size and the interplay between these two forces may be responsible for a finite equilibrium size and thus uniform size of domains.

The interplay between these two forces can also cause transitions between shapes /13,14/. Line tension, being proportional to the boundary

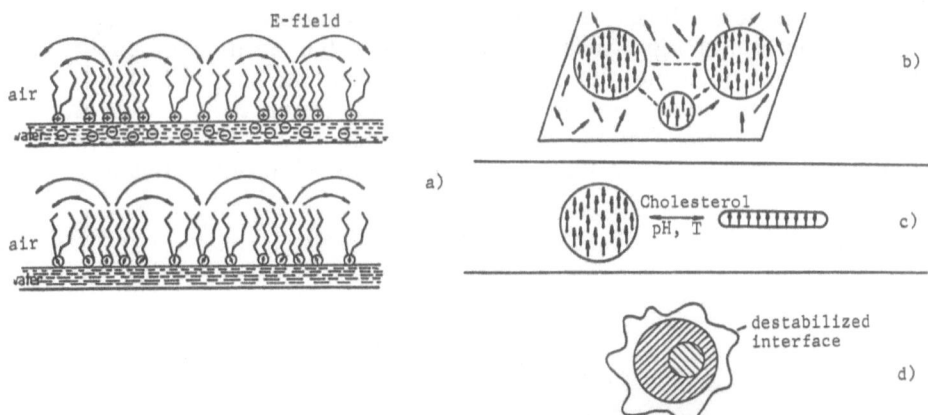

Fig. 4. Sketches of the arrangement of molecular dipoles at the air/water interface. Due to their higher molecular density LC phase domains exhibit an excess dipole density (a). This leads to long range repulsion between domains (b). Repulsive dipolar interaction of molecules within a domain favours elongated domains or destabilized interfaces (c,d).

length, clearly favours circular shapes whereas for a given domain area electrostatic energy would be minimized if the domain would develop into a thin stripe. Hence to enforce a transition from a circle into a stripe electrostatic forces have to be increased and line energy decreased. Comparing the shapes of Figs. 3a and b this has been achieved by increasing the surface charge density via pH and by adding 1mole% of cholesterol as an agent that likes to arrange at the phase boundary. The spiral shapes additionally require lipid chirality, and this can be taken into account by assuming an anisotropic line tension.

These ideas have been elegantly elaborated by Andelman et al. /15,16/ revealing that going along an isotherm in thermodynamic equilibrium there should be transitions between hexagonal, lamellar and inverted hexagonal phases. Also the work of McConnell and co-workers /13,14/ theoretically proved the establishment of the uniform lamellar phase as a result of the counteraction of the above forces. The latter group also considered the shapes of isolated domains and revealed the existence of shape transitions /17,18/. Comparing theory and experiment there are two main problems:

(i) Most experiments are not performed in thermodynamic equilibrium and therefore superlattices determined by nucleation conditions. Only considering the shapes it can be shown for some systems that they are in local equilibrium /9/.

(ii) Estimates of our group revealed that due to the slow (logarithmic) divergence of electrostatic energy it dominates only for systems with very low line tension to influence domain shapes /19,20/. On the other hand there are no data on this energy and it is possible that these experiments may yield practical values.

Extending the ideas on the interplay of forces one may also envision shape transitions during domain growth: Whereas for small sizes the dominating line energy causes a circular shape the electrostatic energy that tends to destabilize the phase boundary governs the shapes at large sizes. Numerical simulations now in progress in our group demonstrate that on increasing the size there will be a sequence of eigenmodes of the twodimensional interface that consecutively develops on domain growth /21/.

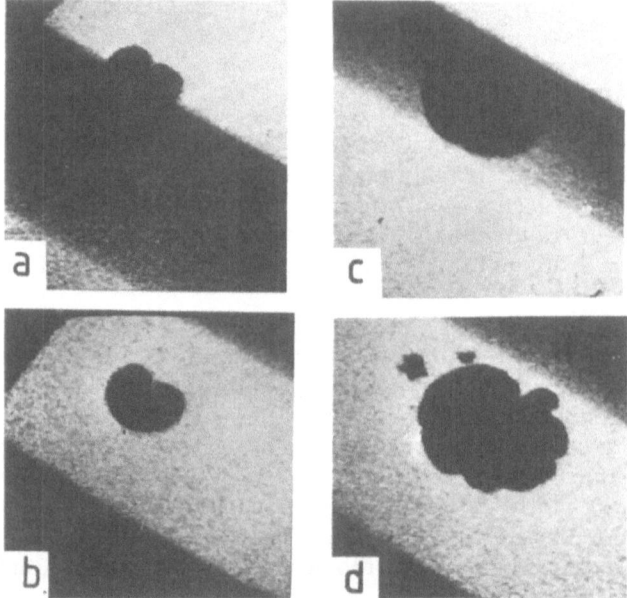

Fig. 5. Development of a domain fixed under an
electrode while repelling neighbours

These transitions can experimentally be observed because it is possible to place peculiar electrodes above the surface /22/. Since a domain behaves like a dipolar disc it can be fixed in an inhomogeneous electric field while repelling all other domains. Now observing the growth Fig. 5 demonstrates that there is a stage (between 5b and 5c) where the boundary becomes unstable.

The peculiarity of Fig. 5d is that the irregular shape exists also under local equilibrium conditions. On the other hand rough interfaces can also be prepared under nonequilibrium conditions. This was demonstrated by quickly varying the surface pressure and observing domain development /23,24/. Thus fractal structures can be prepared and quantitatively be described within the concept of constitutional supercooling /24/. There the diffusion of impurities (impeding domain growth) from the interface is needed for growth to continue. Fig. 6 demonstrates the striking similarity of a typical domain shape with one obtained by computer simulations by Nittman and Stanley /25/ for the case of diffusion limited aggregation.

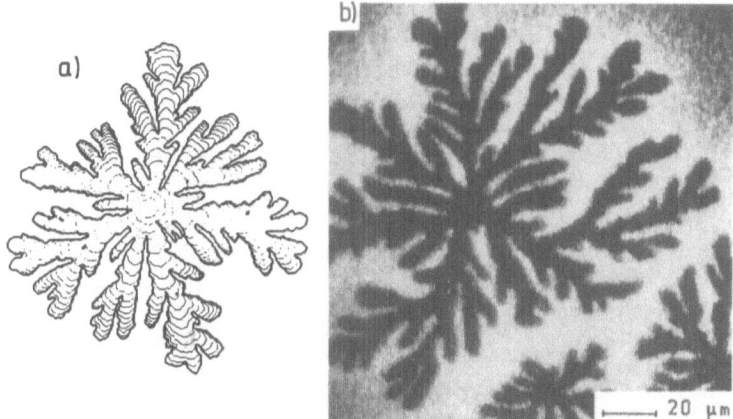

Fig. 6. Domain shape obtained by computer simulation with the "diffusion limited aggregation" model (a), in comparison with experimental observation following a pressure jump (b).

The picture on the formation of a superlattic very much is similar to that of a two-dimensional colloidal crystal /26/. To further quantify this Fig. 7 gives results of digital image analysis of a series of images for a selected system going along an isotherm. To describe the establishment of order we calculate the mean deviation Δf of domain area f from a mean value f, the mean deviation ΔD of the interdomain distance D from a mean value D and a form parameter φ being the average ratio of the boundary length with that expected for a circular domain. On reduction of the molecular area increasing order is reflected in a decrease of ΔD and Δf and in φ tending towards 1. Deviations from perfect order are only about 10%, but there is no abrupt change observable. The increase of φ for low A may be an artefact due to too few data, but it may also be caused by domain deformation on close approach /27/.

3. Structure at molecular level

The analysis of fluorescence micrographs presented in Fig. 8 serves to conclude on the domain structure at the μm as well as nm level. The measurement of the domain number N shows that decreasing the molecular area below A_c N quickly assumes a fixed value which on further compression remains constant. This means there is no further domain fusion or fission or new nucleation, and this holds for most systems encountered hitherto. The absolute value of N depends on nucleation conditions, e.g. on compression speed, and therefore superlattice dimensions are determined kinetically /28/.

Assuming equilibrium between two phases with fixed molecular areas A_{fl} and A'_s the LC phase area fraction ϕ should obeye a lever rule according to

$$\phi = \frac{A_{fl}-A}{A_{fl}-A'_s}$$

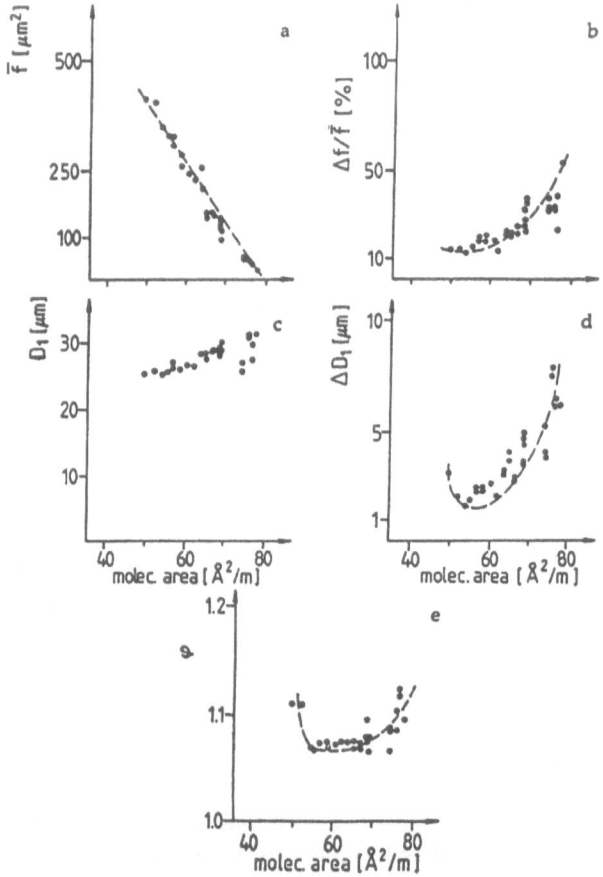

Fig. 7. Average domain area $\bar{f}$ (a), mean deviation Δf of domain areas from the average value (b), interdomain distance D_1(c) and mean deviation (ΔD_1) and form parameter φ, defined as described in the text, for the DMPA monolayer of Fig. 2.

throughout the coexistence range. The upper part of Fig. 8 demonstrates the validity of such a linear relation. As expected, $\phi \to 0$ for $A \to A_c$ since the latter is the area of the fluid phase and the pressure is hardly changed on going through the nearly horizontal part of the isotherm. Extrapolating the linear ϕ versus A plot to $\phi=1$, however, one does not obtain $A'_s=41\text{Å}^2$ but $A'_s=48\pm2\text{Å}^2$. This means that the ordered phase coexisting with the fluid one exhibits a lower density than the solid condensed phase. On the other hand the transition is not terminated on compression below 48Å^2 per molecule. This means that on approaching A_s the LC phase domains change their molecular density.

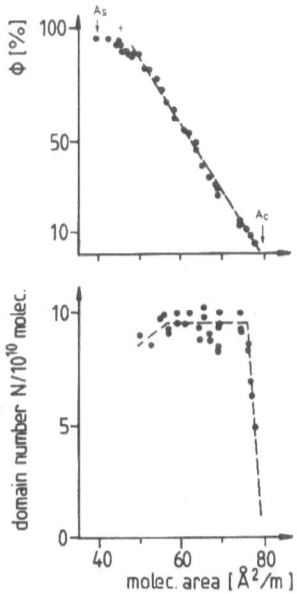

Fig. 8. LC phase area fraction ϕ and domain number N as a function of molecular area for the monolayer of Fig. 2.

A more direct proof of a pressure induced structural change results from X-ray scattering data described in the contribution of Als-Nielsen. Diffraction data taken below and above π_s show a drastic increase in the positional coherence length from about ten to more than 50 lattice spacings and a discontinuity of the lattice compressibility, indicative of a second order transition /29,30/. Reflectivity data indicate a drastic decrease in head group dimensions and increase in electron density above π_s /31/. This can be explained by a dehydration and ordering of head groups observed also for the low temperature phases of bilayer vesicles /32/.

Another strong indication on changes in the head group region results from surface potential data presented in Fig. 9 /33/. There the two lipids L-α-dimyristoylphosphatidylethanolamine (DMPE) and L-α-dilauroyl-phosphatidylethanolamine (DLPE) were chosen. These lipids differ in their chain lengths, and therefore at the same temperature DLPE exhibits a broad LE phase and a high π_c whereas for DMPE the coexistence range is well pronounced. The polar head groups are identical, and as they determine the surface potential ΔV one expects similar values for the latter. This is indeed the case.

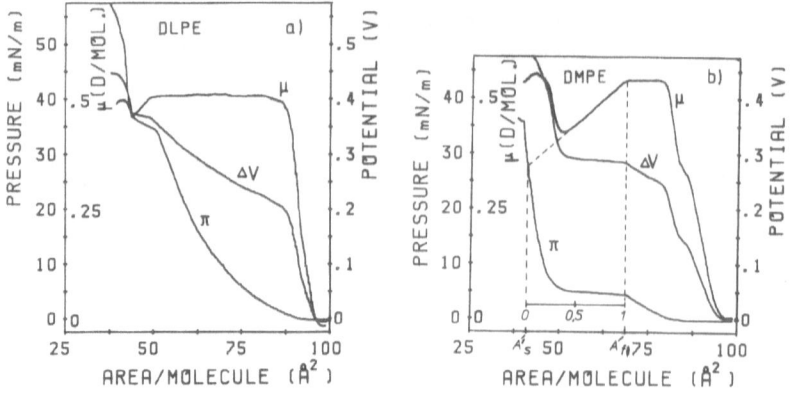

Fig. 9. Surface potential ΔV, perpendicular dipole moment μ and
surface pressure as a function of molecular area for DLPE and
DMPE monolayers.
T=20°C, pH 5.5.

Measuring ΔV on compression the values for zero surface pressure are
meaningless, since they are influenced by large heterogeneities. Throughout the
LE phase ΔV increases proportional to the density $\frac{1}{A}$ of molecules. This is most
clearly demonstrated by recording the effective dipole moment per molecule
$\mu = \varepsilon_0 * A * \Delta V$ The value is independent of A indicating a fixed dipole
orientation in the LE phase. Increasing the pressure above π_c the ΔV versus A
isotherm abruptly changes slope and becomes almost linear. The linear relation
again corresponds to a lever rule where the surface potential is composed of
that of two coexisting phases (ΔV_{fl}, ΔV_s)

$$\Delta V = \frac{A-A_{fl}}{A'_s-A_{fl}} \Delta V_s + \frac{A-A'_s}{A_{fl}-A'_s} \Delta V_{fl} = \phi \Delta V_s - (1-\phi) \Delta V_{fl}$$

ΔV_{fl}, A_{fl} correspond to the values at π_c and if A'_s is known ΔV_s can be
derived. The latter, however is not exactly known. Assuming, derived from
fluorescence micrographs, $A'_s=48\text{Å}^2$ one obtains $\Delta V'_s=290\text{mV}$ for DMPE. The
value would increase to $\Delta V_s=295\text{mV}$ if the corresponding area A'_s would be
41Å². Relevant for the calculation of dipolar forces is the difference in dipole
densities corresponding to the potential differences $\Delta(\Delta V)=\Delta V_s-\Delta V_{fl}$. For the
two situations above one derives 13mV and 18mV, respectively, for DMPE.
Due to the higher molecular density at π_c for DLPE the difference would be
correspondingly smaller.

Inspection of Fig. 9 also demonstrates that the lever rule does not hold
throughout the whole phase coexistence range. Instead a drastic increase in
ΔV is observed on decreasing the molecular area below 50Å² for DMPE as well
as for DLPE. This again indicates a structural change involving a
reorganization of the polar groups contributing to the surface potential.

Relating surface potential data to any molecular groups is rather daring since these groups are at an interface where the dielectric constant ε jumps from 80 to 1. Thus much to our surprise but in accordance with previously published data /34/ the surface potential is positive. Looking at a sketched

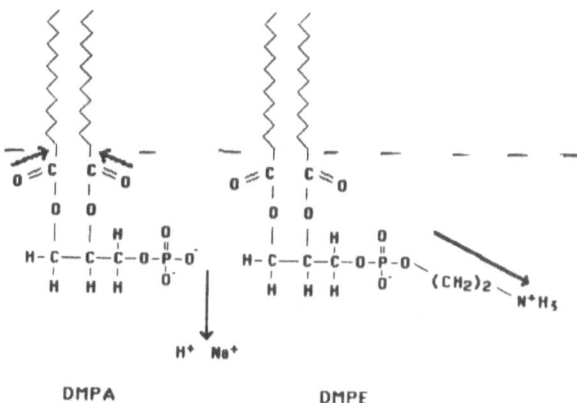

DMPA   DMPE

Fig. 10. Sketch of the phospholipid arrangement at the air water interface (dashed). The dipole moment of the diffuse double layer (for DMPA) or of the phosphatidylethanol-amine head would yield a potential with sign opposite to the findings, whereas the carbonyl groups would yield the correct sign.

arrangement in Fig. 10 this sign is not expected if the potential is due to the vertical component of the P^--N^+ dipole moment. Instead for the probable arrangement of the carbonyl groups as indicated the correct potential sign is obtained. Now assuming that the carbonyl groups determine ΔV and knowing their dipole moment one may estimate the angles of their bond axes with the surface /32/. One obtains a value near 20° for the LE phase which reduces to 10° in the LC phase. Although these values are not unrealistic they should not be taken too serious since they depend on many assumptions. The most important one is that of a local dielectric constant as the discussion below will show.

An explanation for the irrelevance of the P^--N^+ moiety concerning ΔV results from the solution of the Laplace equation for the interface /16/. A polar group in water is not only screened but also exhibits an image group above water with opposite polarity. Hence only groups in the low dielectric region will contribute significantly and a good candidate for this are the carbonyl groups. On the other hand this shows that a change in water structure or water level with respect to the head groups will strongly affect ΔV. Thus the increase in ΔV on decreasing the molecular area below 50Å^2 is consistent with a dehydration and ordering process also indicated from X-ray data.

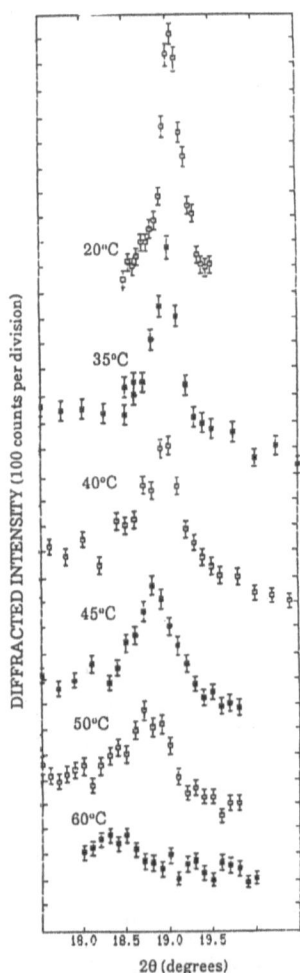

Fig. 11. Diffracted X-ray
intensity as a function
of diffraction angle for
a DMPE monolyer on
a silicon wafer for
various temperatures.
The data a displaced in
vertical direction.

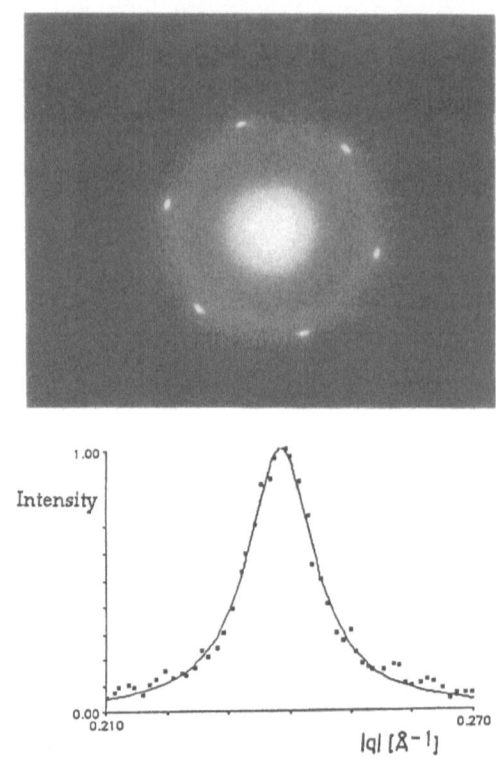

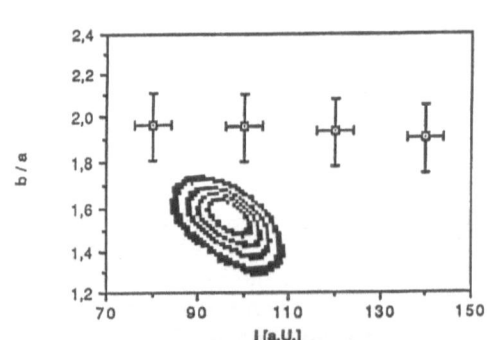

Fig. 12. Electron diffraction pattern of DMPE
monolayer on solid support (top),
radial intensity distribution along one
spot (middle) and ratio of long and
short axes of the contours of equal
intensity (bottom, insert) for various
intensities.

In conclusion optical, X-ray and surface potential data demonstrate that the
transition from LE to S involves an intermediate phase. This phase is more
compressible than the one established at high pressure, exhibits low

positional correlation and undergoes pressure induced structural changes of the head groups. This supports the suggestion /35/ that the head groups order at π_s. The latter is additionally supported by the fact that π_s can be varied by binding of divalent counter ions in a very specific way. On the other hand there are no data on diffusion coefficients or tail orientation in the LC phase. Therefore one can neither proof it to be "liquid" nor "tilted".

Although the molecular interpretation of surface potential data is speculative they can be directly used to calculate dipolar forces. Considering two circular domains with radius R~10μm and a distance D~3R apart the repulsion energy is calculated as $W_{rep} = \dfrac{(\pi R^2 \Delta\mu)^2}{4\pi\varepsilon_0(3R)^3}$.

With an effective dipole density difference $\Delta\mu = \varepsilon_0 A\Delta(\Delta V)$ one obtains $W_{rep} = 0.5eV \approx 20kT$. From this one also estimates an electrostatic energy density $e_{el} \sim \dfrac{1}{2}\dfrac{W_{rep}}{\pi R^2} \sim 10^{-10}N/m$

This shows that, as long as these domains are a distance of some μm apart their interaction does not significantly contribute to the surface pressure. Hence interdomain repulsion is not responsible for the finite isotherm slope in the phase coexistence range.

III. Monolayers on solid support

Monolayers prepared on water can be transferred on solid support, and fluorescence microscopy as well as X-ray data indicate that the structure is not grossly altered /36/. This is not only important for applications but also opens ways for studying phase transitions in layered structures. As an example on this Fig. 11 shows X-ray diffraction data as a function of temperature for a DMPE monolayer on solid support. Increasing the temperature the peaks shift and broaden and disappear above 60°C. Optical and X-ray data (unpublished) prove that the monolayer is stable against desorption up to temperatures well above 100°C. Hence in this case one observes melting of aliphatic tails.
Similar observations can also be performed by electron diffraction (Fig. 12), although in that case the resolution of line width and position is still lower than with X-rays. One unique advantage of this technique is that by help of digital image analysis the broadening of diffraction spots in radial as well as azimuthal direction can be followed. This opens new ways to study correlations between positional and orientational ordering, a theoretical as well as practical important question.

IV. Acknowledgement

This work contains results of the thesis work of my former and present co-workers M. Lösche, A. Miller, W.M. Heckl, H.D. Göbel, M. Flörsheimer, C.A. Helm, P. Tippmann-Krayer and R. Steitz. I am indebted to them as well as to J. Als-Nielsen and K. Kjaer for a fruitful cooperation. The work is supported by the Deutsche Forschungsgemeinschaft, the Bundesministerium für Forschung und Technologie (BMFT), the Stiftung Volkswagenwerk and the Fonds der Chemischen Industrie.

References

/1/   D.A. Cadenhead, F. Müller-Landau and B.M.J. Kellner, Phase transitions in insoluble one and two-component films at the air/water interface, in:"Ordering in Two Dimensions", Elsevier Biomedical Press, Amsterdam (1980)

/2/   O. Albrecht, H. Gruler and E. Sackmann, Polymorphism of phospholipid monolayers, J. Phys., 39:301 (1978)

/3/   G. Gaines, "Insoluble Monolayers at Liquid-Gas Interfaces", John Wiley & Sons, Inc., New York, London, Sidney (1966)

/4/   R. Peters and K. Beck, Translational diffusion in phospholipid monolayers measured by fluorescence microphotolysis, Proc. Natl. Acad. Sci. USA, 80:7183 (1983)

/5/   C.A. Helm, L. Laxhuber, M. Lösche and H. Möhwald, Electrostatic interactions in phospholipid membranes I, influence of monovalent ions, Colloid & Polym. Sci. 264:46 (1986)

/6/   M. Lösche, C.A. Helm, H.D. Mattes and H. Möhwald, Formation of Langmuir-Blodgett Films via Electrostatic Control of the Lipid/Water Interface, Thin Solid Films, 133:51 (1985)

/7/   M. Lösche and H. Möhwald, A fluorescence microscope to observe dynamical processes in monomolecular layers at the air/water interface,
       Rev. Sci. Instrum., 55:1968 (1984)

/8/   R.M. Weiss, and H.M. McConnell, Two-dimensional chiral crystals of phospholipids, Nature (Lond.), 310:5972 (1984)

/9/   W.M. Heckl and H. Möhwald, A narrow window for observation of spiral lipid crystals, Ber. Bunsenges. Phys. Chem., 90:1159 (1986)

/10/   W.M. Heckl, M. Lösche, D.A. Cadenhead and H. Möhwald, Electrostatically Induced Growth of Spiral Lipid Domains in the Presence of Cholesterol, Europ. Biophys. J., 14:11 (1986)

/11/   H.D. Göbel, H.E. Gaub and H. Möhwald, Shape and Microstructure of Crystalline Domains in Polydiacetylene Monolayers, Chem. Phys. Lett., 138:441 (1987)

/12/   A. Fischer, M. Lösche, H. Möhwald and E. Sackmann, On the Nature of the Lipid Monolayer Phase Transition, J. Physique Lett., 45:785 (1984)

/13/   D.J. Keller, H.M. McConnell and V.T. Moy, Theory of Superstructures in Lipid Monolayer Phase Transitions, J. Phys. Chem., 90:2311 (1986)

/14/   R.M. Weis and H.M. McConnell, Cholesterol Stabilizes the Crystal-Liquid Interface in Phospholipid Monolayers, J. Phys. Chem., 89:4453 (1986)

/15/   D. Andelman, F. Brochard, P.-G. de Gennes and J.F. Joanny, Monolayer transitions with polar molecules, C.R. Acad. Sc. Paris, 301:675 (1985)

/16/   D. Andelman, F. Brochard and J.F. Joanny, Phase Transition in Langmuir Monolayers of Polar Molecules, J. Chem. Phys., 86:3673 (1987)

/17/   H.M. MConnell and V.T. Moy, Shapes of Finite Two-Dimensional Lipid Domains, J. Phys. Chem., 92:4250 (1988)

/18/   D.J. Keller, J.P. Korb and H.M. McConnell, Theory of Shape Transitions in Two-Dimensional Phospholipid Domains, J. Phys. Chem., 91:6417 (1987)

/19/   M. Lösche, Das Phasenverhalten eines quasi zwei-dimensionalen Dielektrikums an der Electrolyt/Gas-Grenzfläche, Doctoral thesis, Technische Universität München (1986)

/20/ M. Lösche and H. Möhwald, Electrostatic interactions in phospholipid membranes II: Influence of divalent ions on monolayer structure, J. Coll. Interf. Sci, in press

/21/ T.K. Vanderlick, to be published

/22/ M. Flörsheimer and H. Möhwald, Development of Equilibrium domain shapes in phospholipid monolayers, in Chemistry and Physics of Lipids, in press

/23/ A. Miller, W. Knoll and H. Möhwald, Fractal growth of crystalline phospolipid domains in monomolecular layers, Phys. Rev. Lett, 56:2633 (1986)

/24/ A. Miller and H. Möhwald, Diffusion limited growth of crystalline domains in phospholipid monolayers, J. Chem. Phys. , 86:4258 (1987)

/25/ J. Nittmann and E. Stanley, Tip splitting without interfacial tension and dendritic growth patterns arising from molecular anisotropy, Nature, 321:663 (1986)

/26/ P.A. Forsyth, S. Marcelja, D.J. Mitchell and B.W. Ninham, Phase transition in charged lipid membranes, Biochim. Biophys. Acta, 469:335 (1977)

/27/ M. Lösche and H. Möhwald, Quantitative Analysis of Surface Textures in Phospholipid Monolayer Phase Transitions, J. Coll. Interf. Sci, 126:432 (1988)

/28/ C.A. Helm and H. Möhwald, Equilibrium and non-equilibrium features determining superlattices in phospholipid monolayers, J. Phys. Chem., 92:1262 (1988)

/29/ K. Kjaer, J. Als-Nielsen, C.A. Helm, L.A. Laxhuber and H. Möhwald, Ordering in Lipid Monolayer studied by Synchrotron X-ray Diffraction and Fluorescence Microscopy, Phys. Rev. Lett., 58:2224 (1987)

/30/ C.A. Helm, H. Möhwald, K. Kjaer and J. Als-Nielsen, Phospholipid monolayers between fluid and solid states, Biophys. J., 52:381 (1987)

/31/ J. Als-Nielsen and H. Möhwald, Synchrotron X-ray Scattering Studies of Langmuir Films, in press in "Handbook of Synchrotron Radiation", Vol. 4, eds. S. Ebashi, E. Rubenstein and M. Koch, North Holland 1989

/32/ J.M. Seddon, G. Cevc, R.D. Kaye and D. Marsh, X-ray Diffraction Study of the Polymorphism of Hydrated Diacyl-and Dialkylphosphatidylethanolamines, Biochemistry, 23:2634 (1984)

/33/ A. Miller, C.A. Helm and H. Möhwald, The Colloidal Nature of Phospholipid Monolayers, J. Physique, 48:159 (1987)

/34/ F. Paltauf, H. Hauser and M.C. Phillips, Monolayer characteristics of some 1,2-diacyl, 1-alkyl-2-acyl and 1,2-dialkyl phospholipids at the air/water interface, Biochim. Biophys. Acta, 249:539 (1971)

/35/ E. Sackmann, A. Fischer and W. Frey, Polymorphism of Monolayers of Monomeric and Macromolecular Lipids: On the Defect Structure of Crystalline Phases and the Possibility of Hexatic Order Formation, in: "Physics of Amphiphilic Layers", Springer Proc. Phys. 21:25 (1987) eds. J. Meunier, D. Langevin and N. Boccara

/36/ K. Kjaer, J. Als-Nielsen, C.A. Helm, P. Tippmann-Krayer and H. Möhwald, An X-ray scattering study of lipid monolayers at the air/water interface and on solid supports, Thin Solid Films, 159:17 (1988)

WETTING OF ROUGH SOLID SURFACES BY LIQUIDS

David Andelman[*], Jean-Francois Joanny[@] and Mark O. Robbins[#]

[*]School of Physics and Astronomy, Tel Aviv University, Tel Aviv 69978, Israel. [@]Ecole Normale Superiéure de Lyon, 46, Allée d'Italie, 69364 Lyon cedex 07, France. [#]Department of Physics and Astronomy, Johns Hopkins University, Baltimore, MD 21218, U.S.A.

INTRODUCTION

Recently, newly introduced experimental techniques have advanced our understanding of the phenomena of wetting and spreading of solid surfaces by thin liquid films ranging in thickness between few dozen Angstroms to a fraction of a micron. Among others these include ellipsometry[1]−[3] and grazing incidence X-ray diffraction using a synchrotron source[4][5]. From a theoretical point of view, simple models for profiles of thin liquid films and their relation to the shape of the solid substrate have been proposed for smooth solid surfaces[6], wetting on fibers[7], wetting of porous media and fractals[8]. In this short contribution we focus only on the static behavior of thin liquid films that completely wet a rough solid surface[9]−[11]. Our predictions can be tested using the experiments mentioned above[1]−[5].

When a liquid completely wets a solid surface, its *contact angle* θ_e is zero and the liquid completely covers the solid with a thin layer. The thickness of the liquid layer depends on several factors: *i*) the undersaturation vapor pressure for volatile liquids; *ii*) the height of the horizontal solid surface above a reservoir of liquid; *iii*) the conserved volume of the liquid layer in the case of nonvolatile liquids such as silicone oils (PDMS)[12].

We consider here the structure of such a thin liquid film which completely wets a *rough* solid surface. By rough we mean any surface that is not atomically smooth and that can be described by its height $\zeta_S(\vec{\rho})$ above a two-dimensional reference plane defined by the vector $\vec{\rho}$. The free energy associated with such a thin film depends on the local height of the liquid-vapor interface taken with respect to the same reference plane $\zeta_L(\vec{\rho})$ and can be expressed as

$$\Delta F = \int \{(\gamma_{SL} - \gamma_{SV})\sqrt{1 + |\vec{\nabla}\zeta_S|^2} + \gamma\sqrt{1 + |\vec{\nabla}\zeta_L|^2} + P(\zeta_L) + \Delta\mu(\zeta_L - \zeta_S)\}d^2\vec{x} \quad (1)$$

The first two terms represent the change in interfacial energy with respect to the bare solid surface. γ_{SL}, γ_{SV} and γ are the solid-liquid, solid-vapor and the liquid-vapor surface tensions, respectively. The third term, $P(\zeta_L)$, comes from the long-range interactions such as Van der Waals in nonpolar liquids. The last term is the chemical potential difference between the liquid and the vapor integrated over the volume of the film.

Using the variational principle, an Euler-Lagrange equation can be derived for the liquid-vapor profile $\zeta_L(\vec{x})$

$$\gamma \vec{\nabla} \cdot \{ \frac{\vec{\nabla}\zeta_L(\vec{x})}{\sqrt{1 + |\vec{\nabla}\zeta_L(\vec{x})|^2}} \} = -\Pi_d(\zeta_L(\vec{x})) + \Delta\mu \tag{2}$$

where the disjoining pressure, $\Pi_d = -\delta P/\delta\zeta_L$, is positive for attractive long-range forces. For non-retarded Van der Waals interactions between any two molecules $U_{ij} \sim r_{ij}^{-6}$, the disjoining pressure is calculated to be

$$\Pi_d(\zeta_L) = \frac{3A}{8\pi^2} \int d^2\vec{\rho} \frac{1}{\rho^5} \{ tg^{-1}(\frac{1}{\zeta_{SL}}) - \frac{\zeta_{SL}(5 + 3\zeta_{SL}^2)}{3(1 + \zeta_{SL}^2)^2} \} \tag{3}$$

where $\zeta_{SL}(\vec{x}, \vec{\rho}) = (\zeta_L(\vec{x}) - \zeta_S(\vec{x} + \vec{\rho}))/|\vec{\rho}|$ and A is the Hamaker constant.

Equations (2) and (3) completely specify the liquid-vapor profile ζ_L as a function of the random solid surface ζ_S. In the case of a flat surface $\zeta_S(\vec{x}) = const$, it is easy to verify that $\Pi_d(e) = A/6\pi e^3$, where e is the film thickness. We proceed by presenting the linearized version of (2) and (3). Numerical studies of (2) and (3) for one-dimensional corrugation will be presented elsewhere[13].

LINEAR RESPONSE APPROXIMATION

For weakly fluctuating solid surfaces, (3) can be linearized in ζ_S and $\zeta_L - e$. In addition, keeping only the leading order in $\vec{\nabla}\zeta_L$ in (2) and choosing $\langle\zeta_S\rangle = 0$, we get an equation for the average profile $\langle\zeta_L\rangle = e$, $\Pi_d(e) = \Delta\mu$, and a linear equation for the fluctuation of ζ_L

$$\xi^2\nabla^2\zeta_L(\vec{x}) = (\zeta_L(\vec{x}) - e) - \int d^2\vec{\rho} K(\vec{x} - \vec{\rho})\zeta_S(\vec{\rho}) \tag{4}$$

where the characteristic length in (4)

$$\xi = \sqrt{\frac{Ae^4}{2\pi\gamma}} \tag{5}$$

is called the *healing length* and the kernel in (4) is

$$K(\rho) = \frac{2e^4}{\pi(\rho^2 + e^2)^3} \tag{6}$$

Equation (4) can be solved separately for each Fourier component yielding

$$\tilde{\zeta}_L(q) = \tilde{\zeta}_S(q)\frac{\tilde{K}(q)}{1 + q^2\xi^2} \tag{7}$$

where $\tilde{\zeta}_L, \tilde{\zeta}_S$ and $\tilde{K}(q)$ are the Fourier transforms of ζ_L, ζ_S and $K(\rho)$, respectively.

DISCUSSION

The competition between surface tension and disjoining pressure determines to what extent a thin liquid film follows the undulations of a rough solid substrate and is seen in (7). Long-wavelength fluctuations of the solid surface, $qe \ll 1$ and $q\xi \ll 1$ are followed by the liquid film, $\tilde{\zeta}_L(q) \simeq \tilde{\zeta}_S(q)$ while the short-ones, $q\xi > qe \gg 1$ are strongly damped.

$$\tilde{\zeta}_L(q) \simeq (qe)^{3/2}(q\xi)^{-2}\exp(-qe)\tilde{\zeta}_S(q) \qquad (8)$$

Our linear response theory (7) is a generalization of the so-called Deryagin approximation[14] which amounts to calculating only the local contribution to the disjoining pressure. In our approach this is achieved by replacing the true kernel $K(\rho)$ by a Dirac delta function $K(\vec{\rho}) = \delta(\vec{\rho})$ in (4) or $\tilde{K}(q) = 1$ in (7). The main discrepancy between the improved result (7) and the local approximation occurs in the sort-wavelength limit, $qe \gg 1$. While we get an exponential damping with a characteristic length being the film thickness, $\tilde{\zeta}_L(q)/\tilde{\zeta}_S(q) \simeq (q\xi)^{-2}$ for the Deryagin approximation for any $q\xi \gg 1$. Note that $\xi = e^2/a$, a being a molecular length, is always bigger that e.

A quantitative comparison of our results with experiments should be possible for Van der Waals (non-polar) liquids such as PDMS (silicone oil) spreaded on various preprepared rough surfaces like etched glass, fused silica and mica[1]. In a grazing incidence X-ray diffraction experiment[4][5], the ratio of intensities scattered from the liquid and solid surfaces $I_L(q)/I_S(q)$ is proportional to the ratio of the q-component of the mean-squared height fluctuations of the two surfaces, $\langle \tilde{\zeta}_L^2(q) \rangle / \langle \tilde{\zeta}_S^2(q) \rangle$

$$I_L(q)/I_S(q) \simeq \frac{\tilde{K}^2(q)}{(1 + q^2\xi^2)^2} \qquad (9)$$

In the limit $qe \gg 1$, (9) reduces to $I_L(q)/I_S(q) \simeq (e^3/q\xi^4)\exp(-2qe)$.

We conclude with several remarks on related results and possible extensions of the model presented here. The formalism for non-retarded Van der Waals interactions has been extended to any inverse power law interaction $U_{ij} \sim r_{ij}^{-n}$ and generalized expression for the healing length ξ and the kernel $K(\rho)$ have been found[9]. Numerical studies of (2)–(3) for simple corrugated surfaces like surfaces with square-well grooves[13] confirm qualitatively the damping found within the linear response theory even in the limit where the r.m.s. fluctuation of the roughness is of the same order of magnitude as the film thickness. Complete wetting of chemically heterogeneous surfaces can be modeled by a position dependent Hamaker constant $A(\vec{\rho}, z)$ [15]. Within a linear response theory, a direct analogy exists between roughness and chemical heterogeneities[13].

Height-height correlations in position space, $\langle \zeta_L(0)\zeta_L(\vec{\rho}) \rangle$ can be obtained by an inverse Fourier transform of $\langle \tilde{\zeta}_L^2(q) \rangle$. The height-height correlation of the liquid interface depends on the correlations of the solid surface. We have treated three types of surfaces: (a) solid surfaces with short-range correlations characterized by a Gaussian-like decay $\langle \zeta_S(0)\zeta_S(\vec{\rho}) \rangle \sim \exp(-\rho^2/\sigma^2)$; (b) solid surfaces with algebraically decaying correlations, $\langle \zeta_S(0)\zeta_S(\vec{\rho}) \rangle \sim |\vec{\rho}|^{-\alpha}, \alpha > 0$; and (c) Self-affine solid surfaces where $\langle \zeta_S(0)\zeta_S(\vec{\rho}) \rangle \sim |\vec{\rho}|^{\alpha}, 0 < \alpha < 2$. In all three cases, the correlations of the liquid film depend both on the solid surface structure and on the healing length. In the latter case, for example, the liquid interface is composed of smooth sections up to length scales of order ξ. For length scales bigger than ξ, the liquid film has a self-affine behavior which follows the solid.

163

For a closer comparison with experiments, thermal fluctuations have to be included in the model as an additional source of roughness of the liquid interface[16]. Even in the limit where we predict a smooth and flat liquid interface, e.g. when ξ is big, thermal fluctuations will always cause a r.m.s. roughness of the order of a few Angstroms. They have been found to be about 3Å for thick water films[17].

Finally, we mention two interesting experiments where roughness is superimposed on films whose average thickness varies smoothly with position. The first example is complete wetting of a vertical solid plate. Above a macroscopic miniscus, there is a static Rollin film. For Van der Waals interactions[6][18], the film thickness, $e(h) \sim h^{-1/3}$, and we can define a local healing length using (5), $\xi(h) \simeq (A\gamma^3 h^{-4})^{1/6}$. For solid surfaces with small roughness with typical wavelength λ, there will be a crossover from a relatively smooth liquid film for $\lambda \ll \xi(h)$ to a "wiggly" film for $\lambda \gg \xi(h)$. Thus, there is a characteristic height $h_c \sim (A\gamma^3)^{1/4}\lambda^{-3/2}$ above which the film starts to follow the rough solid.

For a horizontal geometry, a similar prediction holds for the precursor film preceding the macroscopic drop. Since the thickness of the precursor film is found[6] to vary as $e(x) \sim x^{-1}$, x being the distance from the macroscopic edge of the drop, wiggles are expected to be found at the tip of the precursor film where the thickness is smaller than $e_c \simeq \sqrt{a\lambda}$, a being a microscopic length $a \simeq \sqrt{A/2\pi\gamma}$.

Acknowledgements: One of us (D.A.) acknowledges support from the Bat Sheva de Rothschild Foundation and the U.S.-Israel Binational Foundation (BSF).

REFERENCES

1. D. Beaglehole, *J. Phys. Chem.*, to be published (1989).
2. L. Leger, M. Erman, A.M. Guinet-Picard, D. Ausserre and C. Strazielle, *Phys. Rev. Lett.* **60**, 2390 (1988).
3. F. Heslot, A.M. Cazabat and P. Levinson, *Phys. Rev. Lett.* **62**, 1286 (1989).
4. S. Garoff, E. B. Sirota, S. K. Sinha and H. B. Stanley, *J. Chem. Phys.*, to be published (1989).
5. J. Daillant, J.J. Benattar, L. Bosio and L. Leger, *Europhys. Lett.* **6**, 431 (1988).
6. P.G. de Gennes, *Rev. Mod. Phys.* **57**, 827 (1985).
7. F. Brochard, *J. Chem. Phys.* **84**, 4664 (1988).
8. P. G. deGennes, in *Physics of Disordered Materials*, edited by D. Adler, H. Fritzsche and S. R. Ovshinsky (Plenum Press, New York) 1985, p. 227; F. Brochard, *C.R. Acad. Sci. (Paris)* **304** II, 785 (1987).
9. D. Andelman, J.F. Joanny and M.O. Robbins, *Europhys. Lett.* **7**, 731 (1988).
10. Effects of atomic height steps were studied by A.C. Levi and E. Tosatti, *Surf. Sci.* **178**, 425 (1986).
11. P.G. de Gennes, *C.R. Acad. Sci. (Paris)* **300** II, 129 (1985).
12. J. F. Joanny and P. G. de Gennes, *C. R. Acad. Sci. (Paris)* **299** II, 279 (1984).
13. D. Andelman, J.F. Joanny and M.O.Robbins, *to be published*.
14. J.N. Israelachvili, *Intermolecular and Surface Forces* (Academic Press, New York) 1985.
15. M.W. Cole and E. Vittoratos, *J. Low Temp. Phys.* **22**, 1923 (1976).
16. J.L. Harden and R.F. Kayser, *J. Coll. Interface Sci.* **127**, 548 (1989).
17. A. Braslau, M. Deutch, P.S. Pershan, A.W. Weiss, J. Als-Nielsen and J. Bohr, *Phys. Rev. Lett.* **54**, 114 (1985).
18. B. Deryagin, *Kolloidn. Zh.* **57**, 191 (1955).

CRYSTALLINE AND LIQUID CRYSTALLINE ORDER IN CONCENTRATED

COLLOIDAL DISPERSIONS: AN OVERVIEW

H.N.W. Lekkerkerker

Van 't Hoff Laboratorium
Rijksuniversiteit te Utrecht
Postbus 80051, 3508 TB Utrecht
The Netherlands

I. INTRODUCTION

The equilibrium thermodynamic and structural properties of colloidal dispersions may be treated in the same way as in the case of simple liquids by considering the colloidal particles as "supramolecules" dispersed in a continuous (but fluctuating) back-ground. The potential which for the case of fluctuating forces replaces the interaction potential between molecules (in vacuo) is the potential of the average forces which act between the dispersed particles. This effective interaction is the input for statistical mechanical theories. Therefore statistical mechanical theories developed for simple fluids can be applied to colloidal dispersions. The theoretical basis for such a treatment was given by Onsager[1] and Mc Millan and Mayer[2]. In recent years concepts of liquid state theory have been applied successfully to understand the behavior of concentrated colloidal dispersions[3,4].

One of the most remarkable phenomena exhibited by concentrated suspensions of monodisperse colloidal particles is the spontaneous transition from fluid-like structures, in which there are only short-range correlations between the positions and orientations of the particles, to structures which have long-range spatial and/or orientational order. Increasing the concentration of spherical colloidal particles (and/or increasing the range of the repulsive interaction by decreasing the electrolyte concentration of the medium in the case of charged particles) leads to the following progression of phases: colloidal fluid, colloidal fluid and colloidal crystal in coexistence, fully colloidal crystal. Since the interparticle spacing in these ordered structures is of the order of optical wavelengths, light is separated into colors when scattered, giving rise to beautiful iridescence. In Section II I will discuss the nature of the disorder-order transition that results in the formation of colloidal crystals and present examples of colloidal systems that show this transition. In the case of dispersions of sufficiently long rod-like particles transitions from the disordered state to the nematic state and from the nematic state to the smectic state are observed. Dispersions of short rigid rod-like particles transform directly from colloidal fluids to colloidal smectic. These transitions from colloidal fluid to colloidal liquid crystal have a lot in common with the colloidal crystal transition. This will be discussed in Section III.

II. COLLOIDAL CRYSTALS

II.1. **Theoretical aspects**

In the statistical physics of liquids there has been the problem whether any attractive force is necessary for a liquid to transform into a crystal. As early as 1939 Kirkwood[5] speculated that in a system of hard spheres a certain limiting density exists above which a liquid type of distribution and a liquid structure cannot exist. Above this density, only structures with crystalline long range order would then be possible. Kirkwood considered this speculation highly tentative. Early computer simulations by Alder and Wainwright[6] revealed that the equation of state of a hard sphere system has two distinct branches. The low-density branch corresponds to the hard sphere liquid, whereas the high-density branch shows crystalline ordering. Subsequent work of Hoover and Ree[7] established that a hard sphere liquid with volume fraction $\phi = 0.494$ undergoes a transition to a crystalline phase with volume fraction $\phi = 0.545$. This means that when a hard-sphere system is gradually compressed, the system will transform into a state with long-range order long before close packing ($\phi_{cp} = 0.74$) is reached. This transition is sometimes referred to as the Kirkwood-Alder transition[8].

The nature of this transition has been clarified considerably in recent years by the application of density functional theories[9]. In these theories the freezing of hard spheres can be described as a competition between two forms of entropy: a loss in ideal (non-interacting) entropy when the particles go into the ordered state and a gain in packing (excluded volume) entropy resulting from the particle localization. The starting point of density functional theories of freezing is the following exact expression for the free energy difference $\Delta F = F-F_0$ between the free energy F of a solid of density $\rho(r)$ and the free energy F_0 of a fluid of density ρ_0 equal to the average density of the solid

$$\frac{\Delta F}{k_B T} = \int_V d\vec{r}\,\rho(\vec{r}) \ln\left(\frac{\rho(\vec{r})}{\rho_0}\right) - \int_V d\vec{r} \int_V d\vec{r}' \int_0^1 d\lambda (1-\lambda) c(\vec{r},\vec{r}';[\rho_0+\lambda\Delta\rho]) \,\Delta\rho(\vec{r})\Delta\rho(\vec{r}')$$

(1)

where $\Delta\rho(\vec{r}) = \rho(\vec{r})-\rho_0$ and $c(\vec{r},\vec{r}';[\rho_0+\lambda\Delta\rho])$ is the direct correlation function of a system of local density $\rho_0+\lambda\Delta\rho(\vec{r})$. The first term on the right hand side of Eq. (1) represents the ideal (non-interacting) contribution to the free energy difference and the second term represents the packing (excluded volume) contribution to the free energy difference. The first term is positive (corresponding to a loss in entropy) whereas the second term is negative (corresponding to a gain in entropy). In order to perform calculations starting from Eq. (1) one has to know the direct correlation function $c(\vec{r},\vec{r}';[\rho])$. This, however, is usually not the case, and, in practice, it is at this stage that rather drastic approximation have to be introduced. Various approximate treatments of the liquid-solid transition in hard sphere systems, starting from Eq. (1), have been worked out and quite satisfactory agreement with computer simulations has been obtained (see the reviews mentioned under Ref. 9 for a discussion and details).

In the case of hard spheres only the face centred cubic (FCC) structure is stable. However when the interparticle repulsion is sufficiently soft both FCC and body-centred cubic (BCC) structures have a stability range that depends on temperature and density. For the case of inverse power potentials

$$W(r) = \varepsilon(\sigma/r)^n$$

(2)

Hoover, Young and Grover[11] found that for $n \leq 7$ both FCC and BCC have a stability range. More directly relevant for colloidal systems is the recent computer simulation study of Robbins, Kremer and Grest[12] of the phase diagram of particles interacting through a repulsive potential of the Debye-Hückel form

$$W(r) = U_0 \frac{\exp(-\kappa r)}{(r/a_s)} \qquad (3)$$

Here $a_s = (N/V)^{-1/3}$ is a characteristic interparticle spacing. The phase diagram contains both a melting transition and a transition from BCC to FCC. The phase diagram can be expressed in terms of the dimensionless inverse screening length $\lambda = \kappa a_s$ and the dimensionless temperature $k_B T/U_a$, where U_a is the interaction between particles separated by the characteristic length a_s: $U_a = U_0 \exp(-\lambda)$. The result is given in Fig. 1. We note that whereas for hard spheres with density functional theory quite satisfactory agreement with computer simulations has been obtained this is not the case for particles interacting with a long-range repulsion. For that case density functional theory in its present form[13] only predicts a metastable BCC phase and not the stable phase seen in computer simulations. Other analytical treatments in the literature are also not entirely consistent with the computer simulation results[14-16].

II.2. Order-disorder phase separation in dispersions of spherical colloidal particles

II.2.1. Dispersions of (nearly) hard colloidal spheres

Ordering of iso-dimensional colloidal particles in three dimensions was first clearly established in dispersions of Tipula iridescent virus (TIV). Klug et al.[17] observed that the TIV particles are packed in a FCC structure, with an interparticle spacing of 250 nm, a distance nearly twice the diameter of about 130 nm of the frozen-dried particle. These authors therefore concluded that the crystals are probably held together by long-range forces operating at a distance comparable with the size of the particles. With the possibility to prepare stable dispersions of highly monodisperse microspheres of synthetic polymers (latex particles) with concentrations of up to 0.6 in volume fraction a number of studies on the transition behavior from the milky white (disordered) state to the iridescent (ordered) state appeared[18,19]. These first studies were performed on charged latex particles exhibiting long-range Coulombic interactions. The idea that no special long-range interactions are needed to explain the disorder-order transition in colloidal dispersions was first clearly stated by Wadati and Toda[8]. They considered this order-disorder phase separation as an indication for the existence of the Kirkwood-Alder transition in nature.

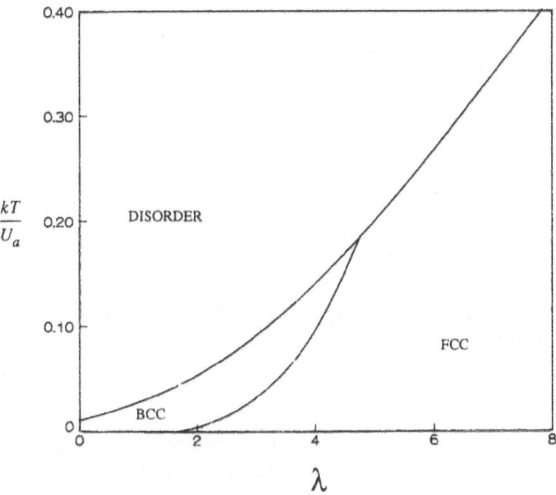

Fig. 1. The phase diagram of particles interacting through a Debye-Hückel potential in terms of the dimensionless temperature $k_B T/U_a$ and the dimensionless inverse screening length. After Ref. 12.

The ordering phenomenon is also observed in dispersions of colloidal particles which interact through a steep repulsive potential. Kose and Hachisu[20] studied polymethylmethacrylate (PMMA) cross-linked with divinylbenzene (DVB) dispersed in benzene. These latex particles were thought to be a good model system for (nearly) hard colloidal spheres because in this non aqueous system both the electric repulsion as well as the Van der Waals attraction do not play a significant role. Spontaneous phase separation was observed in certain particle concentration ranges. The values of the volume fraction at which the phase separation starts, taking into account the swelling of the particles, are very close to 0.5, in good agreement with the value obtained by computer simulations on hard sphere systems[7]. At first sight the results of Kose and Hachisu thus can be said to have substantiated the Kirkwood-Alder transition by experimental evidence. However, although Kose and Hachisu stated that a cross-linked PMMA latex dispersion in benzene can be

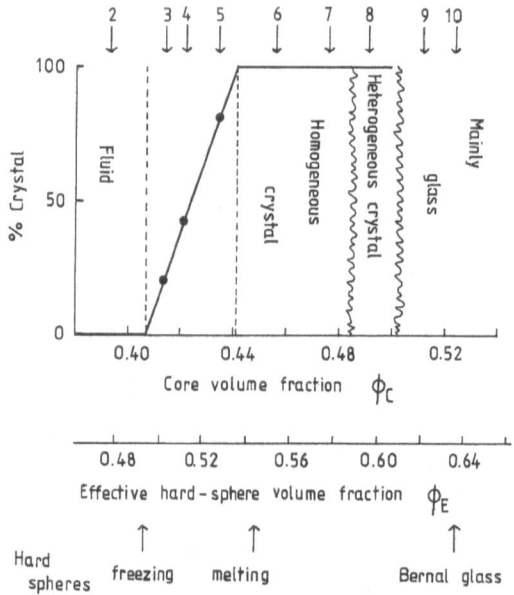

Fig. 2. The phase diagram of PMMA particles stabilized sterically by PHS. After Ref. 23.

considered to be an excellent model for a hard sphere system, static light scattering studies on this system by Nieuwenhuis and Vrij[21] indicate that the interaction between these PMMA particles in addition to a hard core has a soft repulsive tail extending over a considerable distance.

Another system of latex particles which is supposed to be a good model system for nearly hard colloidal spheres is PMMA stabilized sterically by poly-12-hydroxystearic acid (PHS)[22]. The phase behavior of systems of these particles dispersed in a mixture of decalin and carbon disulphide (chosen to match closely the refractive index of the particles) was studied by Pusey and Van Megen[23]. The observed phase behavior is summarized in Fig. 2. Taking into account the stabilizing coating in the effective volume fraction good agreement with the theoretical values for the volume fractions for freezing and melting in an hard sphere system is obtained.

A third system that is claimed to behave as a model hard sphere fluid is a dispersion of colloidal silica spheres sterically stabilized by stearyl chains grafted onto the surface and dispersed in cyclohexane[24]. Experimental studies of both the equilibrium thermodynamic and structural properties (osmotic compressibility and structure factor) as well as the dynamic properties (sedimentation, diffusion and viscosity) established that this system can indeed be described in very good approximation as a hard sphere colloidal dispersion (for a review of these experiments and their interpretation in terms of a hard sphere model see Ref. 4). De Kruif et al.[25] observed that in these lyophilic silica dispersions at volume fractions above 0.5 a transition to an ordered structure occurs. The transition from an initially glass like sediment to the iridescent (ordered) state appears only after weeks or months.

From computer simulations on crystallization in atomic systems it is clear that the harshness of the repulsive part of the pair potential has a pronounced effect on the process of homogeneous nucleation and thereby on the rate of crystal growth[26]. Both the time lapse for the onset of nucleation as well as the time required for the completion of the nucleation process are longer for steeper repulsive potentials. Colloidal systems offer unique possibilities to investigate the influence of the repulsive potential since by suitable modification of the stabilizing coating the range of the repulsive interaction between the particles can be varied more or less continuously. Smits et al.[27] studied the influence of the stabilizing coating of colloidal silica particles on the rate of order formation in dispersions of these particles. They found that with increasing range of the repulsive interaction the rate of crystallization increases dramatically. For example, whereas in dispersions of silica particles stabilized by stearyl chains iridescent crystallites only appear after weeks or months, the same phenomenon takes approximately only one day in dispersions of silica particles coated with terminally attached polyisobutene chains with number averaged molecular weight $M_n = 13000$.

II.2.2. Dispersions of soft colloidal spheres

As mentioned above the disorder-order transition in aqueous dispersions of charged latex particles, in particular well characterized polystyrene latices, has been investigated extensively. These particles interact through screened Coulombic interactions, the range of which depends on the electrolyte concentration in the suspension medium. At very low electrolyte concentration (10^{-6} M or lower) the transition from the milky white (disordered) state to the iridescent (ordered) state may occur at volume fractions below 1%[19]. This means that colloidal spheres in suspension can maintain themselves in a regular lattice structure even when the interparticle spacing is several diameters.

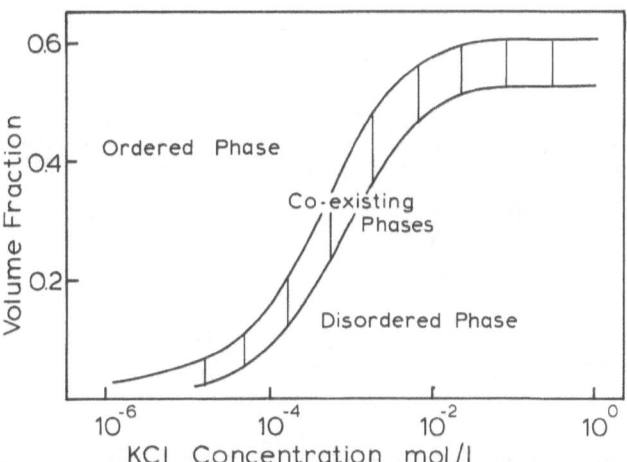

Fig.3. Diagrammatic picture of the phase diagram of charged latex particles a a function of the salt concentrations. After Ref. 28.

Increasing the electrolyte concentration in the suspension medium leads to a reduction of the effective range of the interparticle interaction. At electrolyte concentrations of about 0.1 M the effective range of the interparticle interaction is reduced to a fraction of a particle diameter and the particles behave as nearly hard spheres. The phase transition now occurs at a volume fraction of about 0.5 as in the case of nearly hard colloidal spheres. The phase diagram as function of electrolyte concentration as determined by Hachisu and coworkers[28] is presented in Fig. 3. Unfortunately no simulation studies for the phase diagram of particles of finite size which are interacting with screened coulomb potentials are available to compare the results of Hachisu et al. with.

A further aspect of the disorder-order transition dispersions of charged colloidal particles concerns the question which crystal structure is adopted. The work of Williams and Crandall[29] and Lindsay and Chaikin[30] suggests that the lattice type is BCC at low volume fractions (approximately $\phi < 0.02$) and FCC at high volume fractions (approximately $\phi > 0.03$). Recently Monovoukas and Gast[31] presented a detailed study of the order-disorder and BCC-FCC transitions in a dispersion of charged colloidal polystyrene spheres. They find that at low ionic strength their experimentally determined phase diagram (Fig. 4) good agreement with the simulation results of Robbins et al.[12] can be obtained by using in the interaction potential a renormalized charge[32]. Apparently at low ionic strength, where the range of the interaction is (much) larger than the particle radius, the neglect of the hard core does not appear to be too serious. Nevertheless for a better comparison with experimental data, further simulations intermediate between the Debye-Hückel potential and the hard sphere are needed.

III. COLLOIDAL LIQUID CRYSTALS

III.1. Theoretical aspects

The first theory for the isotropic to nematic phase transition in a solution of rigid rod-like particles that have purely repulsive interactions was developed by Onsager[33] during the 1940's. He modeled the particles as rigid rods and drew an analogy between the dispersion and a dilute gas of rigid rods by formulating a virial series for the free energy. As mentioned in the Introduction such a procedure can be rigorously justified if one uses the potential of the average forces as input for the free energy calculation. Onsager qualitatively showed that for very long slender rods the third virial term is negligible compared to the second virial term and made the plausible assumption that the higher virial coefficients can also be neglected in the same limit.

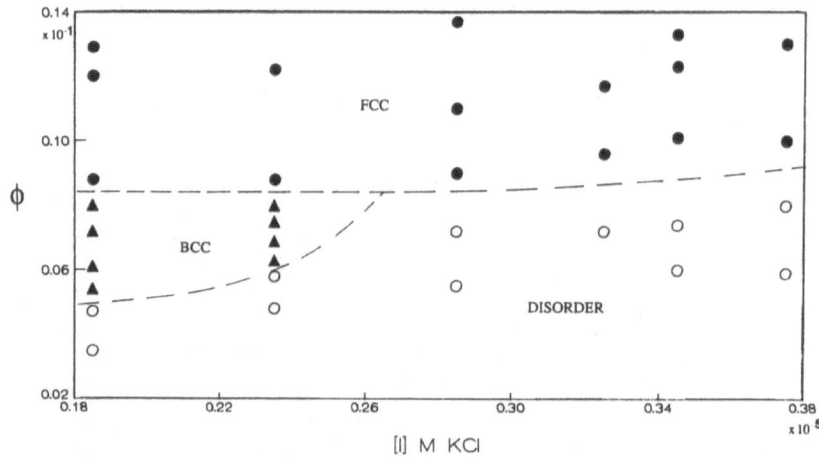

Fig. 4. Experimental phase diagram of charged polystyrene particles at low ionic strengths. After Ref. 31.

Recently Frenkel[34] calculated values for the third to fifth virial coefficients of spherocylinders (cylinders with length L and diameter D, capped at each end with hemispheres of the same diameter) with the Monte Carlo technique and found indeed that for very large L/D ratios the higher order virial coefficients become small compared to the second. Retaining only the first non-ideality correction in the virial expansion of the free energy leads to an elegant and simple theory for the isotropic nematic phase transition that however is only valid for very long rigid rods (L/D > 100 to be quantitatively accurate). Under the above mentioned conditions the free energy for a dispersion of hard spherocylinders can be written as

$$\frac{\Delta F}{N k_B T} = \frac{F(\text{solution}) - F(\text{solvent})}{N k_B T}$$

$$= \frac{\mu^0(T, \mu_0)}{k_B T} - 1 + \ln \rho + \int f(\Omega) \ln [4\pi f(\Omega)] d\Omega$$

$$+ \rho L^2 D \iint \sin \gamma(\Omega, \Omega') f(\Omega) f(\Omega') d\Omega d\Omega' \qquad (4)$$

Here $\mu^0(T, \mu_0)$ represents the standard chemical potential of the particles at temperature T in a solvent with chemical potential μ_0, $\rho = N/V$ is the number density, $\gamma(\Omega, \Omega')$ is the angle between two rods with orientations Ω and Ω' and $f(\Omega)$ is the one-particle orientation distribution function. As far as the isotropic nematic phase transition is concerned the last two terms on the right hand side of Eq. (4) that depend on the orientation distribution function are of importance. Like in the case of the disorder-order transition in a system of hard spheres the isotropic-nematic phase transition in a system of hard rods can be described as a competition between two forms of entropy: a loss in ideal (non-interacting) entropy when the particles go into the orientationally ordered state and a gain in packing (excluded volume) entropy resulting from the particle orientational confinement in the ordered state. By using a clever choice for the orientation distribution function Onsager was able to solve the minimization problem of the free energy and the coexistence conditions for the isotropic and nematic phase in a simple way. He found for the volume fractions in the coexisting isotropic and nematic phase $\phi_i = 3.340$ D/L and $\phi_n = 4.488$ D/L. These approximate results are in good agreement with the accurate numerical solution of the problem[35]: $\phi_i = 3.290$ D/L and $\phi_n = 4.191$ D/L.

Truncating the virial expansion of the free energy after the first term in the density yields an approximate theory for the isotropic-nematic phase transition which should be satisfactory for very long rods (and indeed asymptotically exact for L/D $\rightarrow \infty$) but not for shorter rods. Various statistical mechanical theories for the hard-rod fluid have been developed to deal with the problem such as a scaled particle treatment[36], an integral equation approach[37] and a treatment in which the free energy of a hard-rod fluid is obtained from a hard-sphere fluid by functional scaling[38]. However due to the unavailability of simulation data the predictions for the isotropic-to-nematic transition of these theories have not been tested. The computer simulations of Frenkel, Mulder and Mc Tague[39] on a system of hard ellipsoids of revolution, which was found to exhibit for length-to-breadth ratios a/b > 2.75 (a and b denote the lengths of the major and minor axes of the ellipsoids) an isotropic-liquid to nematic-liquid-crystal transition, have provided an important impetus for the development of analytical theories on this subject. Fair quantitative agreement with the computer simulations has been achieved[40].

Like the fluid to crystal transition in the case of spherical colloidal particles, the isotropic-to-nematic transition in the case of rod-like particles is strongly influenced by the electrostatic repulsion between the particles. Onsager[33] indicated that the effect of the electrostatic repulsion will be equivalent to an increase of the effective diameter of the particles. However, the electrostatic repulsion also depends on orientation and thus the effect of the electrostatic repulsion will be different in the isotropic phase from that in the anisotropic phase. Actually the electrostatic interaction favors perpendicular orientation of the particles. The influence of this twisting effect on the isotropic-to-nematic transition in solutions of rod-like polyelectrolytes was calculated quantitatively by Stroobants, Lekkerkerker and Odijk[41].

While Onsager[33] concentrated in his theory on the isotropic-to-nematic transition he definitely did not exclude the possibility of the occurrence of other types of anisotropic phases such as smectic liquid crystals in a hard rod fluid. Hosino, Nakano and Kimura[42] investigated the nematic-smectic transition in a system of completely aligned hard rods. Even in the second virial approximation this transition is found to occur. The computer simulations of Stroobants et al.[43] revealed that the nematic-smectic transition indeed occurs in a system of hard parallel spherocylinders. For L/D ratios exceeding 0.5, a stable smectic phase is formed at densities well below the thermodynamic melting point. The nematic-to-smectic transition appears to be continuous. This work has inspired a number of theoretical treatments[44-47] all of which are able, with greater of lesser accuracy, to predict the volume fraction at which the nematic phase becomes unstable with respect to smectic perturbations. Recently Frenkel et al.[48] found with computer simulation that hard spherocylinders with both translational and orientational freedom also form a thermodynamically stable smectic phase. It even seems possible that there is a (small) range of L/D ratios where a system of hard spherocylinders has a stable smectic phase but not a stable nematic phase[49]. A tentative picture of the phase diagram is given in Fig. 5. A density-functional theory for nematic and smectic ordering of hard spherocylinders with full translational and orientational degrees of freedom has been developed[50].

So far it has been tacitly assumed that the rod-like particles are completely rigid. Particularly for rod-like polymers this is frequently not the case. For these systems the wormlike chain model, which can be seen as a thin cylinder with an elastic bending modulus provides a better description than the completely rigid rod. This flexibility of the particles has a pronounced effect on the isotropic-nematic phase transition, which has been extensively reviewed by Odijk[51]. In the following I will limit myself to systems that consist of (almost) perfectly rigid rods.

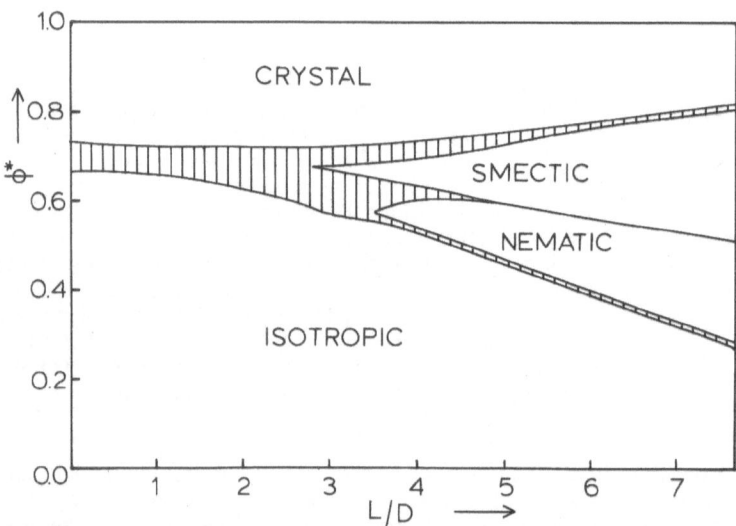

Fig. 5. Tentative picture of the phase diagram of hard spherocylinders with full translational and orientational freedom as a function of the length-to-width ratio L/D.

III.2. Liquid crystal phase transitions in dispersions of rod-like colloidal particles

The formation of a nematic phase in a dispersion of rod-like colloidal particles was first observed and recognized as such by Zocher[52] in 1925. Zocher studied dispersions of colloidal V_2O_5 particles that show a marked streaming birefringence. The time for the disappearance of the birefringence after cessation of flow increases strongly with concentration. In the case of the V_2O_5 dispersion studied by Zocher the birefringence becomes permanent above 1%. Observations with the polarization microscope show that the birefringence is due to spindle shaped birefringent bodies of about 100 μm long and 10 μm wide. These structures were called tactoids by Zocher. Tactoids are droplets of the anisotropic phase, where the particles are lying practically parallel to one another, in the isotropic phase where the particles are oriented in a random way. The concentration of colloidal particles and therefore the density of the nematic droplets is higher than the surrounding isotropic phase and eventually the system separates in two macroscopic phases. The upper phase is isotropic and the bottom layer is birefringent.

This phase separation was subsequently observed in a variety of systems with the common feature that they contain rod-like colloidal particles: dispersions of colloidal β-FeOOH[52,53] and γ-AlOOH[54,55] particles, dispersions of tobacco mosaic virus[56,57] and the rod-like bacterial viruses fd[58,59] and Pf1[59], solutions of deoxygenated sickle-cell haemoglobin[60], dispersions of cellulose microcrystals[61] and quite recently in dispersions of poly(tetrafluoroethylene) microcrystals[62].

The most extensively studied of the above mentioned systems is tobacco mosaic virus (TMV). TMV is a rigid rod of length L=300 nm and diameter D=18 nm. In the pH range of 7-8 TMV has a high negative charge density (about 5e/nm). Like in the case of the disorder-order transition of charged spherical colloidal particles the electrolyt concentration in the suspension medius has a strong effect on the concentration dependence of the isotropic-nematic phase transition. As early as 1939 Best[63] studied the phase diagram of TMV as function of added salt. Without added salt there was coexistence between an isotropic phase of 15 mg/ml and a nematic phase of 23 mg/ml. As the ionic strength increased the transition concentrations of virus and the coexistence range increased but the ratio of concentrations remained constant. Recently Fraden et al[64] measured the concentrations of the co-existing isotropic and nematic phases over a wide range of ionic strength (see Fig. 6). The theoretical results obtained with the Onsager theory including the effect of screened Coulomb interactions lie considerably above the experimental values. This is not surprising as the hard rod dimensions of TMV give a ratio of $L/D \cong 16$, which is certainly not large enough for the Onsager approach to be quantitatively valid.

Systems where one might expect the Onsager theory to be almost quantitatively valid are suspensions of the rod-like bacterial viruses fd (length L=880 nm and diameter D=6 nm, ratio $L/D \cong 150$) and Pf1 (length L=1960 nm and diameter D=6 nm, ratio $L/D \cong 325$). For these long rods fexibility effects may start to play a role however. Maret and Torbet[65] determined the isotropic-nematic transition concentration from the sudden increase of the Cotton-Mouton constant and found in dispersions with 10 mM Tris HCl a critical concentration of 9 mg/ml for fd and 5 mg/ml for Pf1. These values agree reasonably well with the theoretical results obtained with the Onsager theory including the effect of screened Coulomb interactions.

In 1951 Oster[66] reported for the first time the appearance of another phase of TMV. He noticed that from the nematic bottom layer of salt-free purified TMV dispersion an "iridescent gel" formed. The iridescence arises from Bragg diffraction of white light from a structure with a periodicity of the order of optical wavelengths. From one sample Oster measured the distance of the reflecting planes composing the iridescent structure to be 340 nm. Since TMV is 300 nm long he interpreted the structure to be a crystal formed of layers of TMV oriented perpendicular to the scattering planes. Another possibility is that he observed a smectic phase with long range order in one dimension and not a crystalline phase with long range order in three dimensions. In 1980 Kreibig and Wetter[67] extended Oster's work on the iridescent phase of TMV. They identify the iridescent phase of TMV, which they observe in salt free TMV dispersions above 30 mg/ml, as a smectic phase. This smectic phase can be sedimented from the nematic phase. The concentration ratio between the smectic and nematic phase was determined to be $c_s/c_m=1.15$. Zasadzinski et al.[68] studied the iridescent phase of TMV by freeze-fracture electron microscopy. Although no distinct layers were visible, along the virus axis in the freeze-fracture images a sinusoidal density modulation was observed. They conclude that with the experimental evidence at hand it is possible to assign to the iridescent phase either a smectic-B liquid crystal structure, where there is a density modulation along the alignment direction and two dimensional hexagonal packing of the particles within each plane, or a true three dimensional crystal structure.

Very clear evidence for smectic textures in dispersions of the bacterial viruses Pf1 and fd was obtained by Booy and Fowler[69] by freeze-fracture electron microscopy. The liquid-crystalline textures revealed by this technique correlate well with polarizing microscopy of magnetically oriented specimens.

Already in his 1925 paper Zocher[52] described the beautiful iridescence phenomena he observed in the sediments of a number of old (15 years!) β-FeOOH dispersions. He immediately realized that this phenomenon must be due to Bragg diffraction of white light from a structure consisting of regular spaced layers with a distance between the layers of the order of optical wavelengths. He thought that these layers (which he called "Schillerschichten" - glittering layers) were composed of aligned disk-like particles, the large distance between the layers being due to the electrostatic repulsion between the particles. Subsequently Zocher and Heller[53] also observed the iridescence phenomena in dispersions of β-FeOOH freshly prepared by slow (6 months - 1 year) hydrolysis of dilute solutions of $FeCl_3$. In this way monodisperse colloidal β-FeOOH particles are formed, which are, apparently essential for the formation of the Schiller layer structure. For a long time the iridescent phase of β-FeOOH dispersions was not fully understood. The idea of Zocher that the Schiller layers were composed of aligned disk-like particles became untenable when electron microscopy studies of Watson et al.[70] in 1962 revealed that the colloidal β-FeOOH particles forming the Schiller layers were not disks but rods with square cross-section. Quite recently, Maeda and Hachisu[71] using optical and electron microscopy presented clear evidence for the following model for the structure of the iridescent phase of β-FeOOH dispersions.

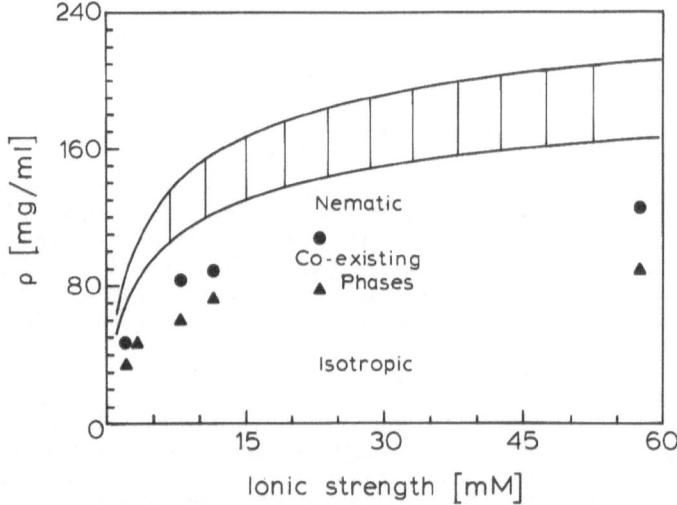

Fig. 6. The concentrations (ρ) of the co-existing isotropic ($\blacktriangle$) and nematic ($\bullet$) phases of TMV as a function of the salt concentration. The drawn lines indicate the coexistence region calculated with the Onsager theory taking into account the electrostatic interaction between the TMV particles assuming a linear charge density of 5e/nm. After Ref. 64.

Rodlike colloidal β-FeOOH gather to form mat-like assemblies in each of which the rods are packed with their axes nearly perpendicular to the mat phase. The mats are superimposed on one another at regular distances which are of the order of the length of the particles. Electron microscopy on dried down samples reveals that there is a high regularity in the packing of the rods in the mat and thus the iridescent phase of β-FeOOH is a smectic-B liquid crystal. There is an interesting difference between the phase behavior of TMV dispersions and β-FeOOH dispersions. Whereas in TMV the smectic phase separates from the nematic phase, in the case of β-FeOOH the smectic phase separates from the isotropic. Maeda and Hachisu note that the fact that β-FeOOH particles are not as elongated as the TMV particles may be the reason why in the case of β-FeOOH dispersions one observes a direct transition from the disordered state to the smectic state. The computer simulations by Frenkel and coworkers lend support tho this idea.

So far I have discussed the occurrence of nematic and smectic phases in dispersions of rod-like colloidal particles. Stroobants et al.[41] observed in their Monte Carlo simulations of systems of hard parallel spherocylinders for $L/D > 3$ a columnar phase intermediate between the smectic and crystalline phases. In their freeze-fracture electron microscopy study of dispersions of the rod like bacterial virus fd, Booy and Fowler[69] may have observed the columnar phase. They report one texture (Figure 2b, in Ref. 69), where the nearly parallel fd particles are arranged randomly in the longitudinal direction but the lateral packing is regular. This is indeed characteristic of a columnar phase.

IV. CONCLUDING REMARKS

I hope that the material presented here will convince you that concentrated dispersions of repulsive spherical and rod-like colloidal particles show interesting phase behavior. The richness of the ordering phenomena in systems of repulsive particles has given rise to interesting questions in the statistical mechanics of phase transitions. Theoretical results, in particular of simulation studies of systems of hard convex bodies, have made it clear that no special attractions are needed to explain the variety of ordered phases that are observed in colloidal dispersions. The fact that it is now possible to prepare a variety of systems of monodisperse spherical particles which interactions can be varied more or less continuously from a steep short-range repulsion to a soft long-range repulsion with concomitant changes in the phase behavior has given a strong boost to the study of colloidal crystals. The situation for rod-like colloidal particles is not nearly as well developed as the related work on spherical colloidal particles. To my knowledge, TMV, the bacterial viruses fd and Pf1 and colloidal β-FeOOH particles are at the present time the only well defined monodisperse rigid rod-like colloidal particles studied. As described the phase behavior observed in these systems closely parallels the theoretical predictions on systems of hard spherocylinders. However the above mentioned particles are charged and although together they span a considerable L/D range there are large gaps in between. What is urgently needed at the present time are methods to prepare in fairly large quantities monodisperse rod-like particles. Ideally it should be possible to vary the L/D ratio of such particles more or less continuously over a considerable range. In addition it should be possible to vary interaction between such particles from a steep short-range repulsion to a soft long-range repulsion. In order to reach this goal I think that further optimization of the preparation of colloidal poly(tetrafluoroethylene) and γ-AlOOH particles should be persued. I am convinced that if systems that satisfy the above mentioned requirements became available many new and interesting phenomena in the field of colloidal liquid crystals could be observed.

Apart from their scientific appeal at the cross roads of statistical mechanics and colloid science the ordering phenomena discussed here may be important in the formation of biological structures and in the manufacture of high-strength fibres and ceramics.

ACKNOWLEDGEMENTS

My views on ordering phenomena in colloidal dispersions presented here have been shaped to a considerable extent by fruitful collaboration and stimulating interactions with

Wim Briels, Jan Dhont, Seth Fraden, Daan Frenkel, Kees de Kruif, Bela Mulder, Theo Odijk, Peter Pusey, Carla Smits, Alain Stroobants and Agienus Vrij. I thank Marina Uit de Bulten for typing this manuscript, Jan den Boesterd for making the photographs and Theo Schroote for drawing the pictures.

REFERENCES

1. L. Onsager, Chem. Rev. 13:73 (1933).
2. W.G. Mc Millan and J.E. Mayer, J. Chem. Phys. 13:276 (1945).
3. A. Vrij, E.A. Nieuwenhuis, H.M. Fijnaut and W.G.M. Agterof, Faraday Discuss. Chem. Soc. 65:7 (1978).
4. C.G. de Kruif, J.W. Jansen and A. Vrij, in "Physics of Complex and Supramolecular Fluids", S.A. Safran and N.A. Clark, eds., Wiley, New York (1987).
5. J.G. Kirkwood, J. Chem. Phys. 7:919 (1939).
6. B.J. Alder and T.E. Wainwright, J. Chem. Phys. 27:1208 (1957).
7. W.G. Hoover and F.H. Ree, J. Chem. Phys. 49:3609 (1968).
8. M. Wadati and M. Toda, J. Phys. Soc. Jpn. 32:1147 (1972).
9. For recent reviews see e.g. A.D.J. Haymet, Ann. Rev. Phys. Chem. 38:89 (1987), M. Baus, J. Stat. Phys. 48:1129 (1987).
10. W.F. Saam and C. Ebner, Phys. Rev. A15:2566 (1977).
11. W.G. Hoover, D.A. Young and R. Grover, J. Chem. Phys. 56:2207 (1972).
12. M.O. Robbins, K. Kremer and G.S. Grest, J. Chem. Phys. 88:3286 (1988).
13. X-G Wu and M. Baus, Mol. Phys. 62:375 (1987).
14. D. Hone, S. Alexander, P.M. Chaikin and P. Pincus, J. Chem. Phys. 79:1474 (1983).
15. W.-H. Shih and D. Stroud, J. Chem. Phys. 79:6254 (1983).
16. W.Y. Shih, I.A. Aksay and R. Kikuchi, J. Chem. Phys. 86:5127 (1987).
17. A. Klug, R.E. Franklin and S.P.F. Humphreys-Owen, Biochem. Biophys. Acta 32:203 (1959).
18. W. Luck, M. Klier and H. Wesslau, Ber. Bunsenges. Phys. Chem. 67:75 (1963).
19. P.A. Hiltner and I.M. Krieger, J. Phys. Chem. 73:2386 (1969).
20. A. Kose and S. Hachisu, J. Colloid Interface Sci. 46:460 (1974).
21. E.A. Nieuwenhuis and A. Vrij, J. Colloid Interface Sci. 72:321 (1979).
22. L. Antl, J.W. Goodwin, R.D. Hill, R.H. Ottewill, S.W. Owens and S. Papworth, Colloids Surfaces 17:67 (1986).
23. P.N. Pusey and W. van Megen, Nature 320:340 (1986).
24. A.K. van Helden, J.W. Jansen and A. Vrij, J. Colloid Interface Sci. 81:354 (1981).
25. C.G. de Kruif, P.W. Rouw, J.W. Jansen and A. Vrij, J. Phys. (Paris) 46:C3-295 (1985).
26. R.D. Mountain and A.C. Brown, J. Chem. Phys. 80:2730 (1984).
27. C. Smits, W.J. Briels, J.K.G. Dhont and H.N.W. Lekkerkerker, Progr. Colloid Polym. Sci. 79: xxx (1989).
28. S. Hachisu, Y. Kobayashi and A. Kose, J. Colloid Interface Sci. 42:342 (1973); 46:470 (1974).
29. R. Williams and R.S. Crandall, Phys. Lett. A48:225 (1975).
30. H.M. Lindsay and P.M. Chaikin, J. Chem. Phys. 76: 3774 (1982).
31. Y. Monovoukas and A.P. Gast, J. Colloid Interface Sci. 128:533 (1989).
32. S. Alexander, P.M. Chaikin, P. Grant, G.J. Morales, P. Pincus and D. Hone, J. Chem. Phys. 80:5776 (1984).
33. L. Onsager, Phys. Rev. 62:558 (1942); Ann. N.Y. Acad. Sci. 51:627 (1949).
34. D. Frenkel, J. Phys. Chem. 91:4912 (1987), 92:5314 (1988).
35. H.N.W. Lekkerkerker, Ph. Coulon, R. Van Der Haegen and R. Deblieck, J. Chem. Phys. 80:3427 (1984).
36. M.A. Cotter, Phys. Rev. A10: 625 (1974).
37. H. Workman and M. Fixman, J. Chem. Phys. 58:5024 (1973).

38. S.D. Lee, J. Chem. Phys. 87:4972 (1987).
39. D. Frenkel, B.M. Mulder and J.P. Mc Tague, Phys. Rev. Lett. 52:287 (1984); D. Frenkel and B.M. Mulder, Mol. Phys. 55: 1171 (1985).
40. J.L. Colot, X.G. Wu, H. Xu and M. Baus, Phys. Rev. A38:2022 (1988).
41. A. Stroobants, H.N.W. Lekkerkerker and Th. Odijk, Macromolecules 19:2232 (1986).
42. M. Hosino, H. Nakano and H. Kimura, J. Phys. Soc. Jpn. 46:1709 (1979).
43. A. Stroobants, H.N.W. Lekkerkerker and D. Frenkel, Phys. Rev. Lett. 57:1482 (1986); Phys. Rev. A36:2929 (1987).
44. B.M. Mulder, Phys. Rev. A35:3095 (1987).
45. X. Wen and R.B. Meyer, Phys. Rev. Lett. 59:1325 (1987).
46. A.M. Somoza and P. Tarazona, Phys. Rev. Lett. 61:2566 (1988).
47. M.P. Taylor, R. Hentschke and J. Herzfeld, Phys. Rev. Lett. 62:800 (1989).
48. D. Frenkel, H.N.W. Lekkerkerker and A. Stroobants, Nature 332:822 (1988).
49. J. Veerman and D. Frenkel, Personal communication.
50. A. Poniewierski and R. Hotyst, Phys. Rev. Lett. 61:2461 (1988).
51. Th. Odijk, Macromolecules 19:2313 (1986).
52. H. Zocher, Z. Anorg. Chem. 147:91 (1925).
53. H. Zocher and W. Heller, Z. Anorg. Chem. 186:75 (1930).
54. H. Zocher and C. Török, Kollod Z. 170:140 (1960); 173:1 (1960); 180:41 (1962).
55. J. Bugosh, J. Phys. Chem. 65:1791 (1961).
56. F.C. Bawden, N.W. Pirie, J.D. Bernal and I. Fankuchen, Nature 138:1051 (1936).
57. F.C. Bawden and N.W. Pirie, Proc. Roy. Soc. B123:274 (1937).
58. J. Lapointe and D.A. Marvin, Mol. Cryst. Liq. Cryst. 19:269 (1973).
59. J. Torbet and G. Maret, Biopolymers 20:2657 (1981).
60. A.C. Allison, Biochem. J. 65:212 (1957).
61. R.H. Marchessault, F.F. Morehead and N.M. Walter, Nature 184:632 (1959).
62. T. Folda, H. Hoffmann, H. Chanzy and P. Smith, Nature 333:55 (1988).
63. R.J. Best, J. Austral. Inst. Agric. Sci. 5:94 (1939).
64. S. Fraden, G. Maret, D.L.D. Caspar and R.B. Meyer, to be published.
65. G. Maret and J. Torbet, unpublished. For an account of their results see G. Maret and K. Dransfeld in "Strong and Ultrastrong Magnetic Fields and Their Applications",pp. 160-167, F. Herlach, ed., Springer, Berlin (1985).
66. G. Oster, J. Gen. Physiol. 33:445 (1950).
67. U. Kreibig and C. Wetter, Z. Naturforsch. 35C:750 (1980).
68. J.A.N. Zasadzinski, M.J. Sammon, R.E. Meyer, M. Cahoon and D.L.D. Caspar, Mol. Cryst. Liq. Cryst. 138:211 (1986).
69. F.P. Booy and A.G. Fowler, Int. J. Biol. Macromol. 7:327 (1985).
70. J.H. Watson, R.R. Cardell and W. Heller, J. Phys. Chem. 66:1757 (1962).
71. Y. Maeda and S. Hachisu, Colloids Surfaces 6:1 (1983); 7:357 (1983).

RANDOM SURFACTANT ASSEMBLIES AND MICROEMULSIONS

M. E. Cates

Cavendish Laboratory

Madingley Road

Cambridge CB3 0HE, UK

INTRODUCTION

Surfactant molecules in dilute solution have a strong tendency to aggregate reversibly into larger structures[1]. While this often involves the formation of simple, roughly spherical micelles, there is now strong experimental and industrial interest in materials whose behaviour is more complicated. The resulting assemblies are often sufficiently large and/or flexible that their spatial configurations are highly entropic. In the one-dimensional case, there is a close analogy with polymers[2]. In what follows, we consider sheet-like bilayer assemblies, for which the corresponding analogy is with the theory of random surfaces, an area in which many fundamental problems remain to be solved. The theory of microemulsions (oil/water surfactant mixtures in which regions of each solvent are separated from one another by a surfactant *monolayer*) can also be described as a random surface problem, at least in the "balanced" case of oil/water symmetry.

SURFACES WITHOUT BOUNDARY

A surfactant bilayer may be modelled as a continuous film of incompressible, two-dimensional fluid. Under most circumstances one can assume that the film has neither torn edges (fig.1a) nor seams (fig.1b), both of which are prohibitively expensive in packing energy – at least for those surfactants which prefer to make bilayer[1]. (These rules might be broken in mixed systems which we do not discuss.) Therefore the important configurations of bilayer-forming surfactants in solution consist of self-and mutually avoiding pieces of surface that are either closed or infinite (*i.e.*, terminating at the edges of the sample). This system of surfaces has a certain number n_c of disjoint pieces or *components* and a certain number n_h of *handles*. (For example, a torus has one component and one handle.)

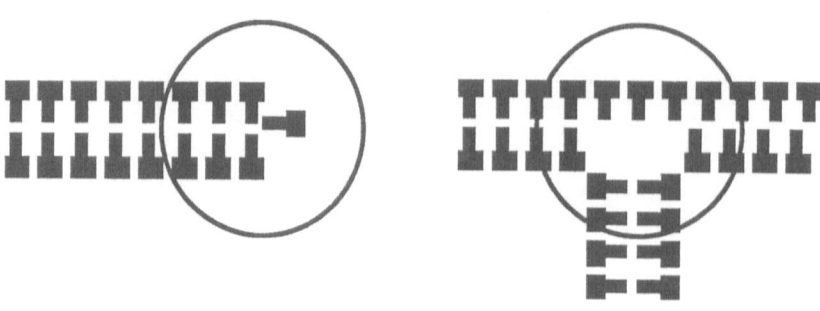

a   b

Fig. 1. Two types of defect: (a) edge; (b) seam.

Notice also that the system of surfaces, even if it has many components, divides space everywhere into an "inside" (I) and an "outside" (O). This definition is unique up to an overall sign; thus, once a given point has been chosen and labelled as I the classification of all other points is determined. If the two sides of the fluid film are equivalent, then the Hamiltonian of the entire system is unaffected by the interchange I ↔ O. This symmetry is exact for bilayers. For microemulsions the two sides of the film are inequivalent, and the I and O regions are filled with different solvents, so the symmetry is an approximate one at best.

CONTINUUM ELASTICITY OF FLUID FILMS

At lowest order in a curvature expansion, the elastic energy of a bilayer film (per unit area) may be written locally as[3,4]

$$H = \frac{K}{2}\left(\frac{1}{R_1} + \frac{1}{R_2}\right)^2 + \frac{\bar{K}}{R_1 R_2} + \sigma \qquad (1)$$

where R_1 and R_2 are the local principle curvature radii of the film. [Note that R_1 and R_2 are coordinate invariant up to an interchange of indices (1 ↔ 2) and an overall sign. Thus the combinations given are invariants.] In a grand canonical ensemble, the term σ is directly proportional to the surfactant chemical potential μ since the film is considered incompressible. (Indeed $\sigma = -\mu a^2$, with a a molecular size.) Often, the σ term is small enough to be ignored when calculating the shape fluctuations of films described by Eq.1; in a thermodynamic calculation this may usually be checked afterwards.

Usually one is interested, not in the bending energy of a certain film configuration that is specified everywhere, but in the free energy of a state with a certain average curvature on a coarse-grained scale. To calculate this, one must allow for the change in the entropy spectrum of short scale undulation modes which results from imposing a large scale "bend", with radius of curvature $\xi \gg a$, say. The calculation of the effective bending constant $K(\xi)$ is quite complicated, but can be set up as a perturbation expansion (about a flat sheet) in the parameter $\tau = T/4\pi K_0$ (where $K_0 \equiv K(a)$ is the "bare" bending constant). The result to first order is[5,6]

$$K(\xi) = K_0 \left(1 - \alpha\tau \log(\xi/a)\right) = K_0 - \frac{\alpha T}{4\pi} \log(\xi/a) \qquad (2)$$

where (according to most authors) $\alpha = 3$ in three dimensions. Thus the effect of thermal fluctuations is to make the bilayer *less rigid* at large distances. Note that this applies only to films with liquid-like in-plane degrees of freedom; films which are internally structured can behave quite differently[6].

The perturbative result, Eq.4, makes sense only for $\xi \leq \xi_K \simeq \exp[4\pi K_0/\alpha T]$. The parameter ξ_K is called the de Gennes–Taupin persistence length; in favourable cases this may be a few hundred Angstroms[7]. For $\xi \leq \xi_K$ an unconstrained film is roughly flat, whereas at larger distances it is some kind of random surface with large overhands and possibly handles; so perturbation theory about a flat sheet breaks down.

We have not yet discussed the gaussian curvature term $\bar{K}/R_1 R_2$ in Eq.1. A well-known theorem (Gauss–Bonnet) tells us that

$$\int \frac{1}{R_1 R_2} d^2 S = 4\pi(n_c - n_h) . \qquad (3)$$

So $\bar{K}$ is like a chemical potential for topology. Notice that if $\bar{K}$ is positive, a fluid film system described by Eq.1 has a ground state instability leading to the formation of a periodic minimal surface which has many handles, only one component, and zero mean curvature (*i.e.*, with $1/R_1 + 1/R_2 = 0$ everywhere)[8]. The energy of such a state with unit cell size l diverges as $-\bar{K}/l^3$. This will eventually be cut off by anharmonic terms, but nonetheless the general expectation is for phase separation to a rather concentrated cubic phase whenever $\bar{K} > 0$. Similarly if $\bar{K}$ is too negative ($\bar{K} \leq -2K_0$) there is an instability to the formation of very small spherical vesicles.

The renormalization of the topological parameter $\bar{K}$ at large length scales is a somewhat delicate issue, but it appears that the first correction term is positive[9]. This raises an interesting possibility that in some systems with small negative $\bar{K}_0$, the effective $\bar{K}$ value changes sign at a length scale $\bar{\xi}$ (small compared to ξ_K). Presumably this could stabilize a cubic structure at the length scale $\bar{\xi}$, which would be a strange example of dilution-induced ordering.

EMPIRICAL MODEL OF RANDOM BILAYER ASSEMBLIES

Let us now estimate the free energy of a homogeneous phase containing bilayer-forming surfactant at volume fraction ϕ. To do this we introduce a coarse grained lattice model as shown in fig.2. [The model originated in the description of microemulsions of Talmon and Prager[10] as modified by de Gennes and coworkers[7], Widom[11] and Safran et al.[12]. It was first applied to bilayer-forming systems, rather than microemulsions, in Ref.13.]

Denote the lattice size by ξ, and let us designate each cell randomly as either inside (I), with probabilty ψ, or outside (O) with probability $1 - \psi$. These cells are then separated by a bilayer whose average curvature radius is of order the cell-size ξ. The volume fraction ϕ is then of order (on a cubic lattice)

$$\phi = 6a\psi(1 - \psi)/\xi . \tag{4}$$

This fixes the cell-size ξ for any given ϕ and ψ.

The entropy of the random assignment of the cells, per unit volume, is given by

$$S = -(1/\xi^3)[\psi \ln(\psi) + (1 - \psi) \ln(1 - \psi)] . \tag{5}$$

We now estimate the free energy of bending the film to wrap around the inside and outside regions. Since the average curvature radius is of order ξ, we have, per unit volume

$$F_{bend} \simeq (1/\xi^3) 8\pi\zeta\psi(1 - \psi)K(\xi) \tag{6}$$

with $K(\xi)$ obeying Eq.2, and ζ a geometrical parameter of order unity. We have set $\bar{K}$ for simplicity. (There is no difficulty estimating $n_c - n_h$ as a function of ψ in the randomly mixed states considered here, and so a term in $\bar{K} < 0$ could be included in principle.)

The total free energy is $F = F_{bend} - TS$. It is convenient to introduce reduced variables $x = \xi/\xi_K$, $f = F\xi_K^3/T$, $\tilde{\phi} = \phi\xi_K/a$ in terms of which one has

$$f = x^{-3}[\psi \ln(\psi) + (1 - \psi) \ln(1 - \psi) - 2\zeta\alpha\psi(1 - \psi) \ln(x)] \tag{7a}$$

$$x = 6\psi(1 - \psi)/\tilde{\phi} \tag{7b}$$

The free energy can now be minimized over the variational parameter ψ with the constraint $\psi \geq \phi/2$ (which ensures $\xi \geq a$). The resulting family of curves $f(\phi)$ is sketched in fig.3 (e.g., with $\zeta\alpha \simeq 1.2$). To the right of the point X the minimization on ψ yields $\psi = 1/2$, to the left $\psi < 1/2$.

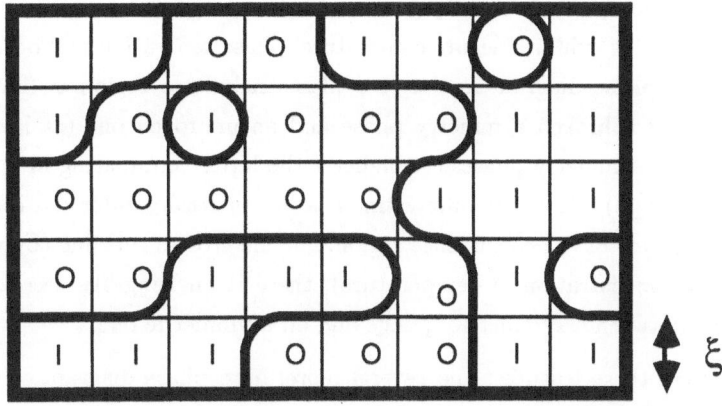

Fig. 2. Lattice model of bilayer sponge, showing inside (I) and outside (O) regions.

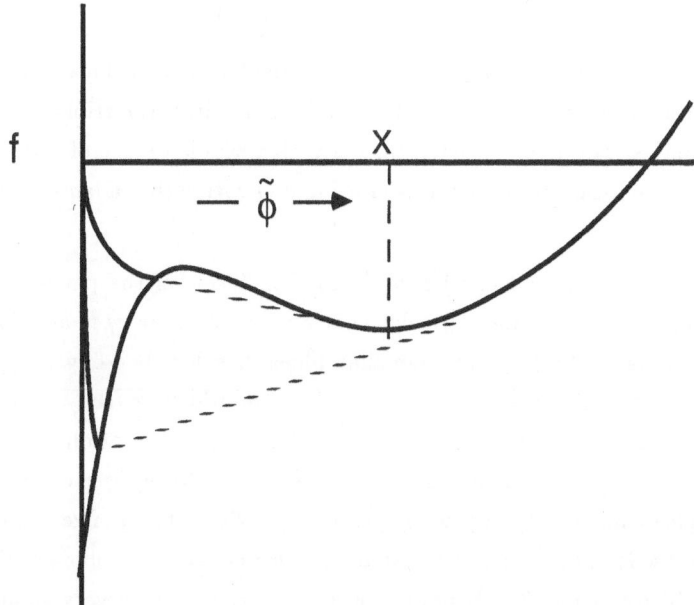

Fig. 3. Schematic curves for reduced free energy. These differ, for different reduced temperatures, only in the left part where the short cutoff violates the scaling of Eq.7. At high temperatures (upper curve) the TR point (X) can lie outside the first order coexistence region.

PHASE EQUILIBRIA

To generate phase equilibria we may perform the usual common tangent construction on $f(\phi)$; when this exhibits negative curvature, phase separation will occur. For $1 \leq \zeta\alpha \leq 1.5$ the phase diagram in the τ, ϕ plane includes a region of first order coexistence between a broken symmetry phase and an unbroken one (at low τ) or between two broken symmetry phases (at higher τ the latter terminating in a critical point of liquid-gas type). Especially interesting is a *line of second order critical points* emerging from the side of the coexistence region. As one crosses this line (by varying either surfactant concentration or temperature), there is an Ising-like second order phase transition between a symmetric sponge and an asymmetric one.

Figure 4 shows these features, and several more, on a phase diagram calculated recently by Golubovic and Lubensky[14], using an improved version of the above model. These authors have added to the above free energy a term involving the entropic repulsion between pieces of film that are too close together:

$$F_{steric} = \text{const.} \frac{T^2}{(\xi - a)^2 K(\xi)} \tag{8}$$

whose form was first predicted by Helfrich[15] for lamellar phases. This term automatically enforces the constraint $\xi \geq a$, and also allows one in a variational approach to incorporate lamellar and cubic ordered phases with variable degree of order. (This is done by biasing the local site probabilities for I and O with either a lamellar or a cubic pattern.)

The basic features of fig.4 can be understood in terms of the persistence length ξ_K. In a smectic phase, the interlamellar spacing d is of order a/ϕ, and for τ small this is large compared to ξ_K. The lamellar phase has low bending energy and is stable. As temperature is raised or d increased (by dilution), entropy becomes more important and bending energy less so (as $K(d)$ falls). At separations of order (but somewhat less than) ξ_K, the lamellar phase melts with a weak first order transition into an isotropic symmetric sponge of length scale $\xi \sim \xi_K$. This sponge can be diluted further (ξ increased) but only a certain amount. This is because a sponge with length scale $\xi \gg \xi_K$ is untenable: the bending constant at this scale is very small and the sponge can increase its entropy by making a shorter length scale structure. One choice is to fragment into a dilute phase of small objects (micelles or vesicles) with a large gain in mixing entropy. (If the change in length scale is large, the entropy shift can overcome the bending energy penalty involved; this happens at low temperatures.) At larger τ, ξ_K is already quite small and the entropy gained by fragmentation is less important. Instead the system makes an asymmetric sponge of $\xi \simeq \xi_K$ which can have a significantly higher entropy than a symmetric one without paying a large bending energy cost.

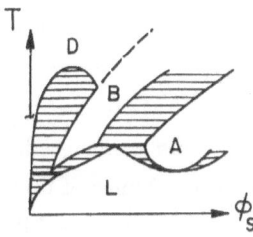

Fig. 4. Phase diagram in reduced-temperature / volume fraction plane. D is phase of broken I/O symmetry, B is symmetric sponge, A cubic and L, lamellar. From Golubovic and Lubensky (Ref.14).

INSIDE–OUTSIDE SYMMETRY BREAKING

The line of second order transitions between a symmetric and an asymmetric sponge, correponding the spontaneous breaking of I/O symmetry, was first predicted by Huse and Leibler[8] who called it the "tense–relaxed" (TR) transition. This transition in bilayer-forming surfactants provides a rather rare example of a second order transition between two isotropic liquids. (^{3}He ^{4}He provides another, quantum mechanical, example.) The I/O symmetry that is being broken is an exact one, at least in the absence of edges and seams in the bilayer. [These could either be relevant, leading to a rounding of the transition very close to the critical line, or irrelevant, in which case a true phase transition would survive until the density of defects becomes large.] In fact, the I/O symmetry of the Hamiltonian is so exact that it is very hard to imagine any way of introducing a field so as to break it "from outside". The symmetry is in this sense "hidden", and some find it difficult to accept that a real phase transition can be associated with its being broken. The order parameter can either be thought of as the difference in volume fractions of I and O, or equivalently as the average mean curvature $\langle 1/R_1 + 1/R_2 \rangle$. This average is well-defined up to an overally sign flip between I and O, and is zero in the symmetric phase.

EXPERIMENTS

Since most experimenters have only recently been made aware of the possibility of a second order TR transition, there are no definitive experimental studies (to my knowledge) that prove its existence. However, a phase diagram that is (topologically at least!) similar to fig.4 is often found, for example in the system of SDS/pentanol bilayers (swollen slightly with water) in dodecane[16]. In this system a maximum of turbidity is measured as one crosses a certain line in the isotropic region[17]. Further experiments are in progress which should confirm whether or not this indeed marks the TR transition line[18]. There is also a chance of finding a similar line of transitions in aqueous systems of lecithin + glycocholate (a bile salt: this system is used in digestion)[19]; it is hoped that further candidates will be forthcoming now that the experimental community is alerted to look for them.

Leaving aside the issue of the second order line, there in any case, quite a large number of systems in which an "anomalous isotropic" phase is found beyond the end of the dilution curve of a highly swollen lamellar phase[16,20,21]. This phase (sometimes called L_3) exhibits strong flow birefringence and abnormal conductivity properties which are consistent[13] with the presence of a percolating "sponge" of bilayer, as the model predicts. (For example, in the SDS system mentioned above, one expects connected water paths across the sample and so a high conductivity even though content is only a few percent; and this is seen[18].)

HOW CUBIC IS THE SPONGE?

The basic proposal then, is that the sponge state is a melted analogue of a lamellar phase, rather in the same way that the well-known viscoelastic phases of entangled, worm-like micelles[2] can be viewed as a melted version of the nearby hexagonal phase of infinite cylinders. On the other hand, something like a sponge could also arise as a partially-melted analogue of a cubic structure[22,23]. There is no experimental evidence for *long range* cubic order in this phase, and presuming this is indeed absent, there is obviously a continuum of possible structures interpolating between those with near perfect short-range cubic order and the completely random sponge described above.

Several authors [22,23] have argued that a low energy for handle formation ($\bar{K}$ high) plays an essential role in stabilizing the L_3 phase. The model above, which takes $\bar{K} = 0$, is quite consistent with this idea: the bicontinuous sponge might well be destabilized by even quite a small negative $\bar{K}$, although this effect remains to be investigated in detail. [In other words, $\bar{K} = 0$ is already a rather high value.] The idea that the phase arises only when the system has a *positive* $\bar{K}$ is more delicate since (as shown following Eq.3 above) this would usually favour a cubic of very small cell-size. Among possible factors giving a stable large cell system when $\bar{K} > 0$ could be (i) Steric repulsion terms (Eq.8 above);[22] (ii) Strong anharmonic corrections to Eq.1[23]; or (iii) The upward renormalization of $\bar{K}$, meaning that this parameter is positive *only* at large distances (see the discussion following Eq.3 above). At present, it seems that the case for a positive $\bar{K}$ in these systems remains open (unfortunately it is not known how to measure this quantity directly in experiments). But, judging from the simple model described above, this does not seem to be essential.

DYNAMICS

The strong flow birefringence of the anomalous isotropic phase is not surprising in view of the large size of the sheet-like assemblies that it contains. One expects a structure like the sponge to be disrupted at quite modest flow rates; hence the system is a good candidate for the observation of flow-induced phase transitions. Preliminary observations[17] suggest that in some cases there may be a transtion from the isotropic to the lamellar phase under flow; a partial theoretical account of this process is given in Ref. 24, at the level of Landau-Ginzburg theory. It is argued there that the main effect of flow is to reduce the amplitude of local lamellar fluctuations in the isotropic phase. The nonlinear effect of these fluctuations is to supress the transition to the ordered phase, and to make it weakly first order (rather than second order as predicted by mean field theory). An imposed flow reduces the effects of the fluctuations and hence pushes the transition temperature upward. Thus in a certain parameter range, the lamellar phase can be induced by flow[24]. The details of this prediction remain to be tested experimentally. However, the basic message is that the presence of large connected pieces of bilayer in these systems can render them highly susceptible under shear. A detailed model of how the pieces of bilayer move (this must involve mechanisms for changing the topology, for example) is lacking at present.

MICROEMULSIONS

In a microemulsion[25], the inside component (say) of our ensemble of surfaces is water and the outside component oil. The surfactant film is now a monolayer with a definite direction and there is no inside-outside symmetry. However, this symmetry can be approximately restored by (i) tuning parameters such as the concentration of added salt and alcohol, so as to avoid a strong preferred curvature of the monolayer towards one solvent rather than the other; and then (ii) using equal amounts of oil and water. Under these conditions, the phase diagram (fig.4) for the bilayer model can be reinterpreted as that of a "balanced" microemulsion system.

The "variational parameter" ψ is now the volume fraction of water[12]. The main difference of interpretation is that any broken-symmetry state ($\psi \neq 1/2$) must be interpreted, not as a single phase, but a coexisting symmetric pair of states related by the interchange of oil and water. Thus the region of coexistence between the symmetric sponge phase and a dilute phase (small vesicles or micelles) becomes a coexistence between a symmetric, bicontinuous, microemulsion and *a pair* of dilute phases of surfactant in water or surfactant in oil. Since the microemulsion has a density less than that of the oil-rich phase and more than that of water, it floats between the two in a test-tube and is referred to as a "middle phase microemulsion". At higher temperatures, the second order TR transition is now between one symmetric micromulsion and a symmetric pair of asymmetric ones. Somehow the transition seems much more normal because the two broken symmetry phases are observably different (which is the same as to say that the symmetry is only exact at a special point in parameter space). The corresponding symmetry in the bilayer case holds everywhere, and the "symmetric pair of coexisting states" are not merely related by symmetry, but are the same state.

In microemulsions, as indicated already, it is extremely easy to break the oil/water symmetry of the hamiltonian, either by detuning parameters so that the monolayer has a *spontaneous curvature* toward water or oil, or merely using unequal amounts of water and oil as ingredients. These effects can quite easily be incorporated into models of the kind described above, and of course yield asymmetric phase behaviour[12]. The various phase equilibria (both experimentally and in theory) can become quite complicated and will not be discussed further here.

In any case, for several practical purposes, the "balanced" case is among the most interesting. For example, a balanced middle phase microemulsion is found to have extremely low interfacial tension with both of the coexisting dilute phases[25]. This is presumably because the middle phase "sponge" only has to make entropic rearrangements of the monolayer on the scale of $\xi \simeq \xi_K$ to seal off the oil cells (say) and thus make a low energy interface with water. A thin slab of microemulsion between water and oil can thus provide an effective interfacial tension between these two phases which is very low (much lower than obtainable simply by dissolving a lot of surfactant in the water phase). For this reason, attempts have been made to use microemulsions in oil recovery, to help displace oil ganglia trapped by capillary forces in water-flooded wells[26]. At present, the procedure appears to be too expensive for widespread use. There are many other uses for microemulsions, for example in lubrication applications where a high thermal or electrical conductivity is required (provided by the water), in dry cleaning, wax polishes, *etc.* [25,26].

OTHER APPROACHES

Simplistic models of the type sketched above are able to reproduce, qualitatively, some of the main experimental results. (If one goes further than the random mixing approximation to study local correlations, then these include the characteristic peaked $S(q)$ seen for microemulsions[27].) Certainly these models can help us decide what are the important interactions to include in a more complete approach; also they can exhibit interesting features (like the TR transition) that turn out to be far more general. However many of the assumptions and parameters in the models are rather arbitrary; for example, a completely different phase diagram from fig.4 is obtained if the parameter ζ is chosen so that $\zeta\alpha \geq 1.5$. There are other artefacts as well: in particular, the use of a single length-scale (ξ) to model a sponge phase prohibits the small micelles (with which it is predicted to coexist) from "leaking" into the interior of the sponge. This is surely unrealistic, and it is hard to quantify the importance of such approximations.

Obviously a more fundamental approach is needed. One line of development is that of microscopic lattice modelling[28], in which surfactants and solvent molecules are represented as monomers and dimers (for example) on a lattice. This is a reasonable route, but perhaps even more difficult to purge of artefacts than the coarse-grained models described here. Another important, more empirical, approach is that of Landau Ginzburg theory[29,24] though it is hard to find an expansion that works in more than a small part of the phase diagram at a time. A third good idea concerns off-lattice variational models, in which the random surface is represented as an "contour" of a three-dimensional random field[30]. The method has been used to model interfaces in spinodal decomposition[31], but to apply it in a thermodynamic context one must estimate of the entropy of surfaces made in this way, which is very difficult. Discretized surfaces, represented as assemblies plaquettes (either on- or off-lattice) have been studied extensively[32]; because of the difficulty of treating curvature terms, these are perhaps most useful to describe structures on scales larger than ξ_K.

FIELD THEORY

Finally, the most fundamental approach available is that based on the continuum field theory of random surfaces, also known as string field theory. [An excellent review is Ref.6; see also Ref.33.] As outlined briefly below, this theory can become extremely complicated for even (say) an isolated piece of surface having the topology of a sphere, without self-avoidance effects. Thus a convincing description of a sponge phase (say) in this language remains a long way off. For the moment, perhaps the main use of the field theory approach lies in identifying unambiguously different possible classes of behaviour (though it has also provided some results for parameters like α in Eq.2).

One question of interest concerns the asymptotic behaviour of a large, isolated vesicle (size $\gg \xi_K$) that (for simplicity) is not self-avoiding but is controlled by the bending energy Hamiltonian, Eq.1. This is certainly an academic question, since such a surface would prefer to break up into small pieces, but it is a well-posed one. Can this surface be represented as a set of plaquettes of size ξ_K, glued together in a completely random fashion subject to the overall constraints on topology? If so, an old argument[34,35] suggests that the asymptotic structure is the same as a branched polymer[36] "coated" with plaquettes of size ξ_K. A demonstration that the large isolated vesicle was indeed a branched polymer would provides support for the empirical procedure (used above) of modelling a sponge as pieces of surface of size $\xi \simeq \xi_K$ glued together randomly at a certain overall density. If, on the other hand, the structure of an isolated vesicle is different, then this way of modelling the sponge is strongly called into question.

Describe the surface by a spatial coordinate $\mathbf{r}(s)$ where $\mathbf{r}$ is a vector in d–space and $s = (s_a, s_b)$ is a coordinate system on a two dimensional parameter space such as the unit square. Then the bending energy may be written

$$H = \int d^2s\, g^{1/2}(K_0 c^2/2 + \sigma) \tag{9a}$$

$$g_{ab} = \partial_a \mathbf{r} \cdot \partial_b \mathbf{r} \tag{9b}$$

$$c = 2g^{-3/2}\partial_a g^{1/2} g^{ab} \partial_b \mathbf{r} . \tag{9c}$$

Here g is the induced metric and c is the scalar Laplacian of $\mathbf{r}$ in the metric g. The partition function of the vesicle is

$$Z = \int e^{-H/T} \tag{10}$$

with the integral (formally) over all maps $\mathbf{r}(s)$ that correspond to *distinct* spatial loci of the surface: note that reparametrizations $s \to s'$, $\mathbf{r} \to \mathbf{r}'$, with $\mathbf{r}(s) = \mathbf{r}'(s')$ are not to be counted as distinct. To eliminate Eq.9b introduce a lagrange multiplier $\lambda^{ab}(s)$ so that

$$Z = \int d[\mathbf{r}(s)]\, d[g_{ab}(s)]\, d[\lambda^{ab}(s)] \exp\left[-H/T - \int d^2s\, \left(\lambda^{ab}(\partial_a \mathbf{r} \cdot \partial_b \mathbf{r} - g_{ab})\right)\right] . \tag{11}$$

From this it is possible to extract an "effective action" that describes the behaviour of the surface at large distances[33,6]:

$$Z = \int d[\mathbf{r}(s)]\, d[g_{ab}(s)] e^{-F/T} \tag{12a}$$

$$F/T = \bar\lambda \int g^{1/2} \left(g^{ab}\partial_a \mathbf{r} \cdot \partial_b \mathbf{r} + \sigma\right) d^2s \tag{12b}$$

$$\bar\lambda = 1/\xi_K^2 , \quad \xi_K \simeq a \exp[4\pi K_0/dT] . \tag{12c}$$

This is known to be the appropriate continuum action for random fluid surfaces *without bending energy*[37], but with an effective short cutoff ξ_K rather than a (the molecular scale). This development looks quite consistent with the idea that the vesicle is completely random at a scale above ξ_K.

To complete the argument, one must still show that the ensemble of surfaces defined by Eq.12 is equivalent to branched polymers at large distances. Some progress can be

190

made by choosing coordinates so that $g_{ab}(s) = \delta_{ab} e^{\phi(s)}$. It is then possible to integrate out the spatial coordinate $\mathbf{r}$, leaving behind an effective theory[36,33] for ϕ,

$$Z = \int d[\phi] \, e^{-F(\phi)} \tag{13a}$$

$$F(\phi) = \frac{25 - d}{48\pi} \int \left(\partial_a \phi \partial_a \phi + \sigma e^\phi \right) d^2 s \,. \tag{13b}$$

This is called the Liouville field theory. For $\sigma \to 0$ it looks like a free field theory, but in fact is nothing of the kind, because there is supposed to be a short cutoff fixed at ξ_K on the surface as actually embedded in real space. In terms of the s coordinates this translates into a cutoff length $\zeta(s) \simeq \xi_K e^{\phi(s)}$ that depends locally on the value of the field ϕ that is being cut off! This can lead to very complicated behaviour, and it has not yet known whether the internal connectivity of the surface (which is completely specified by ϕ) is the same as that of branched polymers.

What is known, of course, is how to solve the free field theory obtained by setting $\zeta = $ constant. This is quickly found to be totally *inconsistent* with branched polymers, instead corresponding to an almost "tethered" surface[35,38,39] whose internal connectivity remains that of a flat sheet at all distances. If this result were to carry over to the variable cutoff version of the theory, it would probably spell disaster for all the empirical models discussed earlier. Fortunately, the self-interacting cutoff may provide an escape route by permitting the occurence of localized "spikes" in the ϕ field[39]. In real space these correspond to cylindrical protrusions of the surface, which may signal the onset of a branched polymer instability. Although much work remains before this is firmly established, it looks as if our heuristic assumptions about the structure of fluid films beyond ξ_K will probably remain intact.

CONCLUSION AND OUTLOOK

Random surface models, both for bilayer-forming surfactant solutions and for microemulsions, play a useful role in helping us to understand the experimental properties of these systems. It appears that many such properties can be understood in terms of the interplay between entropy and bending energy. One interesting prediction is that of a line of second order Ising-like critical points corresponding to the breaking of I/0 symmetry (the tense-relaxed transition[8]). Experimental confirmation of this prediction is eagerly awaited.

Some possible refinements to the basic approach, which nonetheless retain the same picture of the surfactant layer as a semiflexible incompressible film, were discussed above. Another way in which the models could be refined is to include a more detailed description of the various interactions (van der Waals, electrostatic, hydrophobic...)[1].

Most local terms can of course be lumped into the elastic constants of the film, but the longer-range contributions of these interactions must be treated separately, and may play an important role in many cases. (In particular, attractions between sheets can lead to "binding" of lamellar phases[4]; a tacit assumption in all the above is that there are no strong binding effects.) It is quite possible that some observed phase equilibria are correctly predicted by the random surface models only as a fortuitous consequences of the approximations made: in particular it remains an open question whether the familiar three-phase equilibrium of microemulsions can really arise purely from the competition of bending energy and entropy, or whether some attractive interactions between pieces of sheet are also required. To study these problems further, one would like an approach somewhere between the empirical lattice models and string field theory, in terms of both complexity and rigour.

Acknowledgements: I am indebted to numerous colleagues for discussions on this subject, and I apologize to them for the places above where their ideas are regurgitated without proper attribution. These colleagues include especially David Andelman, Francois David, Leonardo Golubovic, Jacob Israelachvili, Stanislas Leibler, Tom Lubensky, Scott Milner, Fyl Pincus, Gregoire Porte, Didier Roux, and Sam Safran.

REFERENCES

1. See, *e.g.*, J. N. Israelachvili, *Intermolecular and Surface Forces*, Academic Press, London (1985).

2. See, *e.g.*, M. E. Cates, J. Phys. Paris 49:1593 (1988) and references therein.

3. P. B. Canham, J. Theor. Biol. 26:61 (1970); W. Helfrich, Z. Naturforsch. 28C:693 (1973).

4. See the review by S. Leibler in *Statistical Mechanics of Membranes and Surfaces*, editors D. R. Nelson, T. Piran and S. Weinberg, World Scientific, Singapore (1989).

5. W. Helfrich, J. Phys. Paris 46:1263 (1985); L. Peliti and S. Leibler, Phys. Rev. Lett. 54:1690 (1985); D. Foerster, Phys. Lett. A114:115 (1986).

6. F. David, in *Statistical Mechanics of Membranes and Surfaces*, editors D. R. Nelson, T. Piran and S. Weinberg, World Scientific, Singapore (1989).

7. P. G. de Gennes and C. Taupin, J. Phys. Chem. 86:2294 (1982); J. Jouffroy, P. Levinson and P. G. de Gennes, J. Phys. Paris 43:1241 (1982).

8. D. A. Huse and S. Leibler, J. Phys. Paris 49:605 (1988).

9. H. Kleinert, Phys. Lett. 114A:263 (1986); E. Guitter, unpublished.

10. Y. Talmon and S. Prager, J. Chem. Phys. 69:2984 (1978); 76:1535 (1982).

11. B. Widom, J. Chem. Phys. 81:1030 (1984).

12. S. A. Safran, D. Roux, M. E. Cates and D. Andelman, Phys. Rev. Lett. 57:491 (1986); D. Andelman, M. E. Cates, D. Roux and S. A. Safran, J. Chem. Phys. 87:7229 (1987); M. E. Cates, D. Andelman, S. A. Safran, and D. Roux, Langmuir 4:802 (1988).

13. M. E. Cates, D. Roux, D. Andelman, S. T. Milner and S. A. Safran, Europhysics Lett. 5:733 (1988), erratum: *ibid* 7:94; also, unpublished calculations for the case $\zeta \alpha \neq 1$.

14. L. Golubovic and T. C. Lubensky, preprint (1989).

15. W. Helfrich , Z. Naturforsch. 33A:305 (1978).

16. C. R. Safinya, D. Roux, G. S. Smith, S. K. Sinha, P. Dimon, N. A. Clark and A. M. Bellocq, Phys. Rev. Lett. 57:2718 (1986).

17. D. Roux and C. M. Knobler, Phys. Rev. Lett. 60:373 (1988).

18. D. Roux, private communication.

19. N. A. Mazer, G. B. Benedek and M. C. Carey, Biochemistry, 19:601 (1980); P. Schurtenberger, N. A. Mazer and W. Kaenzig, J. Phys. Chem. 89:1042 (1985).

20. F. Harusawa, S. Nakumara and T. Mitsui, Colloid Polym. Sci. 252:613 (1974); D. J. Mitchell, G. J. Tiddy, L. Waring, T. Bostock and M. P. McDonald, J. Chem. Soc. Faraday Trans. I, 79:1975 (1983); R. G. Laughlin, in *Advances in Liquid Crystals*, editor G. H. Brown, Academic Press, New York (1978) pp 44 and 91; P. G. Nilsson and B. Lindman, J. Phys. Chem. 88:4764 (1984); W. J. Benton and C. J. Miller, J. Phys. Chem. 87:4981 (1983).

21. J. C. Lang and R. D. Morgan, J. Chem. Phys. 73:5849 (1980).

22. G. Porte, J. Marignan, P. Bassereau and R. May, J. Phys. Paris 49:511 (1988).

23. A detailed mechanism, that involves strong nonlinear corrections to Eq.1, is suggested by D. Anderson, H. Wennerstrom and U. Olsson, to be published.

24. M. E. Cates and S. T. Milner, Phys. Rev. Lett. 62:1856 (1989).

25. See, *e.g., Surfactants in Solution*, editors K. Mittal and B. Lindman, Plenum, New York (1984, 1987).

26. See, *e.g., Surfactants*, editor T. F. Tadros, Academic Press, London (1984), especially the article on oil recovery by E. Neustadter.

27. S. T. Milner, S. A. Safran, D. Andelman, M. E. Cates and D. Roux, J. Phys. Paris 49:1065 (1988).

28. B. Widom, J. Chem. Phys. 84:6943 (1986); B. Widom, K. A. Dawson and M. D. Lipkin, Physica 140A:26 (1986), M. Schick and W. H. Shih, Phys. Rev. B 34:1797 (1986), K. Chen, C. Ebner, C. Jayaprakash and R. Pandit, J. Phys. C 20:L361 (1987).

29. M. Teubner and R. Strey, J. Chem. Phys. 87:3195 (1987), and references therein.

30. N. F. Berck, Phys. Rev. Lett. 58:2718 (1987); M. Teubner, unpublished.

31. J. W. Cahn, J. Chem. Phys. 42:93 (1965).

32. See J. Froehlich, in Lecture Notes in Physics Vol. 216, *Applications of Field Theory to Statistical Mechanics*, editor L. Garrido, Springer, Berlin (1985); also A. Krzywicki, J. Ambjorn, V. A. Kazakov and I. K. Kostov, in *Field Theory on the Lattice*, Nucl. Phys. B (Proc. Suppl.) 4 (1988).

33. A. M. Polyakov, *Gauge Fields and Strings*, Harwood Academic Publishers, London (1987).

34. B. Durhuus, J. Froehlich and T. Jonsson, Nucl. Phys. B240:453 (1984); G. Parisi, Phys. Lett. B 81:357 (1979); J. M. Drouffe, G. Parisi and N. Sourlas, Nucl. Phys. B 161:397 (1980); F. David and E. Guitter, Europhysics Lett. 3:1169 (1987), Nucl. Phys. B 295:332 (1988).

35. M. E. Cates, Phys. Lett. B 161:363 (1985).

36. T. C. Lubensky and J. Isaacson, Phys. Rev. A 20:2130 (1979).

37. A. M. Polyakov, Phys. Lett. B 103:207 (1981).

38. Y. Kantor, M. Kardar and D. R. Nelson, Phys. Rev. Lett. 57:791 (1986).

39. M. E. Cates, Europhysics Lett. 8:719 (1988).

OBSERVATIONS OF SPHERE TO ROD TRANSITION IN A

THREE-COMPONENT MICROEMULSION

J. Samseth*, S.-H. Chen**, J.D. Litster**, and J.S. Huang***

* Institute for Energy Technology, N-2007 Kjeller, Norway
** Mass. Institute of Technology, Cambridge, MA 02139, USA
*** Exxon Research & Engineering, Annandale, NJ 08801, USA

Microemulsions are thermodynamically stable mixtures of three components, water, oil and a surfactant. Sometimes a cosurfactant is needed
for the microemulsion to form. In the system studied here, a surfactant
didodecyldimetyl ammonium bromide (DDAB) is used[1]. This surfactant has a
polar ammonium head group and two hydrocarbon tails of twelve carbons.
The oil is hexane and the water is either D_2O or H_2O. The droplet nature
of the aggregates have been indicated by previous studies[2,3].

In this paper we first report viscosity data performed at the phase
boundaries on both sides of the one-phase region. We then use neutron
scattering data along the same lines and a series of additional neutron
data at a constant surfactant concentration of 5% across the one phase
region (see the solid line in the diagram) to support the viscosity
interpretation.

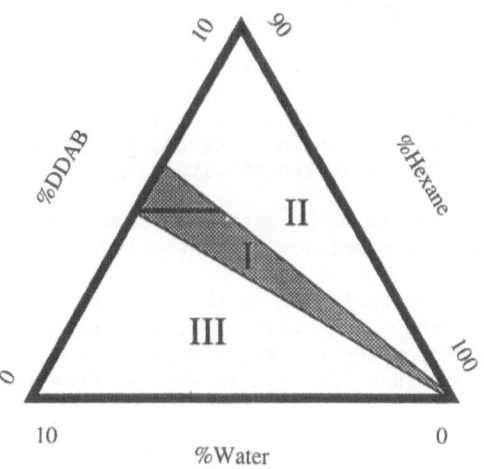

Fig. 1

PHASE DIAGRAM

The phase diagram near the hexane corner is shown in Figure 1. It has 3 distinct regions that can be described as follows:

    I one phase region: transparent microemulsion
   II two phases region: the lower phase is a transparent
                         microemulsion while the upper phase
                         is predominantly hexane
  III two phases region: the upper phase is a transparent
                         microemulsion while the lower
                         phase is a soapy opaque substance

The phase boundaries appear at a water-to-surfactant ratio (w/s) of 0.6 between region I and II and at 1.0 between region I and III.

VISCOSITY

The viscosity measurements were done using a Haake Rotoviscometer. By keeping the water-to-surfactant ratio constant in a water-in-oil microemulsion, the only parameter that is changing is the particle concentration.

If the microemulsion droplets were spheres we expect the viscosity to follow the Einstein relation

$$\eta = \eta_0 \ (1 + 2.5 \ \varphi) \tag{1}$$

where η_0 is the viscosity of the solvent, in our case the solvent is hexane and φ is the volume fraction of the particles, namely the micro-emulsion droplets.

Bedeaux has exended eq. (1) further to include the hard-sphere interactions to the third order. At high sheer rate the viscosity is

$$\eta = \eta_0 \ \frac{1 + 3/2\varphi(1 + S(\varphi))}{1 - \varphi(1 + S(\varphi))} \tag{2}$$

where $S(\varphi) = 3.08 \ \varphi + 3.15 \ \varphi^2$.

Figure 2 shows the viscosity along the lower phase boundary where the water-to-surfactant ratio is 1.0. The straight line in the figure correspond to the Einstein relation, eq. (1). In the limit $\varphi \to 0$ we see that that the Einstein relation is observed. At higher volume fractions Bedeaux's form is in better agreement with data.

Based on this analysis we suggest that along this dilution line, where w/s = 1.0, the microemulsion droplets have a spherical shape.

If the shape of the particles in the solution are not sphereical, but have a different shape like rods or ellipsoids of revolution, a possible way to check for the non-sphericity is to use Simha's factor v. The viscosity at low volume fraction is then written as

$$\eta = \eta_0 (1 + v\varphi) \tag{3}$$

For spheres this reduces to the Einstein form with v = 2.5.

Using this relation in Figure 3, which gives the viscosity along the upper phase boundary, $\omega/s = 0.6$, we find $\nu = 7$. This correspond to an ellipsoid of revolusion with an axial ratio of 6. Along this dilution line the result clearly shows that the microemulsion droplets are not spherical, but the stimated axial ratio can be to high because of interactions between particles.

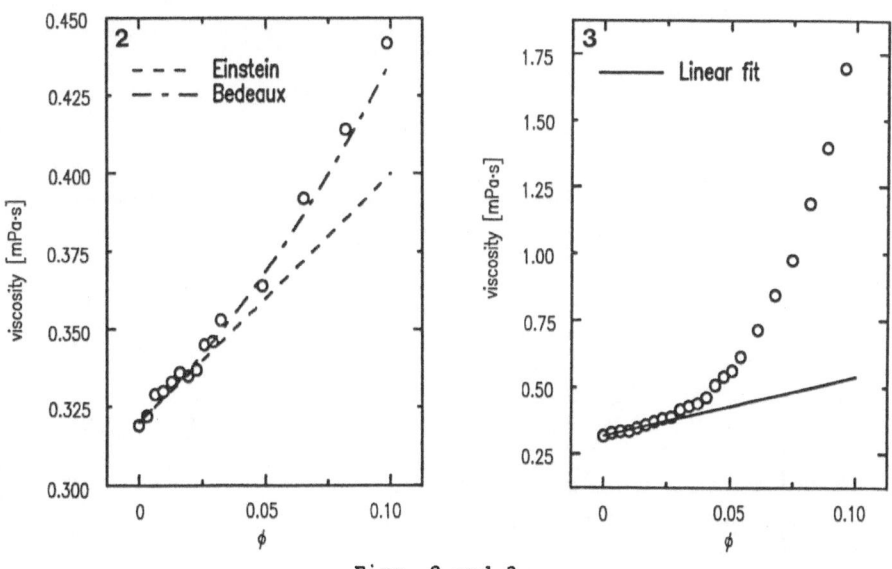

Figs. 2 and 3

NEUTRON SCATTERING

Small angle neutron scattering measurements were made on samples with D_2O. When H_2O is replaced by D_2O, the lower phase boundary changes slightly from $\omega/s = 1.0$ to ω/s 0.95. The neutron scattering cross section can in the dilute case be written as

$$\frac{d\Sigma}{d\Omega} = n_p P(q) \tag{4}$$

where n_p is the particle number density and $P(q) = \langle |F(q)|^2 \rangle$ is the particle structure factor. For a single sphere the form factor, $F(q)$, is

$$F(q) = v(\varrho_1 - \varrho_s) \frac{3j_1(qR)}{qR} \tag{5}$$

where

197

v   is the volume of the sphere
ϱ_1 is the scattering length density for the particle
ϱ_s is the scattering length density for the solvent
R is the radius of the sphere
$j_1(x)$ is the first order spherical Bessel function

If we fit some SANS data along the dilution line where $w/s = 0.95$, we find that a model of poly-disperse spheres fits the data best. By using a gaussian size distribution

$$f(r) = \frac{1}{\sqrt{2\pi}\ \sigma}\ e^{-\ (r-R)^2/\sigma^2} \qquad\qquad (6)$$

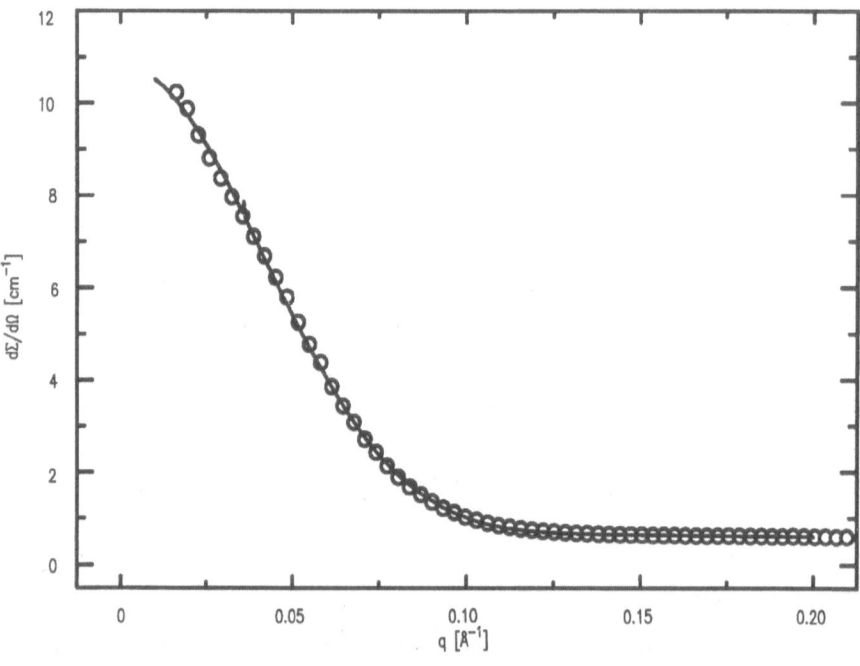

Fig. 4.

we have fitted the data as shown in Figure 4 (3% DDAB). The radius of the water core comes out to be 2.9 nm. When we relate this result to the viscosity measurement, we conclude that at the lower phase boundary the shape of the microemulsion droplets is spherical.

TABLE 1
Results from fits to a Gaussian distribution of non-interacting spheres

Percent DDAB	R (nm)	Polydispersity σ/R
1	2.5	0.29
2	2.8	0.30
3	2.9	0.30
4	2.8	0.30
5	2.9	0.30

On the upper phase boundary, where $\omega/s = 0.6$, spherical object does not fit SANS data.

In the most dilute sample, 0.5 % DDAB, single ellipsoids model will fit. However, at 1.0 and 1.5 % a pronounced flatness appears in the spectra at about $q = 0.1 \ \mathring{A}^{-1}$. One way to explain this is by introducing dimers of microemulsion droplets. Let us assume a dimer where two ellipsoidal droplets are joined together along their major axis. The form factor for such a configuration is

$$F(q,\mu)_{dimer} = F(q,\mu)_{monomer}(1 + e^{-i2\mu qt}) \qquad (7)$$

where 2t is the distance between the centers and μ is the direction cosine describing the orientation. The form factor is then squared and averaged over μ to get the one particle structure factor.

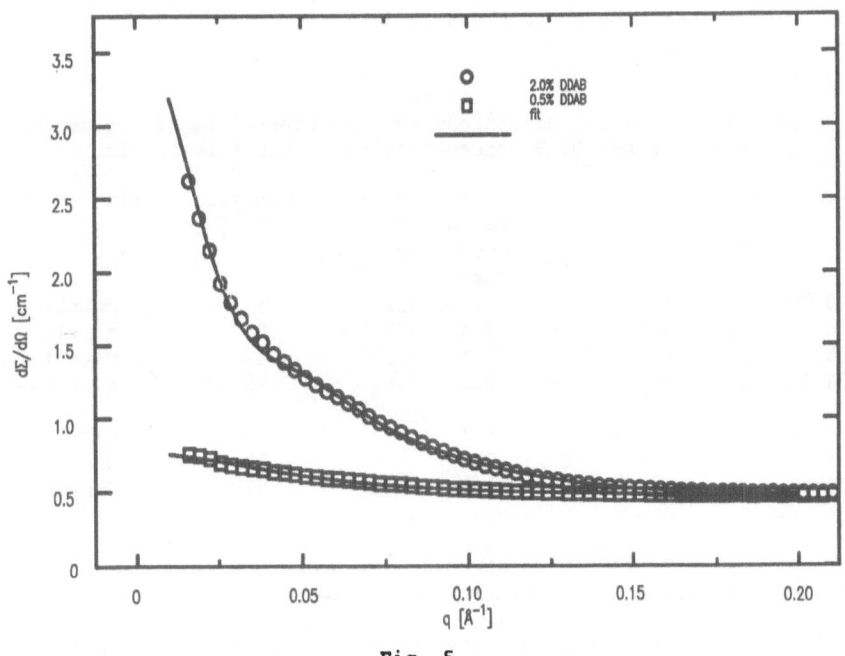

Fig. 5.

Using this P(Q) to fit the data we see that the flat portion of the
scattering curve can be accounted for. This can be seen in Fig. 5. It
should be noted that if we take a model where the two ellipsoidal droplets were aligned along their minor axis, this model will not fit the
data. As indicated by the viscosity data, the shape of the microemulsion
droplets along the w/s = 0.6 line is not spherical, but rather ellipsoidal or elongated. The neutron data show that a single ellipsoid model
fit the very dilute region (0.5% DDAB). At higher concentrations these
may form dimers. The results from the best fits are summarized in Table
2. From this we see that the minor axis remains constant at about 2.0 nm,
whereas the axial ratio is about 2.0. The separation distance is about
1.0 nm.

TABLE 2
Best fits for water to surfactant ratio of 0.6

DDAB	minor axis (nm)	major axis (nm)	axial ratio monomer droplet	configuration
0.5	1.7	6.8	2.5	monomer
1.0	1.8	7.4	2.7	monomer
1.5	2.0	5.6	2.0	dimer
2.0	2.0	5.4	2.0	dimer
2.5	2.0	5.4	2.0	dimer

In order to establish where the sphere-to-ellipsoid transition takes
place a concentration scan with constant surfactant concentration was
performed. The surfactant concentration was 5%. This correspond to the
horizontal line in the phase diagram Figure 1. The neutron data were
fitted to both spherical and ellisoidal models. The results from the fit
to the ellipsoidal models are shown in Table 3.

TABLE 3
Best fit to single ellipsiods and two connecting ellipsoids
for constant DDAB concenttration of 5.0 % by weight

hexane	a(nm)	b(nm)	t(nm)	axial ratio water core	outer ratio droplet	headarea (nm^2)	fit
90.00	3.4	4.3	-	1.3	1.2	-	monell
90.25	3.2	4.4	-	1.4	1.3	.55	monell
90.50	2.9	4.4	-	1.5	1.3	.42	monell
91.00	2.4	5.5	-	2.2	1.8	.55	monell
91.50	2.1	5.6	6.2	2.7	-	.58	dimell
91.80	2.1	5.7	6.2	2.8	-	.52	dimell
92.00	2.1	5.7	6.2	2.8	-	.51	dimell

Ellipsoids of revolution with an axial ratio of less than 2.0 cannot be
separated from polydisperse spheres using SANS.

From the table we see that the sphere-to-ellipsoidal transition
takes place between a hexane concentration of 90.5 and 91. This

correspond to a ω/s-ration of 0.8, which is at the center of the phase region.

DISCUSSION

A simple explanation for the "sphere-to-rod" transition may be put forward in terms of the bending curvature of the surfactant layer[4]. The surfactant layer has a natural radius of curvature say ϱ_0. This is what is measured in the case where we have spheres. As we move towards the ellipsoidal regime, we are effectively taking out water from the water core without reducing the surfactant concentration. Since the surfactant layer has a natural curvature it cannnot reduce its diameter without paying to much energy cost in the terms of the surface tension ($1/\varrho^2$ - $1/\varrho^2_0$). Instead it takes on an elongated shape. Because of the elongated shape, the surfactant tails close to the minor axis are much more densely packed than at the ends. These ellipsoids that meet end-to-end can get very close to each other.

REFERENCES
1. D.F. Evans, D.J. Mitchell and B.W. Ninham, J. Phys. Chem. 90, 2817 (1986)
2. W. Jahn and R. Strey, J. Phys. Chem. 92, 2294 (1988).
3. J. Samseth, S.-H. Cgen, J. Litster and J.S. Huang, J. Appl. Cryst. 21, 835 (1988).
4. S.A. Safran, L.A. Turkevich and P. Pincus, J. Physique L45, L-89 (1984)

FIELD-INDUCED PERCOLATION

IN MICROEMULSIONS

Marc Aertsens[1] and Jan Naudts[2]

[1]Limburgs Universitair Centrum, Universitaire Campus

3610 Diepenbeek, Belgium

[2]Departement Fysica, Universiteit Antwerpen (UIA)

2610 Antwerpen, Belgium

INTRODUCTION

Microemulsions of water in oil exhibit an insulator-conductor transition: a slight increase of the concentration of nanodroplets increases the conductivity with several orders of magnitude[1-6]. The basic phenomenon[7-9] is percolation of clusters of water droplets throughout the system, although the volume concentration is often too low for continuum percolation. Safran, Webman and Grest[8] have shown that the weak mutual attraction of the droplets lowers the density treshold for percolation. There is experimental evidence that in presence of a strong electric field the electric conductivity is enhanced. Indeed, recent experiments[5] have shown that the electric current in the insulator phase consists of two contributions: a fast response due to diffusion of charged droplets, and a slow response due to changes in the shape of the clusters induced by the electric field.

An explanation for these field-induced effects is found in ref. 10 and 11. The latter proposes a microscopic lattice hamiltonian which reproduces partly the experimental behaviour. Simulation results for this hamiltonian are discussed below.

THE MODEL

The water in oil microemulsion is described by a diluted ising model on a simple cubic lattice. All water droplets are assumed to be on lattice sites. Each occupied site has a spin which is either $+1$ or -1 depending on the charge of the droplet. In the experimental system most droplets are electrically neutral. Hence, charges larger than $\pm e$ are very unlikely[12]. For efficiency reasons we simulate a system where all particles have charge either $+e$ or $-e$. The total charge of the system is zero.

The particles interact according to the hamiltonian

$$H = -4J \sum_{|i-j|=1} t_i t_j - E \sum_i t_i s_i z_i$$

The variable t_i is one when a particle is present at site i, zero when the site is empty. The variable s_i is the spin variable and gives the charge of the particle at site i. The parameter $4J$ is the nearest neighbor attraction between the particles. The electrical field has been taken parallel to the z-axis (z_i is the z-coordinate at site i). In absence of the electric field the hamiltonian reduces to the usual lattice gas hamiltonian.

The particles move according to Kawasaki dynamics[12]. During one Monte-Carlo step each particle tries exactly once to move to a site, randomly chosen from 18 neighboring positions (nearest and next to nearest neighbor positions). If the chosen site is occupied the particle does not move. Otherwise the change in energy is calculated and the move is accepted using the Metropolis criterium[13]. Due to the particle-hole exchange the individual particles diffuse and the cluster configuration changes.

In addition a mechanism for charge exchange between particles on neighboring sites is introduced. The algorithm for charge exchange is executed a large number of times N after each completion of the particle-hole exchange algorithm. In the present simulations $N = 50$ was used. The charge exchange algorithm selects for each particle a random site out of 18 neighboring sites. If it is occupied the exchange of charges between both particles is accepted using the Metropolis criterium. Annihilation of charges cannot occur because neutral charges are not included.

An analoguous hamiltonian using only one kind of charges has been studied extensively[14-20]. We believe that in the latter model steric hindrance effects are important. In order to avoid such effects we use a coordination number of 18 in the dynamics of the present model.

SIMULATIONS

We simulate a simple cubic lattice with size $15 \times 15 \times 20$ and periodic boundary conditions in all directions. Because the electric potential is different at the $z = 0$ and $z = 20$ electrodes we cannot take the periodic boundary conditions literally. We have to imagine that the system is infinite in the z-direction and consists of an infinite repetition of identical configurations.

The measured quantities are the electric current i and the probabilities n_x and n_z that a particle has a neighbor in the x- resp. the z-direction. For the calculation of the current all motions of the charges are taken into account. In order to avoid wild fluctuations the current is averaged over 1000 MCS.

The particle density is $\rho = 0.08$. From ref. 21 we know that in absence of an electric field the percolation treshold ρ_c increases from 0.16 at zero temperature to 0.31 at infinite temperature. At the density $\rho = 0.08$ the system does not percolate in absence of the electric field. In the related work of ref. 14-20 the cluster shape changes drastically in the presence of a strong electric field. Particle-rich and particle-poor strips are formed along the direction of the field at low enough temperature. In this way percolation at very low densities becomes possible.

For simplicity we take $4J = 1$ everywhere. In absence of the electric field and

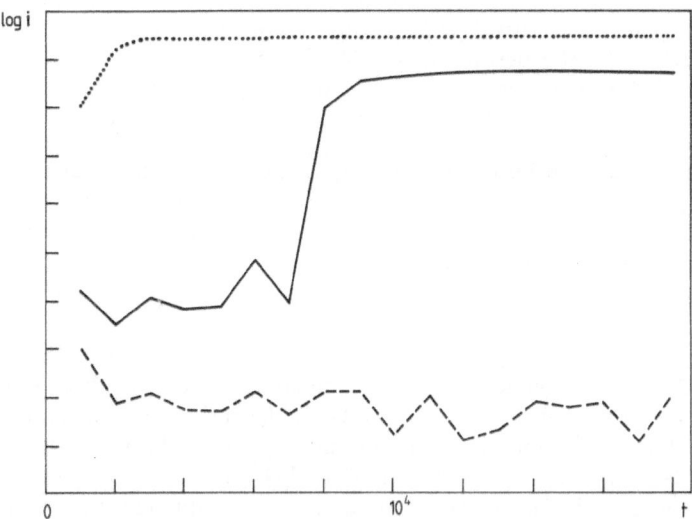

Figure 1. Electric current versus time for different values of the temperature T and the electric field E: $k_BT = 0.45$ and $E = 0.025$ (dashed), $k_BT = 0.40$ and $E = 0.125$ (full), $k_BT = 0.40$ and $E = 0.40$ (dotted).

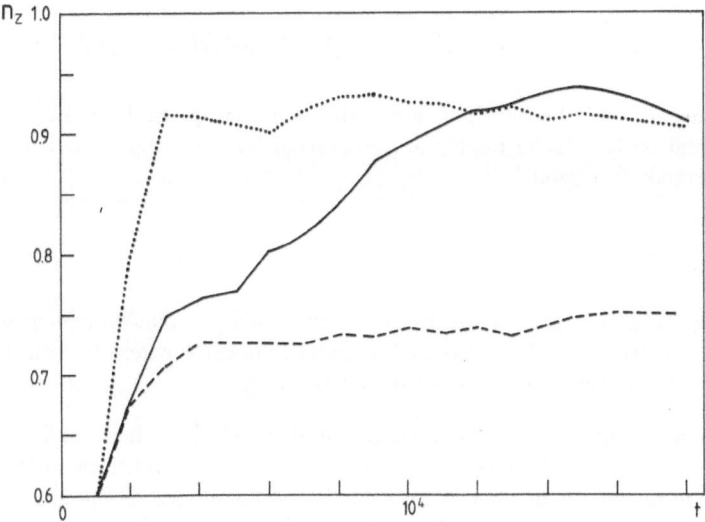

Figure 2. Probability n_z that a particle has a neighbor in the positive z-direction. The dashed, full and dotted lines are as in fig. 1. The curves for n_x and n_y coincide within errorflags with the dashed line.

at density $\rho = 0.5$ the critical temperature below which phase separation occurs is $k_B T_c \simeq 1.12$. At the density $\rho = 0.08$ the critical temperature is lowered to $k_B T_c \simeq 0.90$. But in presence of the electric field the critical temperature rises again[16,20]. Up to now we have investigated the range of temperatures 0.2 to 0.7.

Our simulations are implemented as quenches from high to low temperature. Initially the particles are distributed randomly on the lattice. In principle we should first thermalise before switching on the external electric field. The quenching procedure was chosen in order to save computer time.

RESULTS

Fig. 1 shows the electric current as a function of time for three different field strengths at about the same temperature. For sufficiently large fields a large current is observed almost immediately. For intermediate fields the increase in current is delayed. Indeed, we see from fig. 2 that in this case it takes more time for the clusters to elongate. For small fields the current is completely due to the diffusion of very few isolated particles. The percolation transition cannot occur because the system remains nearly isotropic. A snapshot of the particle configuration is shown in fig. 3. Note that at these temperatures almost all particles are in the solid phase and the clusters are very compact.

At higher temperatures the clusters become smaller and more ramified. The percolating cluster can be destroyed temporarily by thermal motion. Indeed, at $k_B T = 0.65$ we observe intermittent behaviour (see fig. 4). Fig. 5 shows the conductivity as a function of the electric field at the latter temperature. The measured points are fitted with

$$\sigma = i/E = \sigma_0 + \frac{1}{2}(\sigma_\infty - \sigma_0)\left(1 + \tanh c(E^2 - E_0^2)\right)$$

where σ_0 and σ_∞ are the ohmic conductivity at low resp. high fields. One has constant percolation for strong fields, no percolation for weak fields, and intermittency in the intermediate region.

CONCLUSIONS

During the simulations of our model we observe a percolation transition induced by a strong electric field. It occurs at low particle density where percolation cannot be explained by mutual attraction of particles alone.

We are aware of the fact that the simulated system is rather small and that our results are size-dependent. The use of periodic boundary conditions in the direction of the field is rather artificial, but because of the small size of the system it is the only way to avoid dominant boundary effects.

One of us (M.A.) wishes to thank the Belgian *Nationaal Fonds voor Wetenschappelijk Onderzoek* and the *Vlaamse Wetenschappelijke Stichting* for financial support.

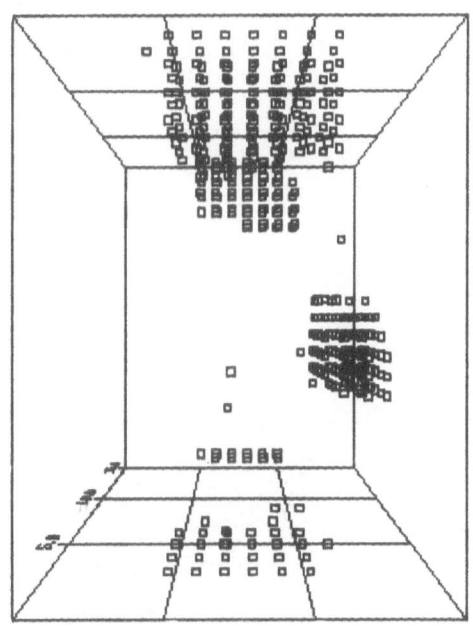

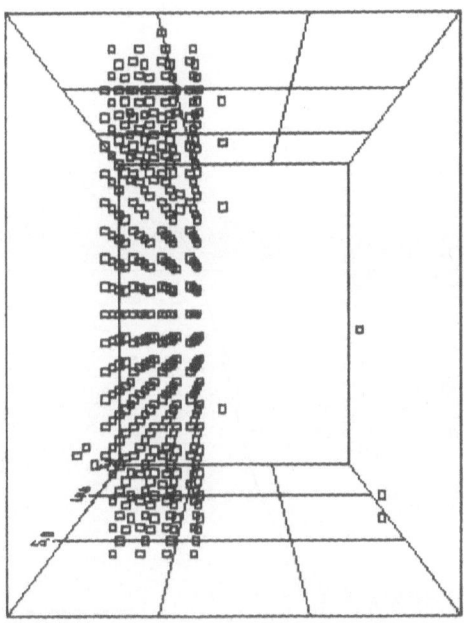

Figure 3. Three-dimensional view of a particle configuration at $t = 18750$ MCS, $E = 0.025$, and $k_BT = 0.45$ resp. $t = 21750$ MCS, $E = 0.40$, and $k_BT = 0.40$.

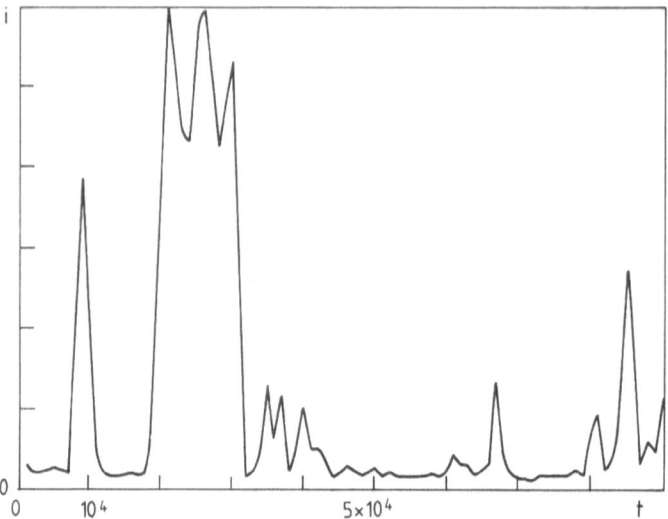

Figure 4. Current versus time at $k_B T = 0.65$ and $E = 0.11$.

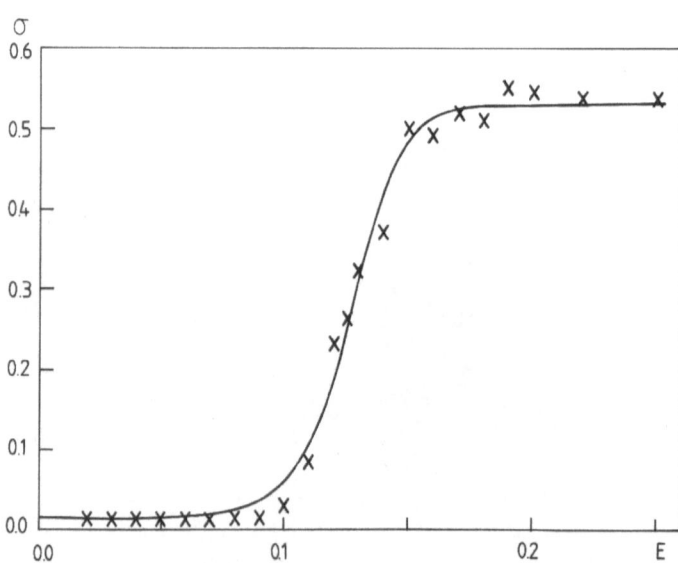

Figure 5. Conductivity versus field at $k_B T = 0.65$. The crosses are the measured points, after averaging. The dotted curve is the result of a fit with the formula discussed in the text.

REFERENCES

1. M. Lagues, J. Phys. (Paris) Lett. **40**, L331 (1979)
2. H.-F. Eicke, R. Hilfiker, M. Holz, Helv. Chim. Acta **67**, 361 (1984)
3. M. A. Van Dijk, Phys. Rev. Lett. **55**, 1003 (1985)
4. S. Bhattacharya, J. Stokes, M. W. Kim, J.S. Huang, Phys. Rev. Lett. **55**, 1884 (1985)
5. H.-F. Eicke, R. Hilfiker, H. Thomas, Chem. Phys. Lett. **125**, 295 (1986)
6. J. Samseth, this volume, and Ph.D. thesis, MIT Boston, 1987.
7. A. Safran, L.A. Turkevich, Phys. Rev. Lett. **50**, 1930 (1983)
8. S.A. Safran, I. Webman, G.S. Grest, Phys. Rev **A32**, 506 (1985)
9. G.S. Grest, I. Webman, S.A. Safran, A.L.R. Bug, Phys. Rev. **A33**, 2842 (1986)
10. H.-F. Eicke, J. Naudts, J. Chem. Phys. Lett. **142**, 106 (1987)
11. M. Aertsens, H.-F. Eicke, J. Naudts, and H. Thomas, *to be published*
12. K. Kawasaki, in *Phase transitions and critical phenomena*, ed. C. Domb and M. Green (Academic Press, London, 1972), vol. II, p. 443.
13. N. Metropolis, A. Rosenbluth, M. Rosenbluth, A. Teller, and E. Teller, J. Chem. Phys. **21**, 1087 (1953)
14. J. Lebowitz, H. Spohn, J. Stat. Phys. **28**, 539 (1982)
15. H. van Beijeren, L.S. Schulman, Phys. Rev. Lett. **53**, 806 (1984)
16. S. Katz, J.L. Lebowitz, H. Spohn, J. Stat. Phys. **34**, 497 (1984)
17. J. Marro, J.L. Lebowitz, H. Spohn, M.H. Kalos, J. Stat. Phys. **38**, 725 (1985)
18. J. Krug, J.L. Lebowitz, H. Spohn, M.Q. Zhang, J. Stat. Phys. **44**, 535 (1986)
19. J.L. Vallés, J. Marro, J. Stat. Phys. **49**, 121 (1987)
20. J. Marro, J.L. Vallés, J. Stat. Phys. **49**, 121 (1987)
21. S. Hayward, D.W. Heerman, K. Binder, J. Stat. Phys. **49**, 1053 (1987)

NEW SOURCE OF CORRECTIONS TO SCALING FOR MICELLAR SOLUTION CRITICAL BEHAVIOR

G. Martínez-Mekler,[*][+] G.F. Al-Noaimi[**] and A. Robledo[+]

[*]Istituto Nazionale di Fisica Nucleare, L. Enrico Fermi 3,
Firenze, Italia
[**]Department of Physics, University of Qatar, Doha, Qatar.
[+]Instituto de Fisica, UNAM, Apdo. Postal 20-364, Mexico, D.F.

INTRODUCTION

The work here presented originated as an attempt to clarify the apparent "non-universality" of critical exponents measured in micellar solutions of non-ionic surfactants[1,2,3] and in microemulsions.[4] The controversial issue in these systems is the existence of continuously variable exponents which are solvent dependent and may assume values below mean-field predictions. Our approach is based on a lattice model[5] for the solution which is mapped into an Ising spin- 1/2 model.[6] The mapping generates a parameter dependent function that modifies the Ising critical exponents. The original system and the Ising image are shown to have different reduced temperature scales which may be related, for certain values of the parameters, by a non-analytic transformation. As the critical point is approached universality is preserved, however some conjectures related to the anomalous behavior of the effective exponents follow from the formalism presented. Though the full understanding of the problem remains open, we should stress that the mechanism that produces the corrections to scaling is of a very general nature and should by taken into account for a large class of statistical mechanics models.[6]

MICELLAR SOLUTION MODEL[5]

Consider a two component mixture of species that completely occupy the sites on a regular lattice with coordination number q. Species 1 represents solvent molecules while species 2 represents small empty micelles or small permanent aggregates with many internal degrees of freedom. The two species interact by pairs only through nearest-neighbor interactions. Among the solvent molecules (A) the interaction is isotropic with an energy parameter ε_{AA} The second species is formed by q molecules oriented towards the bonds of the lattice, each with two active ends a and b, one at the lattice site the other at the bond center. These micelles have $n=2^q$ possible internal configurations Micelles interact amongst themselves and with the solution only through their outermost ends via the energy parameters ε_{aa}, ε_{bb}, ε_{Aa} and ε_{Ab}.

Let N_{ij} be the number of nearest neighboring pairs of species (i,j=1, 2) and P_{kl} denote the number of encounters at bond centers of species with active ends k,l=A,a,b. Then the energy E of a mixture configuration will be the sum of the pair interaction energy E_{pair} and the internal energy E_{int}

given by

$$E_{pair} = \sum_{i,j=A,a,b} e_{ij}P_{ij} = e_{AA}N_{11} + e_{Aa}P_{Aa} + e_{Ab}(N_{12}-P_{Aa})$$

$$+ e_{aa}P_{aa} + e_{ab}P_{ab} + e_{bb}(N_{22}-P_{aa}-P_{ab}), \qquad (1)$$

$$E_{int} = \sum_{k=1}^{n} N_2^k \alpha_k , \qquad (2)$$

where α_k is the energy parameter of the kth micellar configuration and N_2^k is the number of micelles in each kth internal state in a given configuration of the mixture.

If we now give equal weights to all the internal micellar configurations. i.e. α_k=const, then the partition function for the mixture will be given by:

$$\Xi = Q_{11}^{qN/2} \sum_{N_2=0}^{N} \sum_{N_{12}=0}^{qN/2} \Omega_0(N,N_2,N_{12}) K^{N_{12}} \lambda^{N_2} \qquad (3)$$

where $Q_{11}=4 \exp(-\beta e_{AA})$, $\Omega_0(N,N_2,N_{12})$ is the number of configurations with N sites, N_2 micelles and N_{12} encounters,

$$K \equiv \exp(-\beta_{Aa})[1+\exp(-\beta\delta_{12})]D_{12}^{-1/2} \qquad (4)$$

$\lambda \equiv 4\exp(q\beta_{Aa})\zeta(D_{12})^{-q/2}$, with $D_{12}=1+\exp(-\beta d_{ab})+2\exp(-\beta(\Delta_{ab}+ 1/2 \ d_{ab}))$, $\Delta_{Aa}=e_{Aa}-1/2(e_{AA}+e_{aa})$, $\Delta_{ab}=e_{ab}-1/2(e_{aa}+e_{bb})$, $\delta_{12}=e_{Ab}-e_{Aa}$, $d_{ab}=e_{bb}-e_{aa}$, $d_{Aa}=e_{AA}-e_{aa}$, and ζ the activity ratio of the two species. If we now make the identifications:

$$K= \exp(-2\beta^i J) \quad and \quad \lambda=\exp(2\beta^i H) \qquad (5)$$

then, except for a proportionality constant, equation (3) is the partition function for a nearest-neighbor spin-1/2 Ising model with N sites, at a temperature β^i, with coupling J and external magnetic field H (N_2 would then be the number of spins up in a configuration and N_{12} the number of spin-up spin-down encounters).

ISING MODEL WITH TEMPERATURE DEPENDENT COUPLING PARAMETERS

If we assume in equations (5) that the Ising image temperature is the same as the mixture temperature, i.e. $\beta=\beta^i$, then the coupling parameters J and K are temperature dependent (the alternative assumption gives the same final results). From equation (4) we have that depending on the mixture energy parameters, the associated Ising model may be of a ferro or an antiferromagnetic nature. Given a set of values of the mixture energy parameters $ep\equiv\{d_{ab},d_{Aa},\Delta_{Aa},\Delta_{Ab},\delta_{12}\}$, if $K(\beta,ep)$ is less than some critical value K_c, then there occurs a phase transition between isotropic (uniform) phases in the Ising image as H passes through zero, with inverse critical temperature β_c satisfying $K(\beta_c)=K_c$. The existence of a minimum of K at β_0 with $K(\beta_0)<K_c$ is a requisite for the occurrence of close-loop coexistence curves. Anisotropic (nonuniform) phases occurring at low temperatures appear if $K(\beta)$ can reach values larger than K_c^{-1}. In this case the mixture undergoes sub-lattice ordering transitions that correspond to antiferro-magnetic transitions in the Ising image; this ordering is the model's representation of liquid

212

crystalline phases. The conditions $K(\beta_{LC})=\bar{K}_c^i$ and $K(\beta\pm)=1$, define respectively liquid-crystal critical temperatures and the mixture temperatures at which the Ising image ferro/antiferro-magnetic switch takes place.

Let us now look into the case when there is a solubility loop and focus our attention on the behavior near the lower critical point with temperature β_c. For the usual Ising model the critical temperature $(T_c^i=1/\beta_c^i)$ is dependent on the lattice geometry and the model interactions, i.e., $\exp(-2\beta_c^iJ)=K_c$, with K_c fixed by the lattice geometry, determines β_c^i. Since in our model, we have from (4) and (5) that $J(\beta)=-\ln K(\beta,ep)/2\beta$, it follows that

$$\beta_c^i(\beta)=\beta \ln K_c/\ln K(\beta,ep). \tag{6}$$

A crucial consequence of equation (6) is that the distance to the critical temperature in the Ising image $\varepsilon_i=|T-T_c^i(T)|/|T_c^i(T)|$, will differ from the measured one $\varepsilon=|T-T_c|/|T_c|$. More explicitly using (6) we obtain the following non-trivial relation between ε_i and ε:

$$\varepsilon_i(\varepsilon,ep)=\left|1-\frac{\ln K_c(\beta_c,ep)}{\ln K(\varepsilon,ep)}\right|. \tag{7}$$

EFFECTIVE EXPONENTS AND NEW SOURCE OF CORRECTIONS TO SCALING

Consider now the micellar solution correlation length ξ_m, near the critical point, the following expansion should hold:

$$\xi_m=\xi_{0m}\varepsilon^{-\nu_m}[1+ a(T)\varepsilon^{wa}+ b(T)\varepsilon^{wb}+ \cdots] \tag{8}$$

where the leading singularity comes from the exponent ν_m, and the terms within the brackets are the usual corrections to scaling contributions with wa, wb,... the crossover exponents. The effective exponent for the mixture measured at a value $\varepsilon=\varepsilon^*$ will be given by the logarithmic derivative:

$$\nu_{m,eff}(\varepsilon^*) = -\left.\frac{d \ln \xi_m}{d \ln \varepsilon}\right|_{\varepsilon=\varepsilon^*} \tag{9}$$

For the Ising image we have the equivalent expressions for ξ_i, in terms of ε_i, ξ_{0i} and ν_i. Since the value of the correlation length at a given temperature should be the same in both descriptions of the system, we have that $\xi_i(\varepsilon_i(\varepsilon,ep))=\xi_m(\varepsilon)$. With this last expression, (8), (9), and the corresponding Ising image counterparts we arrive by a simple chain rule at:

$$\nu_{m,eff}(\varepsilon^*,ep)=\nu_{i,eff}(\varepsilon_i(\varepsilon^*,ep)) f(\varepsilon^*,ep), \tag{10}$$

$$f(\varepsilon^*,ep) \equiv \left.\frac{d \ln [\varepsilon_i(\varepsilon,ep)]}{d \ln \varepsilon}\right|_{\varepsilon^*}. \tag{11}$$

Expressions (10) and (11) are the main result of this paper, exhibiting explicitly, through the parameter dependent function f, that for the mixture a new source of corrections to the scaling of the Ising image, produced by the mapping into the spin model, should be taken into account when comparing with experiment.

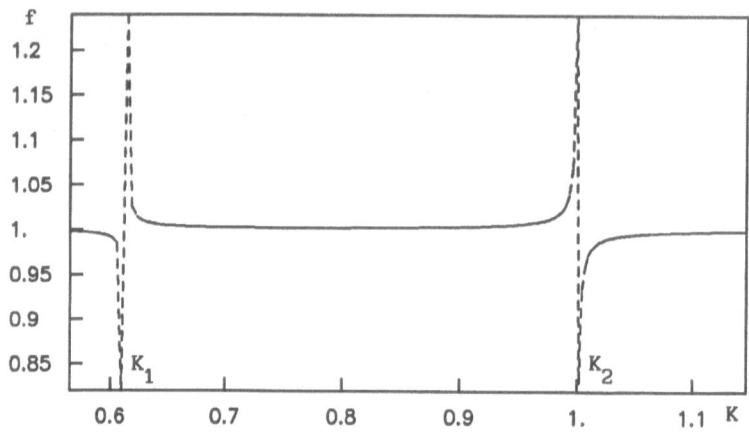

Fig. 1. Graph of f(K) defined by equation (11) as we vary d_{ab}
from -100 to 1900, leaving $\delta_{12}=1000$, $\Delta_{Aa}=-50$, $\Delta_{ab}=0$,
(all in ε_{AA} units) $\beta_m^c=0.0029$ [$^\circ K^{-1}k_B$] and $\varepsilon_m=0.001$ fixed.

Our experience is that f is almost always very close to 1, i.e., in general $\varepsilon_1(\varepsilon)$ is a linear relation, however when a critical point of the micellar solution is in the proximity of another critical point, f may give a considerable contribution. In fig. 1 we show a graph of f vs K with parameter values in the range considered experimentally[1,2]. Two singular points appear at the values K_1 and K_2 which correspond respectively to the values of ep such that $K(\beta,ep)=K_c(ep)$ and $K(\beta,ep)=1$ (i.e., $\beta=\beta\pm(ep)$). For $K=K_1$ we are already looking at the upper critical temperature point of the solubility loop, and for $K>K_2$ we are looking at the behavior of a ferromagnetic critical point at a temperature within the antiferro-magnetic type interaction region (this is the cause for the negative sign of f). In this last case, if we send ε to zero, the ferro-antiferromagnetic critical points collapse, while in the former case the solubility loop is lost at a value slightly above K_1 where $K_c(ep)=K(\beta_0)$ and beyond which we have switched from the lower critical temperature point to the upper critical temperature point.

Numerically we have seen that as ε approaches zero the region for which $f\neq1$ diminishes, moreover a straightforward application of d'Hopital's rule to equation (11) gives the following exact results: $\lim \varepsilon\to0 \; f(\varepsilon)= 1$, whenever $K_1<K<K_2$ and $\lim \varepsilon\to0 \; f(\varepsilon)= 2$, for $K_c=K_1$. The former limit is a formal statement of the persistence of universality, while the latter one shows that the critical exponent doubles at the merger point of two ferromagnetic type critical points. This has been observed[7] and is a known result, however, within our formalism the calculation is trivial and the generalization to the merger of more critical points is immediate, namely if n critical points of this type coalesce, the critical exponent is multiplied by n. We have also obtained[8] a series of additional exact results concerning scaling relations and various limiting behaviors.

DISCUSSION

The purpose of this work is to raise a note of caution when comparing theoretical predictions and experimental data related to critical phenomena whenever the solution of the theoretical modelling involves the type of mapping described above. We have shown that these mappings, that lead to effective Hamiltonians with thermodynamically dependent interaction parameters, can induce in the original system corrections to the scaling behavior of the image system. The generality of this result should be stressed, since it may be appliable to a variety of statistical mechanical models. The following comments summarize our findings:

On Universality. Within our formalism *universality is not violated.* Both numerically and analytically we have shown that as $\varepsilon \to 0$, the unusual corrections to scaling disappear. Notice that this result is parameter independent. If ε is sufficiently small the usual Ising results should be retrieved, however, values of ε which usually fall within the scaling region may not be small enough in the Ising image, leading thus to unusual, parameter dependent, exponent values.

On the value of the critical exponents: We have shown that for finite ε, when in the physical system under study an antiferro- and ferro-magnetic type critical point are close to each other, the critical region is drastically *contracted.* In the lattice model this occurs when $K_c \to 1$, i.e. the coordination number is high and the micelles have many components. Under these circumstances, $\nu_{1,eff}$ should be close to its mean field value and the effect of the function f would be to increase the value towards the usual Ising result for full criticality. A detailed calculation on these lines would involve a careful consideration of the usual corrections to scaling and possibly of additional contributions due to a crossover of the type suggested by Fisher[8], if lower than mean-field values are to be retrieved. Before entering such a calculation it is advisable to extend the modelling here considered in order to be able to obtain asymmetric coexistence loops since this is the experimental case. This effort is underway with positive preliminary results. The role of the internal configurations which in our model are related to the energy parameters α_k may turn out to be essential.

Alternatively,[6] if the experimental uncertainty is interpreted as a coarse graining in a $\log(\xi_m)$ vs $\log(\varepsilon)$ plot, then the determination of the exponents would be closer to the experimental one, if a limiting quotient definition of the critical exponent is used instead of the logarithmic derivative of equation (9). In this case f can take values in the range (0,1) for a wide set of *ep* values, thus lowering the Ising exponents. A more thorough examination of this situation is called for.

On the generality of our findings. Our formalism suggests that other critical points of the system under study such as the upper critical point may also show unexpected behaviors. For example, if $K > K_2$ then the liquid-crystal critical point shows a pathological behavior as $K \to K_2$ due again to the proximity of the lower solubility loop critical point. Recently, through an alternative treatment, Gazeau[9] has argued that the lower than mean-field exponents measured in microemulsions are due to the proximity of another critical point. Similar conclusions have been drawn by Vives[10] with regard to Monte Carlo calculations for phase transitions in liquid crystals. A reexamination of these critical points in the perspective of the formalism here introduced may contribute to interesting theoretical and experimental developments.

REFERENCES

1. M. Corti and V. Degiorgio, Phys. Rev. Lett. **55**, 2005 (1985).
2. G. Dietler and D.S. Cannell, Phys. Rev. Lett. **60**, 1852 (1988).
3. J.P. Wilcoxon and E.W. Kalor, J. Chem. Phys. **86**, 4684 (1987).
4. A.M. Bellocq, P. Honorat, and D. Roux, J. Phys. (Paris) **46**, 743 (1985)
5. A. Robledo, C. Varea and G. Martinez-Mekler (unpublished).
6. G. Martinez-Mekler, G.F. Al-Noaimi and A. Robledo (unpublished).
7. R. Johnston, N Clark, P Wiltzus, D Cannell, Phys. Rev. Lett. **54**, 49 (1985)
8. M.E. Fisher, Phys. Rev. Lett. **57**, 1911 (1986).
9. D. Gazeau, Doctoral Thesis, University of Bordeaux I (1989)
10. E. Vives (private communication).

TWO-DIMENSIONAL PHASE TRANSITIONS OF A SURFACTANT MONOLAYER

Mahn Won Kim[1], Th. Rasing[2,*], and Y.R. Shen[2]

[1]Exxon Research and Engineering Company
 Corporate Research, Annandale, N.J. 08801
[2]Dept. of Physics, Univ. of California and
 Center of Advanced Materials, LBL, Berkeley, CA 94720

The phenomenon of phase transitions in two dimensions is of great fundamental interest and has therefore drawn a considerable amount of attention.[1,2] Insoluble monomolecular layers at a water-air interface provide a quite ideal two-dimensional model system with an isotropic substrate and an easily controllable density of molecules. At low densities they often exhibit a two-dimensional gas behavior,[3] whereas at higher densities transitions to liquid and solid states can be found. In many systems, the liquid phase is further divided into the so-called liquid-expanded (LE) and liquid-condensed (LC) phases.[4] Though observed and intensively studied, the nature of the LE-LC phase transition is still controversial.

In this paper we will demonstrate how we can use ellipsometry and optical second harmonic generation (SH) to study the structure of a monolayer of pentadecanoic acid (PDA) [$CH_3(CH_2)_{13}COOH$] near its LE-LC transition at a water-air interface. By simultaneously measuring the surface pressure versus surface molecular area we can show that this LE-LC transition is accompanied by a reorientation of the molecules and that the two phases are separated by an inhomogeneous coexistence phase.

A monolayer at an interface between air and water can be considered as an infinitely thin dielectric sheet represented by an induced polarization, which is usually a complicated nonlinear function of the electric field (E) of the incident laser beam. In the linear case, the induced polarization is a simple linearized form, $\vec{P} = \chi_s \vec{E}$ where χ_s is the surface susceptibility $\chi_s = \chi_s \delta(z)$, for simplicity here taken as a scalar, i.e. the monolayer is assumed to be isotropic and z is the direction normal to the surface. With the light incident from air, the phase retardation[5] between the reflected S and P waves ($\Delta\phi = \phi_s - \phi_p$), which is measured by ellipsometry, is linearly proportional to the surface susceptibility χ_s. For a monomolecular layer in the absence of local field effects χ_s can be related to the molecular polarizability $\alpha^{(1)}$ by $\chi_s = N_s <\alpha^{(1)}>$, where N_s is the surface density of molecules and the angular brackets indicate an orientational average. Thus, $\Delta\phi$ is linearly proportional to N_s.

In the nonlinear cases, the surface nonlinear susceptibility (SNS) arising from a monolayer of adsorbates is linearly proportional to the surface density of the molecules and the second order molecular polarizability (SMP) averaged over the molecular orientational distribution. Since the SH output in the reflected direction[6] is proportional to the square of SNS the intensity is proportional to the square of N_S. Furthermore, with input and output beam polarizations and geometry properly chosen, the various elements of SNS can be selectively measured by the SH. Consequently in the case that SMP is dominated by a single component along a molecular axis ξ one can then determine the orientation of this axis.[7] Furthermore, for a constant orientational distribution, the SH output can be nulled[8] by properly chosen optical polarization and geometry.

Monolayers of PDA were prepared by spreading a solution of PDA in petroleum ether or hexane on a thoroughly cleaned water surface (pH = 2) in a glass trough with paraffin coated edges. The surface density of molecules was controlled by a movable Teflon barrier and the surface tension was measured by a Wilhelmy plate[8]. For the ellipsometry measurements we used a weak He-Ne laser at 632.8 nm. The phase shift $\Delta\phi$ was measured by a standard high resolution ellipsometry technique.[5] The accuracy in $\Delta\phi$ was ~ 10^{-4} rad.

Figure 1 shows typical ellipsometry data obtained at three different parts of the phase diagram of a monolayer of PDA on

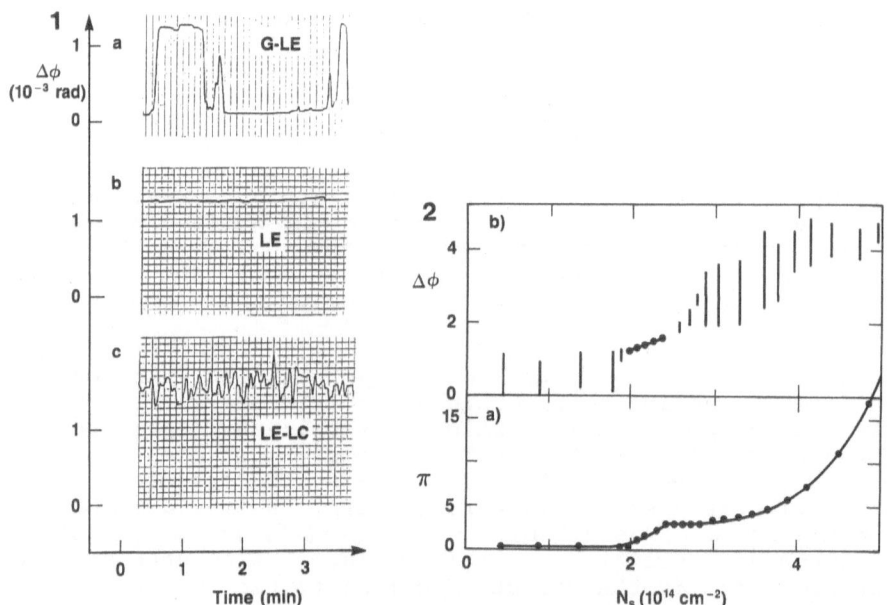

Figure 1. Some typical ellipsometry data for three different parts of the phase diagram of a PDA monolayer on water.

Figure 2. (a) Surface pressure, π, (dynes/cm) and (b) observed phase shift, $\Delta\phi$, (10^{-3} rad) for a PDA monolayer on water at 21°C.

water at 21°C. Figure 2 summarizes the results of both surface pressure and ellipsometry measurements. For $0.13 < N_S < 1.9$ x 10^{14} cm^{-2} the horizontal plateau in the π-N_S curve indicates the coexistence region between the G and LE phases. The corresponding measured $\Delta\phi$ appears to fluctuate very strongly (see Fig. 1a). Plotted in Fig. 2b are the peak-to-peak amplitudes of the fluctuating ellipsometry signal at a given density (with the present sensitivity of ~ 10^{-4} rad the signal in the pure G phase for $N_S < .13$ x 10^{14} cm^{-2} was too small to be detected). For $1.9 < N_S < 2.5$ x 10^{14} cm^{-2} the monolayer is in the pure LE phase, characterized by an increasing surface pressure and a stable $\Delta\phi$ (see Fig. 1b), that is linearly increasing with N_S. For $N_S > 2.5$ x 10^{14} cm^{-2}, the LE-LC transition point as indicated by a kink in the surface pressure curve, the fluctuations appear again (see Fig. 1c). When the density reaches a full monolayer at $N_S = N_O = 5$ x 10^{14} cm^{-2}, the amplitude of the fluctuations decreases but remains finite.

For the SHG measurements we used the frequency-doubled output of a Q-switched neodymium-doped yttrium aluminum garnet laser at 532 nm with a 7-nsec pulse width as the pump beam.

Figure 3 shows the results of both surface pressure and SHG intensity fluctuations ($\Delta I = |I- <I>|/<I>$) as a function of surface area per molecule. Again the ΔI appears large in the LE → LC transition regime. Furthermore, the SH intensity fluctuates significantly at the concentration regime which the surface pressure increases. Since the SH output is sensitive to the molecular orientation (see ref. 7), we can determine the origin of the SH intensity fluctuations as due to either concentration and/or orientation of adsorbates. For this purpose, the null technique was used. If the phases in a coexistence regime have the same orientation, the zero output of SH can then be obtained by rotating the output polarization for the given optical geometry and input polarization.[8] Figure 4 shows the null angle for the LE phase (solid circles with a solid line). However, the output of SH (open circles with a dashed line) at the surface concentration in the coexistence regime cannot be nulled, as shown in Figure 4. This indicates that the orientation of the two phases in a coexistence phase is not the same, consistent with our orientational measurements reported in ref. 7.

Knowing the size of the probe area (~ 0.5 mm^2), an estimate of the domain size can be made by looking at the $\Delta\phi$ data (Figs. 1 and 2b). In the G-LE region, the signal fluctuates more or less step-wise even at very low N_S, i.e. $\Delta\phi$ is either at its maximum or minimum value. This indicates that this region of the phase diagram is characterized by a few, but quite large (>> .5 mm^2), domains. On the other hand, in the LE-LC region $\Delta\phi$ fluctuates much more rapidly and with a relatively smaller amplitude indicating the presence of many smaller domains (< .5 mm^2).

In conclusion, by using highly sensitive ellipsometry and an optical second harmonic generation technique together with a surface balance we have been able to study the appearance and average growth of domains in the G-LE and LE-LC coexistence regions for a monolayer of PDA on water. The G-LE region is characterized by fairly large (>> .5 mm^2) domains that merge into a homogeneous LE phase. In the LE-LC region, the domains

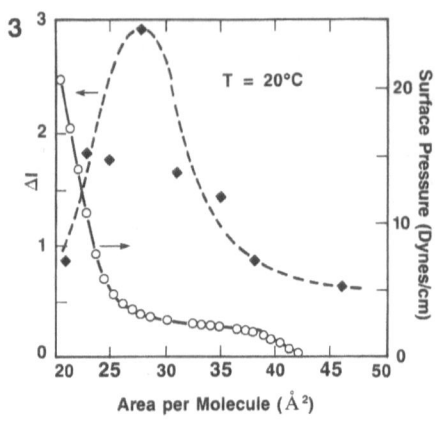

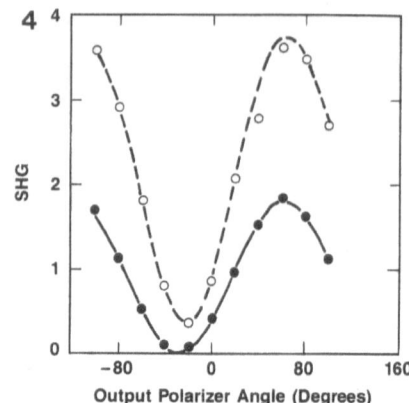

Figure 3. Surface pressure (0, solid line) and observed SH intensity fluctuation (♦ , dashed line) vs. specific surface area per PDA molecule.

Figure 4. SH intensity at LE phase (● solid line) and at a phase transition regime (o dashed line) of a PDA monolayer on water as a function of an output polarizer angle.

are smaller and do not merge into a uniform phase, but rather form an aggregate. The results do not support the distinction between a LC and S phase.

ACKNOWLEDGEMENTS

This work was partially supported by the Director, Office of Energy Research, Office of basic Energy Sciences, Materials Sciences Division of the U.S. Department of Energy under Contract No. DE-AC03-76SF00098.
* Present address: Univ. of Nijmegen, Toernooiveld, 6525 ED Nijmegen, The Netherlands.

REFERENCES

1. J. M. Kosterlitz and D. J. Thouless, J. Phys. C6, 1181 (1973); B. I. Halperin and D. R. Nelson, Phys. Rev. Lett. 41, 121 (1978); D. R. Nelson and B. I. Halperin, Phys. Rev. B19, 2457 (1979).
2. See, for instance, V. I. Imry, in Chemistry and Physics of Solid Surface IV, edited by R. Vanselow and R. Howe (Springer-Verlag, N.Y. 1982), p. 461 and references therein.
3. M. W. Kim and D. S. Cannell, Phys. Rev. Lett. 35, 889 (1975).
4. For a review, see, e.g., G. M. Bell, L. L. Coombs and L. J. Dunne, Chem. Rev. 81, 15 (1981).
5. Th. Rasing, H. Hsiung, Y. R. Shen and M. W. Kim, Phys. Rev. (A) 37, 2732 (1988).
6. Y. R. Shen, The principles of nonlinear optics (Wiley, N.Y. 1984).
7. Th. Rasing, Y. R. Shen, M. W. Kim and S. Grubb, Phys. Rev. Lett. 55, 2903 (1985).
8. T. F. Heinz, thesis, University of California, Berkeley, 1982 (unpublished).
9. G. E. Gaines, Jr., Insoluble Monolayers at Liquid Gas Interfaces (Wiley, N.Y. 1966).

DYNAMICS OF WETTING

Jean-François Joanny

E.N.S.   46Allée d'Italie
69364 Lyon Cedex 07 France

I INTRODUCTION

Wetting and spreading phenomena are classical areas which have known a renewed interest over the last few years due both to their important industrial applications and to some recent theoretical and experimental developments.

The classical analysis of Young and Laplace[1] of static wetting problems rests on the characterization of each interface by a macroscopic surface tension. At the intersection of three bulk phases, the three phase contact line is at rest only if the capillary forces represented by these surface tensions balance. When the three phases are a solid substrate S, a wetting liquid L and a vapor V, the mechanical equilibrium condition parallel to the solid gives the Young-Dupré equation for the contact angle θ_0

$$\gamma_{SV} - \gamma_{SL} = \gamma_{LV} \cos \theta_0 \qquad (1)$$

where γ_{ij} is the interfacial tension between phases i and j.

This construction however requires a partial wetting condition that can be expressed on the sign of the spreading power which is the difference of interfacial tensions

$$S = \gamma_{SV} - \gamma_{SL} - \gamma_{LV} \qquad (2)$$

The contact angle given by the Young construction exists only if S is negative.

If the spreading power S is positive, the wetting liquid spreads onto the substrate and forms a thin film of microscopic thickness. The so-called Antonov rule states then that the spreading power never is strictly positive and thus vanishes in a complete wetting situation. This is certainly true for volatile liquids with which one cannot make a solid-vapor interface without having a liquid film wetting the interface by recondensation of the vapor, leading to S=0. For non volatile liquids the vapor is inert and cannot recondense on the solid substrate ; a finite volume of liquid after spreading covers only a finite area of the solid. The spreading power is here well defined and in general has a positive value. In the following we limit our study to non volatile liquids, typical examples being heavy oils or polymeric liquids although polymers sometimes have a peculiar hydrodynamic behavior.

The thickness of the wetting liquid film (which has sometimes been called a pancake) results from a competition between the capillary forces (spreading power) which tends to spread the film and the disjoining pressure characterizing the long range character of the molecular interactions which in general tend to thicken the film.

Several other factors are known to modify this ideal wetting picture. Heterogeneities of the solid surface, such as roughness or chemical contamination lead to contact angle hysteresis in incomplete wetting situations : the measured contact angle is not the Young angle but depends on the history of the liquid. A larger angle is measured in an advancing experiment and a smaller angle is measured in a receding experiment. This phenomenon is also associated with a wiggly contact line. In the complete wetting situation, the heterogeneities may lead to the formation of holes in the wetting film.

Hydrodynamic motions of the contact line also have a strong influence on the wetting behavior. Close to the contact line, the liquid thickness is very small and the viscous dissipation is singular : viscous forces must be taken into account in a force balance on the contact line as for instance given by the Young equation. This has remained an unsolved theoretical problem for a long time because of the apparent contradiction between the advancing motion of the contact line and the no slip hydrodynamic boundary condition for the liquid at the solid surface : this paradox is solved by a rolling motion of the spreading liquid on the solid surface[2].

Even in a complete wetting situation, the macroscopic dynamic contact angle has then a finite value which has been studied in great details both theoretically and experimentally. The properties of the liquid in the vicinity of the contact line are not however well described by macroscopic concepts, it is now well established that an advancing liquid front is preceded by a microscopic precursor film. The existence of this film has been explained first by deGennes[3] and Teletzke et al.[4] by the thickenning effect of the long range molecular interactions (disjoining pressure).

In an incomplete wetting situation the disjoining pressure is important in the very vicinity of the contact line (within a few angstöms) and as in the static case, the major effect is the importance of the heterogeneities of the solid surface. The motion of a contact line on a heterogeneous surface has been studied recently and is only partially understood.

In this short review, we discuss recent theoretical developments of these dynamic wetting phenomena. The paper is divided into two main sections :
- the first section presents the basic mechanisms which govern the spreading of a liquid on a solid surface both in situations of complete and incomplete wetting. In the complete wetting situation we emphasize the role of long range forces and the disjoining pressure, in the incomplete wetting situation, we emphasize the role of the heterogeneities of the solid surface.
-the second section presents some generalizations and applications of these basic principles. We first discuss the role of the viscous dissipation in the vapor (or when the vapor is replaced by an immiscible viscous liquid) and the spreading on a liquid substrate. Finally we study three situations where the spreading is unstable and leads to hydrodynamic instabilities, a Hele-Shaw cell with open boundaries, a rotating drop spreading under the action of the centrifugal force and a film falling down a plane under the action of gravity.

Several very detailed reviews on dynamic wetting phenomena[5,6,7,8,9] have been published over the last few years, and we will give here only a summary of the results and the main physical ideas. The reader is referred to these longer reviews for more detailed calculations and for a complete description of the recent experimental results.

II SPREADING KINETICS

1-Complete wetting

a-Disjoining pressure, pancake thickness

The free energy of a macroscopic liquid film on a solid surface is the sum of the two interfacial tensions $\gamma_{SL} + \gamma_{LV}$. The molecular interactions being long ranged, the physical properties of microscopic films cannot be described only in terms of macroscopic quantities such as the surface tensions. The free energy per unit surface of a macroscopic film varies with its thickness h and is equal to $\gamma_{SL} + \gamma_{LV} + P(h)$ where P is the component of the free energy due to the long range forces. The microscopic film has a higher energy and tends to thicken if $P(h)>0$.

Rather than the surface free energy $P(h)$ it is common to use the disjoining pressure Π defined as

$$\Pi(h) = -\frac{dP}{dh} \tag{3}$$

The disjoining pressure of various thin liquid films has been studied quite extensively in particular by the Russian school of Derjaguin[10]. Several contributions are in general separated, the Van der Waals component, the electrostatic component and the structural component. In most of the following we will take as our example Van der Waals liquids where the disjoining presure decays as

$$\Pi(h) = \frac{A}{6\pi h^3} \tag{4}$$

A is the negative of the usual Hamaker constant and will be chosen positive : the disjoining pressure has a thickening effect.

A non volatile liquid completely wetting a solid substrate forms a very thin pancake covering an area A with a uniform thickness e[11]. The energy of the pancake is the sum of the capillary and disjoining pressure contributions

$$F(e) = A\{-S + P(e)\} \tag{5}$$

The equilibrium thickness is obtained by minimizing this energy while keeping the total liquid volume Ae constant, it is given by

$$S = P(e) + e\,\Pi(e) \tag{6}$$

For Van der Waals liquids

$$e = (\frac{A}{4\pi S})^{1/2} \tag{7}$$

In usual situations this thickness is of the order of a molecular size. It may be larger in the vicinity of a wetting transition where the spreading power S vanishes.

b-Spreading kinetics, steady state motion

We discuss first a steady state experiment where the three phase contact line moves at a constant velocity U on the solid substrate. Whenever the capillary number $\frac{\eta|U|}{\gamma_{LV}}$ is small (η is the liquid viscosity), the hydrodynamic motion of the liquid may be studied within the framework of the so-called lubrication approximation[12]. The equation of motion is then of the Darcy form : the advancing velocity is proportional to the pressure gradient

$$\eta U = -\frac{1}{3} h^2 \nabla P \tag{8}$$

h(x) being the local thickness of the liquid film.

As in the static case, the pressure has two contributions, the Laplace pressure and the disjoining pressure

$$P = P_V - \gamma_{LV} \nabla^2 h - \Pi(h) \qquad (9)$$

where P_V is the pressure in the vapor phase.

i-Advancing liquid[3]

We first assume that the liquid is advancing to the left (negative x) with a velocity $U = -|U|$, the liquid front may then be divided into three regions : a macroscopic wedge at large liquid thicknesses where the disjoining pressure is small compared to the Laplace pressure, a precursor film at smaller thicknesses where the disjoining pressure dominates over the Laplace pressure and a static region at the very tip of the precursor film where viscous effects are negligible.

The macroscopic region is with a very good approximation a wedge characterized by a well defined dynamic contact angle $\theta_d = \dfrac{dh}{dx}$ that can be determined from the equation of motion (8) neglecting the disjoining pressure

$$\theta_d = (9 \frac{\eta |U|}{\gamma_{LV}})^{1/3} \text{Log}^{1/3} k x \qquad (10)$$

where k is a short distance cutoff.

This law often referred to as Tanner law[13] seems well verified experimentally[5]. One should notice that the macroscopic contact angle θ_d does not depend on the spreading power S and is thus independent of the precize state of the solid surface (as soon as there is complete wetting) ; it is in particular not affected by heterogeneities of the solid surface.

In the precursor film, the Laplace pressure may be neglected and for Van der Waals liquids, the film thickness decays smoothly as

$$h = - \frac{A}{6\pi\eta|U|(x-x_0)} \qquad (11)$$

A matching of this profile with the wedge determines the crossover point between the precursor film and the macroscopic region. This crossover occurs at a thickness $h_0 \sim a/\theta_d$ and $x_0 \sim k^{-1} \sim a/\theta_d^2$ (a is a microscopic length, $a^2 = \dfrac{A}{6\pi\gamma_{LV}}$). The maximum thickness of the precursor film h_0 is larger than a molecular thickness only if $\theta_d \ll 1$. Precursor films only exist for small dynamic contact angles.

The precursor film cannot be thinner than the equilibrium static value e(S) given by equation (6) ; when the thickness reaches this value, it decays rapidly to zero at the nominal contact line. This leads to a length l of the precursor film increasing with the spreading power as

$$l \sim \frac{S \, e(S)}{\eta|U|} \qquad (12)$$

Although the existence of precursor films has been known experimentally for a long time[14] and precursor films have now been studied experimentally quite precizely[15], there does not seem to be any clear experimental evidence for a thickness decaying as predicted by equation (11).

ii-Receding liquid

We now assume that the liquid is receding towards positive x values and has a velocity U= IUI. As in a static experiment, the contact angle is exactly zero and the liquid leaves behind a film with a macroscopic thickness d. The value of d explicitly depends on the macroscopic length scale of the flow such as a gravity capillary length or a capillary radius that imposes the curvature b^{-1} of the liquid-vapor interface far away from the contact line. The film thickness d has been calculated by Landau and Levich[16] for a liquid moving vertically along a plate starting from the equation of motion (8) and including gravity

$$d \sim b \left(\frac{\eta |U|}{\gamma_{LV}}\right)^{2/3} \qquad (13)$$

At low velocities, the film drains out and the thickness is very small, if the velocity is too small, the film becomes microscopic and the effect of the disjoining pressure should be included.

c-Droplet spreading

Many of the spreading experiments are not made in the steady state geometry but study the spreading kinetics of droplets on solid surfaces. The early stages of the spreading are very fast and dominated by inertial effects. At later times, viscous effects become dominant and the equations governing the spreading are the Darcy law (8) and the conservation equation for the liquid

$$\frac{\partial h}{\partial t} + \nabla hU = 0 \qquad (14)$$

where U is now the local velocity.

As for the steady state experiment, the liquid profile may be divided into a macroscopic drop where the disjoining pressure has a small effect and a precursor film where capillary effects are negligible.

The macroscopic droplet may be studied quantitatively by looking for a self-similar solution of the equations of motion. This profile has one single characteristic length scale along the solid surface, the apparent radius R(t). Close to the edge the macroscopic drop has a well-defined contact angle which is related to the local velocity $\frac{dR}{dt}$ by Tanner law (10)

$$\theta_d = (9 \eta \frac{dR}{dt}/\gamma_L V)^{1/3} \qquad (15)$$

A second relation between the contact angle and the radius R is obtained by noting that the drop is in a first approximation spherical (in the macroscopic region the flow is easy and the pressure equilibrates rapidly) ; the volume Ω is such that

$$\Omega \sim \theta_d R^3 \qquad (16)$$

This leads to a radius increasing with time

$$R(t) \sim \Omega^{3/10} \left(\frac{\gamma_L Vt}{\eta}\right)^{1/10} \qquad (17)$$

This result is in good agreement with many experimental results[5].

In the early stages, the precursor follows film adiabatically the edge of the drop and its properties are given by the steady state results (11-12) with a velocity $\frac{dR}{dt}$. Its length in this adiabatic regime increases much faster than the radius and rapidly becomes larger than $R(t)$. The adiabatic approximation then fails and one should use the full equations of motion to study the precursor film. Substitution of the equation of motion (8) where we neglect the Laplace pressure into the conservation equation (14) leads to a non linear diffusion equation[17] for the thickness h with a diffusion constant

$$D(h) = \frac{A}{6\pi\eta h} \qquad (18)$$

The length of the precursor film has thus a diffusive growth with time with the fastest possible diffusion constant corresponding to the minimum thickness $e(S)$

$$l^2 = D(e)\, t \qquad (19)$$

The fraction of the liquid volume that is in the precursor film becomes larger as the spreading proceeds, the macroscopic drop empties in the film to reach its final stage, the static pancake.

2-Incomplete wetting

a-Contact angle hysteresis

Contact angle hysteresis is due to heterogeneities of the solid surface which in general have two origins : roughness and chemical contamination.

The early studies of contact angle hysteresis were concerned with periodic roughness[18]. On a rough surface, a contact line has several equilibrium positions corresponding to places where the angle between the interface and the local solid surface is given by Young law. These various positions correspond to different macroscopic contact angles (defined with respect to the average planar solid surface). For stability reasons the macroscopic contact angle must increase if the liquid advances and the advancing angle θ_A is the largest possible angle ; correspondingly, the receding angle θ_R is the lowest possible angle.

We consider here chemical heterogeneities that can be modelled by local variations of the spreading power $S = S_0 + H(x)$. To explain contact angle hysteresis, two models have been proposed : in the first model the heterogeneity is represented by strong localized defects in the dilute limit[19] ; in the second model the heterogeneity is random[20-21].

A localized defect pins the contact line wich locally deforms in the vicinity of the defect. When the contact line is moved away from the defect, if the deformation is too large, it depins from the defect and the capillary energy stored in the defect W_d is dissipated in viscous energy.

The combined effect of all defects leads to a contact angle hysteresis

$$\gamma_{LV}\cos\theta_R - \gamma_{LV}\cos\theta_A = nW_d \qquad (20)$$

where n is the defect density.

On a random surface, the contact line distorts in order to avoid the less wetting regions and has a large number of metastable positions. Wetting of random surfaces turns out to be a problem quite similar to so-called random field problems in the stastical mechanics of disordered systems (random field Ising model, charged density waves pinned by impurities, vortices in type II superconductors...). The contact angle hysteresis may be analysed following the ideas introduced by Grinstein and Ma[22] for the random field Ising model and is proportional to the square of the heterogeneity.

b-Spreading kinetics on a perfect solid surface
We consider an incomplete wetting situation in the limit where the static contact angle θ_0 is small. The lubrication approximation may still be used in this limit and the equation of motion of the liquid in a steady state is given by a Darcy law (equation 8) with a pressure gradient given by (9).

If the capillary number is very small, both for an advancing and a receding liquid, the profile of the liquid-vapor interface may be expanded around its static value : $h = \theta_0 x + h_1(x)$

There is no precursor film in this situation of incomplete wetting and we separate the profile into two different regions a macroscopic region where the disjoining pressure is small and a static region where the viscous friction is small.

The macroscopic region is a wedge with a dynamic contact angle $\theta_d = \theta_0 + \delta\theta$ (where $\delta\theta = \dfrac{dh_1}{dx}$) varying linearly with the velocity

$$\delta\theta = -\frac{3\eta U}{\gamma_{LV}\theta_0^2} \text{Log } x/x_0 \qquad (21)$$

The static region has a size x_0 of order a/θ_0^2 over which the liquid thickness decays to zero.

When the capillary number $\dfrac{\eta|U|}{\gamma_{LV}}$ increases and becomes larger than θ_0^3 the correction to the static profile is large and should not be treated as a perturbation. For an advancing liquid the dynamic contact angle increases with the capillary number and is given by the same Tanner law as in a complete wetting situation (equation (10)) with the major difference that there is no precursor film in a case of incomplete wetting. For a receding liquid, the contact angle decreases and jumps discontinuously to zero at a finite value of the capillary number of order θ_0^3.

c-Spreading kinetics on a heterogeneous surface

The motion of a contact line on a heterogeneous solid has not been studied theoretically till recently and is at the present time very partially understood ; only very specific types of heterogeneities have been considered : a capillary tube with a single isolated chemical defect[23] and a plate dipping into a liquid with a periodic chemical heterogeneity parallel to the contact line[24] i.e. a spreading power varying as $S=S_0 + H \cos qx$ where the x axis is along the motion. In both these geometries a qualitative difference has been predicted between experiments where the liquid advances at constant velocity or under a constant force.

In a constant force experiment, the important feature is the slowing down of the contact line when it hits regions with a lower spreading power (less wetting); this leads to a non linear relation between the applied force $F = \gamma_{LV}$ $\cos\theta_d$ and the average velocity

$$F - \gamma_{LV} \cos\theta_A \sim \gamma_{LV} \left(\frac{\eta U}{\gamma_{LV}}\right)^2 \qquad (22)$$

In a constant velocity experiment, the contact line remains pinned in regions with a lower spreading power, the important feature is the depinning mechanism ; for strong heterogeneities, this leads to a relation between the average force and the imposed velocity of the following form

$$F - \gamma_{LV} \cos\theta_A \sim \gamma_{LV} \left(\frac{\eta U}{\gamma_{LV}}\right)^{2/3} \tag{23}$$

These results are very similar to those obtained by Fisher[25] in a mean field theory for another dynamic random field problem, the depinning of charge density waves. One should notice however that the two geometries considered are rather artificial and that although such solid surfaces may certainly be realized experimentally they do not correspond to usual heterogeneous surfaces where the heterogeneity is rather randomly distributed. In such a case the effect of neighboring impurities should average   and one expects that the constant force and the constant velocity experiments are equivalent. The precize relation between force and velocity has not however been worked out in details theoretically.

III- GENERALIZATIONS AND APPLICATIONS

1- Solid-liquid-liquid contact line

The general principles of the previous section were obtained by assuming that the vapor is inert and thus does not recondense on the solid surface but also that it is a non viscous fluid that does not contribute to the spreading hydrodynamics. In many of the practical wetting problems, the external phase is not a vapor but a liquid immiscible with the spreading liquid such as oil and water for instance. The hydrodynamic study must then take into account viscous dissipation in the external phase[26-27].

The solution of the complete hydrodynamic equations is a rather formidable task that has been only very partially achieved. In the limit of small dynamic contact angles, it can be shown however that this viscous dissipation in the external liquid is small and that the contact angle is still given by Tanner law. The basic idea is that there is no characteristic lengthscale in the external liquid and that at a distance x from the contact line, the viscous stress in the external liquid is of the order of

$$\sigma_{ex} \sim \frac{\eta_{ex} U}{x} \tag{24}$$

where η_{ex} is the viscosity of the external liquid.

This is to be compared with the viscous stress in the spreading liquid on the solid surface

$$\sigma = 3\frac{\eta U}{h} \sim \frac{\eta U}{\theta_d x} \tag{25}$$

As soon as the viscosity ratio $M = \dfrac{\eta}{\eta_{ex}}$ is larger than the dynamic contact angle, $\sigma_{ex} \ll \sigma$ and the viscous dissipation in the external liquid is small. If the dynamic contact angle is not small or if the viscosity of the external phase is extremely large (for polymeric liquids), this is not true any more and the dynamic contact angle is a function not only of the capillary number $\dfrac{\eta U}{\gamma_{LV}}$ but also of the viscosity ratio M.

2- Liquid-liquid vapor contact line

Another necessary generalization of the spreading problem is to the case where the substrate is not a solid but an immiscible liquid. Two new features must be taken into account in this geometry [28]:

-the interface between the substrate and the spreading liquid is not flat but adjusts itself in order to minimize its interfacial energy

-the spreading liquid induces a flow in the substrate and the velocity of the liquid-liquid interface does not vanish but has a finite value u_S which must be determined

This modifies the spreading equation that becomes

$$\eta(U-u_S) = \frac{h^2}{3} [\gamma \nabla \nabla^2 h + \nabla \Pi(h)] \qquad (26)$$

where h is the total thickness of the spreading liquid and γ an effective surface tension

$$\gamma^{-1} = \gamma_{LV}^{-1} + \gamma_{SL}^{-1} \qquad (27)$$

the substrate liquid and liquid vapor interfacial tensions act thus in parallel for the spreading kinetics. In the following we discuss only macroscopic properties and thus ignore the disjoining pressure term in equation (26)

If the liquid substrate S has a viscosity η_S much larger than the spreading liquid viscosity η, it may in a first approximation be considered as a solid and the spreading laws are the same as on a solid substrate. We consider here the other limit when $\eta_S \ll \eta$. No steady state could be found for this problem and we thus focus on droplet spreading. For simplicity, we first assume a one-dimensional geometry. The results can then be used for the more realistic two-dimensional problem.

A precize determination of the interface velocity u_S requires the determination of the flow field in the substrate and the equality of the tangential stress in the two phases. This has been done only partially   the discussion here will be limited to scaling laws.

If the viscosity η is very large, the viscous stress in the spreading liquid $\frac{3h(U-u_S)}{h}$ remains finite only if $U=u_S$ which implies that the flow in the spreading liquid is in a first approximation a plug flow. Since the viscous stresses at the S-L interface are equal in both liquids, we determine them in the substrate, and distinguish two different situations.

If the liquid substrate is thin (thinner than the drop radius R) the velocity of the substrate varies over a distance d equal to the substrate thickness. The order of magnitude of the viscous stress is :

$$\sigma_{xz} = \frac{\eta_S U}{d} \qquad (28)$$

The total viscous force on a half drop obtained by integration of this stress must balance the capillary force

$$F_{visc} = \frac{\eta_S U}{d} R = 1/2 \gamma \ \theta_d^2 \qquad (29)$$

This is quite different from Tanner's law. For a real drop of volume Ω, the radius increases with time during the spreading as

$$R(t) \sim (\frac{\gamma \Omega^2_d}{\eta S} t)^{1/8} \tag{30}$$

Notice that this depends only a the substrate viscosity ηS (it obviously is a first order term in an expansion in $\frac{\eta S}{\eta}$ <<1) and that it has the same scaling behavior as that obtained for the spreading polymer melts by Brochard and de Gennes[29]. The substrate allows a finite slip of the spreading liquid on the solid surface. These results seem in reasonably good agreement with recent experimental data[30].

If the liquid substrate is thick, the only characteristic length in the substrate is the drop radius R and the velocity decays over a length R. The viscous stress is of the order of

$$\sigma_{xz} = \frac{\eta S U}{R} \tag{31}$$

The force balance leads to

$$Fvisc = \eta S U = 1/2\gamma \; \theta_d^2 \tag{32}$$

As in Tanner's law, this gives a relation between the local contact angle and the capillary number $\frac{\eta S \; U}{\gamma}$ (calculated also here with the substrate viscosity) but the power law is different.

The radius of a spreading drop scales as :

$$R(t) \sim (\frac{\gamma \Omega^2}{\eta S} t)^{1/7} \tag{33}$$

3-Hele-Shaw cell with open boundaries

We now study the Rayleigh-Taylor instability in a Hele-Shaw cell with open boundaries where the general wetting results introduced above play an important role.

The cell is vertical and the distance betwen the two parallel plates is b. It is filled with a horizontal slab of liquid of height L (L>>b). In contrast to usual Hele-Shaw experiments, both the upper and lower menisci are in contact with air at the the atmospheric pressure[31].

In the lubrication approximation, the falling velocity U is related to the pressure gradient in the liquid by a Darcy law

$$\eta U = -\frac{1}{12} b^2 \; [\frac{dP}{dz} + \rho g] \tag{34}$$

For a steady state motion where both the lower and upper menisci remain flat, the pressure gradient along the vertical z axis is constant. If the contact angles of these two menisci are equal, the pressures at the top and the bottom of the liquid slab are equal and the pressure gradient vanishes.

However, if the liquid partially wets the walls of the cell which are always slightly heterogeneous, one meniscus is advancing and the other is receding, the contact angle hysteresis creates a negative pressure gradient

$$\frac{dP}{dz} = \frac{2\gamma_{LV}}{bL} (\cos\theta_A - \cos\theta_R) \tag{35'}$$

(we have neglected here the variation of the contact angle due to the velocity discussed above).

If the liquid completely wets the wall, the front angle θ_d is the dynamic contact angle given by Tanner law (10) ; the back angle vanishes and the liquid leaves a small film behind of thickness d given by equation (13), the upper meniscus curvature is thus smaller than b ; these two contributions lead to a pressure gradient

$$\frac{dP}{dz} = -\alpha \frac{\gamma_{LV}}{bL} \left(\frac{\eta|U|}{\gamma_{LV}}\right)^{2/3} \tag{35''}$$

where α is a numerical constant of order unity that also includes the logarithmic corrections.

In general, the pressure gradient is a small contribution to the velocity given equation (34) and $\eta U = -\frac{1}{12} b^2 \rho g$. However this term is destabilizing and provokes a Rayleigh-Taylor instability of the lower meniscus.

The linear stability analysis follows the same lines as for the usual closed cell[32]. We study a front with a periodic undulation of wave vector q , $\delta z(y) = z_0$ cosqy exp s(q)t, and calculate the growth rate s(q).

In a situation of incomplete wetting, we obtain

$$s(q) = \frac{|q|\gamma_{LV}}{12\eta} \left[\frac{2b}{L}(\cos\theta_R - \cos\theta_A) - (qb)^2\right] \tag{36}$$

all the modes with small wave vectors are unstable and the fastest growing mode has a wavevector

$$q_c = \frac{2^{1/2}(\cos\theta_R - \cos\theta_A)^{1/2}}{(3Lb)^{1/2}} \tag{37}$$

in a situation of complete wetting

$$s(q) = \frac{|q|\gamma_{LV}}{12\eta} \left[\frac{b}{L}\alpha\left(\frac{\eta|U|}{\gamma_{LV}}\right)^{2/3} - (qb)^2\right] \tag{36'}$$

where we have assumed that $qb/\theta_d \ll 1$; and

$$q_c = \left(\frac{\alpha}{3Lb}\right)^{1/2} \left(\frac{\eta|U|}{\gamma_{LV}}\right)^{1/3} \tag{37'}$$

The important result is that contrary to the usual closed cell experiment, the most unstable wavelength is not proportional to the capillary length $\left(\frac{\gamma_{LV}}{\rho g}\right)^{1/2}$ but explicitly depends on the height of liquid L. With reasonnable orders of magnitude (L=10cm, b=1mm, ρ=1g/cm^3) we obtain a wavelength of the order of 1cm i.e.larger than the capillary length. The difference between the instabilities in an open cell and a closed cell should thus be observed easily experimentally.

4-Spreading driven by body forces

In all the above examples, the spreading of the liquid is driven by capillary forces. Some body forces may however considerably accelerate the spreading : two examples are now discussed, gravity down an inclined plane[33-34] and the centrifugal force due to the spinning of a drop[35-36].

We first study a liquid film spreading down a plane inclined from the horizontal by an angle α. In the lubrication approximation, the equation of motion in the macroscopic region is obtained by adding the gravity contribution to the Darcy law (8)

$$\eta U = \frac{h^2}{3} \left[\rho g \sin\alpha + \gamma_{LV} \nabla \nabla^2 h \right] \tag{38}$$

We have here neglected the hydrostatic pressure contribution and ρ is the fluid density. This equation must be supplemented by the conservation equation (14).

Away from the contact line, the gravitational force dominates over the capillary force and may be neglected. The one dimensional flow has a self-similar solution for the film thickness corresponding to an infinite film perpendicular to the motion

$$h = (\eta/\rho g \ \sin\alpha)^{1/2} x^{1/2} t^{-1/2} \tag{39}$$

This profile ends abruptly at $x = x_N$ which is determined from volume conservation as

$$x_N = (9A^2 \rho g \ \sin\alpha/\eta)^{1/3} t^{1/3} \tag{40}$$

where A is the cross sectional area (the volume of the film per unit length). The thickness at x_N is $h_N = 3A/2x_N$

The real profile does not end abruptly but is smoothed by surface tension. In the very vicinity of the contact line, both surface tension and gravity are important. In this region[37] we take the origin at the contact line and scale the profile as

$$h(t) = h_N z (\xi) \tag{41}$$

where $\xi = \dfrac{x}{l(t)}$ and the unit length l is chosen as

$$l(t) = h_N \left(3 \frac{\eta U}{\gamma_{LV}} \right)^{-1/3} \tag{42}$$

The relevant velocity is the velocity at the edge $U = \dfrac{dx_N}{dt}$

With this scaling and the assumption that $l(t) \ll x_N$ the dimensionless profile z is a solution of

$$z^2 \left(1 - \frac{d^3 z}{d\xi^3} \right) = 0 \tag{43}$$

This profile must match with the gravity region in the limit of large ξ. The limit of z as ξ goes to infinity is thus one.

In the very vicinity of the contact line ($\xi \to 0$) the gravity term is negligible and the profile is a wedge with a contact angle given by Tanner law (10). At intermediate distances $\xi \sim 1$ the profile has a strong maximum and then decays exponentially to 1 in an oscillatory manner. The existence of this bump is of major importance because it is at the origin of a fingering instability of the contact line. Its height depends explicitly on the macroscopic cutoff of the logarithm in Tanner law. A very rough estimate leads to

$$z_{max} \sim Log^{1/2} kl \tag{44}$$

Experimentally this one dimensional profile is not stable and shows a fingering instability. In order to show that this profile is indeed linearly unstable, we perturb the contact line in the direction y perpendicular to the motion by a value $\xi_q = \xi_0(t) \cos q\zeta$ where we use the same scaling along the x and y direction : $\zeta = y/l$. In a first approximation the liquid profile in the vicinity of the contact line is the one-dimensional profile but with a shifted contact line

$$z'(\xi,\zeta) = z\{ \xi - \xi_q(t)) \} \tag{45}$$

The time evolution of this profile is obtained by writing the conservation of the liquid volume in a slice of the film of infinitesimal width dy

$$\frac{d}{dt} \int h \, dx = - \int dx \, \frac{dj_y}{dy} \tag{46}$$

where h is the profile in the physical units and $j_y = U_y h$ is the lateral flux of liquid (perpendicular to the motion). In the limit of low wavevectors (q small), this y component is the main contribution to the liquid flux.

The flux in the lateral direction is obtained from the equation of motion ; at low wave vectors $Q = q/l$

$$\frac{dj_y}{dy} = - \frac{h^3 \gamma_L V Q^2}{3\eta} \frac{d^2 h}{dx^2} \tag{47}$$

Substituting this in the conservation equation with the profile given by equation (45), we obtain an equation for the growth of the contact line distortion

$$\frac{d\xi_q}{d\tau} = \xi_q \, sq^2 \tag{48}$$

We have introduced here the dimensionless time $\tau = \int \frac{dx_N}{dt} \frac{dt}{l}$ and s is a function of the one-dimensional self-similar profile $z(\xi)$

$$s = \int_0^\infty z(z^2 - 1) \, d\xi \tag{49}$$

The sign of the integral s determines the stability of the profile. If s is negative the perturbation decays, if s is positive the perturbation grows and the profile is unstable. Whenever the bump in the profile z_{max} is much larger than one, the integral is dominated by the region around the maximum of z, s is positive and scales approximately as

$$s \sim \text{Log}^{5/3} kl \qquad (50)$$

This shows that in general the one dimensional profile is unstable to perturbations of the contact line. A more detailed stability analysis has been performed in reference (37) which shows that the wavelength of the fastest growing mode is of the order of the size of the bump l and also varies slightly with the microscopic cutoff k of Tanner law (the thickness of the precursor film).

The mechanism of the fingering instability is thus related to the lateral flux of liquid from the bumpy edge into the fingers that feeds these fingers and make them grow.

One should notice however that these results were obtained under the two important assumptions that the contact angle is small and that the size l of the region around the contact line where the capillary effects are important is small ($l \ll x_N$). If l is larger than the macroscopic lengthscale x_N, only capillary forces are important and the spreading is obviously stable ; this suggests a threshold for the instability when $l \sim x_N$. The instability has been studied experimentally by several authors[33-34], but in none of these experiments is the dynamic contact angle small ; it is not clear whether the mechanism invoked here applies to these cases.

A very similar analysis can be carried out for a drop spreading on a rotating surface. In general the Coriolis force is negligible and the centrifugal force must be added to the spreading equation(10) which becomes

$$\eta U = \frac{h^2}{3} [\rho \omega^2 r + \gamma_{LV} \nabla \nabla^2 h] \qquad (51)$$

A self-similar profile is found where the radius of the drop increases with time as

$$R(t) \sim (\frac{\rho \omega^2 \Omega^2}{3\eta} t)^{1/4} \qquad (52)$$

Ω being the radius of the drop. The profile in the central region is flat

(h=constant). Around the central region there is a small rim of size

$l \sim (\frac{\Omega \gamma_{LV}}{\rho \omega^2 R^3})^{1/3}$ where the capillary pressure is important. Experimentally a

fingering of the contact line is observed. The linear stability analysis is very similar to that of the gravity problem and leads to a most unstable wavelength of the order of l.

AKNOWLEDGEMENTS

I am deeply indebted to D.ANDELMAN, F.BROCHARD, P.G.deGENNES, E.HERBOLZHEIMER, F.MELO, M.ROBBINS, S.SAFRAN, S.TROIAN. Various parts of the work presented here have been done in collaboration with them.

REFERENCES

[1]-J.R.ROWLINSON, B.WIDOM Molecular theory of capillarity (Oxford University Press 1982)

[2]-E.DUSSAN, S.DAVIS Journal of fluid Mechanics 65 71 (1974)

[3]-P.G.deGENNES Reviews of Modern Physics 57 827 (1985)

[4]-G.TELETZKE Thesis University of Minnesota (1982)

[5]-Les Houches Summer school "Liquids at interfaces" to be published 1989 see the lectures by P.G.deGENNES and A.M.CAZABAT

[6]-Revue de Physique Appliquée special issue on "Wetting " June 1988

[7]-J.F.JOANNY Journal of theoretical and applied Mechanics 23 249 (1986)

[8]-A.MARMUR Advances in Colloid and Interface Science 19 75 (1983)

[9]-E.DUSSAN Annual Review of Fluid Mechanics 371 (1979)

[10]-N.CHURAEV in ref 5 page 975, J.N.ISRAELACHVILI Intemolecular and surface forces (Academic Press New-York 1985)

[11]-J.F.JOANNY, P.G.deGENNES C.R.A.S. 299II 279 (1984)

[12]-G.K.BATCHELOR Introduction to fluid dynamics (Cambridge University press 1971)

[13]-L.TANNER J.Physics D 12 1473 (1979)

[14]-H.W.HARDY Phil.Mag. 38 49 (1919)

[15]-L.LEGER et al. Revue de Physique Appliquée 23 1047 (1988)

[16]-L.LANDAU, B.LEVICH Acta Physicochemica USSR 17 42 (1942)

[17]-J.F.JOANNY, P.G.deGENNES J.Physique 47 121 (1986)

[18]-R.COX Journal of Fluid Mechanics 131 1 (1983)

[19]-J.F.JOANNY, P.G.deGENNES J.Chem.Phys. 81 552 (1984)

[20]-M.O.ROBBINS, J.F.JOANNY Europhysics letters 3 729 (1987)

[21]-Y.POMEAU, J.VANNIMENUS J.Colloid and Interface Sci. 104 477 (1985)

[22]-G.GRINSTEIN, S.MA Phys.Rev.B 28 2588 (1983)

[23]-E.RAPHAEL, P.G.deGENNES preprint 1989

[24]-M.O.ROBBINS, J.F.JOANNY in preparation

[25]-D.S.FISHER Phys.Rev.B 31 1396 (1985)

[26]-C.HUH, L.SCRIVEN J.Colloid and Interface Sci. 35 85 (1971)

[27]-J.F.JOANNY, D.ANDELMAN J.Colloid and Interface Sci. 119 451 (1987)

[28]-J.F.JOANNY PhysicoChemicalHydrodynamics 9 183(1987)

[29]-F.BROCHARD, P.G.deGENNES J.Phys.Lett. 45 L479 (1984)

[30]-J.FRAAIJE, A.M.CAZABAT preprint 1989

[31]-K.M.JANSONS J.Fluid Mechanics 163 59 (1986)

[32]-P.PELCE in reference 5

[33]-H.HUPPERT Nature 300 427 (1982)

[34]-N.SILVI, E.DUSSAN Phys.Fluids 28 5 (1985)

[35]-L.TANNER La Recherche 17 184 (1986)

[36]-F.MELO in preparation

[37]-S.TROIAN, S.SAFRAN, E.HERBOLZHEIMER, J.F.JOANNY submitted to Europhysics letters

ESTIMATION OF WETTABILITY IN A COMPLEX POROUS NETWORK

Anatol Winter

Geological Survey of Denmark
Thoravej 8
DK-2400 Copenhagen NV, Denmark

INTRODUCTION

Wettability of a solid-liquid-fluid or solid-liquid-liquid system is defined as the ability of one fluid, in the presence of another, to spread on a solid surface. The degree of wetting of a solid by liquids can be quantified, for instance, by measuring the angle of contact that a liquid-fluid interface makes with a solid. However, such measurements are extremely sensitive to the degree of smoothness and composition of the underlying substrate (cf. Chattoraj & Birdi, 1984; de Gennes, 1985). Consequently, contact angles are inefficient as a tool for evaluations of wettability in porous media suffering both from geometrical chaoticity (pore surfaces are a random object) and from chemical disorder (wetting properties may vary from one point to another).

The strategy for evaluation of wettability in a hydrocarbon reservoir described in this paper is based on the assumption that reservoir rocks are either predominantly water-wet or oil-wet (cf. Brown et al., 1956; Salathiel, 1973; Melrose, 1982).

The suggested evaluation procedure consists of two stages:

a. Estimation of wettability at the level of a single pore.

b. Extension of (a) to the level of a network of capillaries.

At the level of a single pore the strategy introduces a concept of the state of the aqueous phase in a water-filled pore. More specifically, it utilizes the fact that for capillary tubings of a polygonal cross-section there exists a threshold value of the capillary pressure (and, consequently, also of the pore size) separating two different wetting regimes (cf. Joanny, 1985): for the capillary pressure weaker than the threshold value, all the pore surface

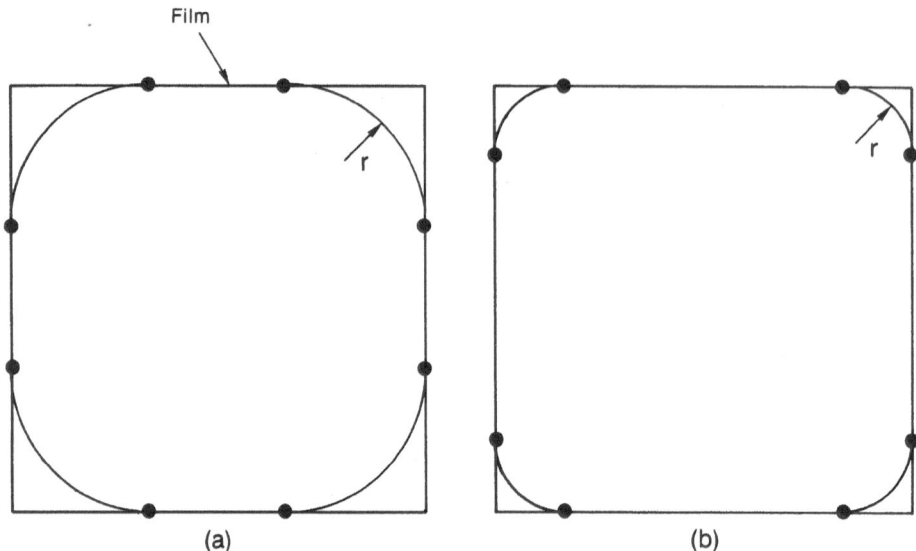

Figure 1. Behaviour of the wetting fluid in a square-sectional capillary: (a) The case of a weak capillary
pressure. All the solid surface is wetted. (b) The
case of a strong capillary pressure. Only corners
of the capillary are wetted.

is wetted by a thin film of the aqueous phase. On the other
hand, below the threshold only corners are wetted (see
Figure 1). It has been shown by Frenkel (1955) that the
wetting transition between the two regimes is controlled by
a stability condition imposed on the aqueous phase: the
gradient of the net force tending to separate the interfaces
between the film and neighbouring phases must be negative in
sign.

At the level of a network of capillaries wettability is
controlled by the pore size distribution function and initial water saturation in the porous network (this is a
fraction of the pore space occuppied by the water phase
prior to the start of hydrocarbon recovery). A method for
determination of the distribution function will be given in
the sequel. The initial water content is usually known from
measurements performed inside drilled wells. For a description of the interplay of physical, physicochemical and
geological mechanisms responsible for establishing the
initial water saturation in a hydrocarbon reservoir the
reader is referred to Winter (1987).

ESTIMATION OF WETTABILITY AT THE LEVEL OF A SINGLE PORE

At the level of a single pore wettability is estimated
by investigating an oil droplet situated in a square-sectional tubing and surrounded by a thin film of the aqueous
phase (see Figure 2).
The wetting film is assumed to be in hydrostatic and chemi-

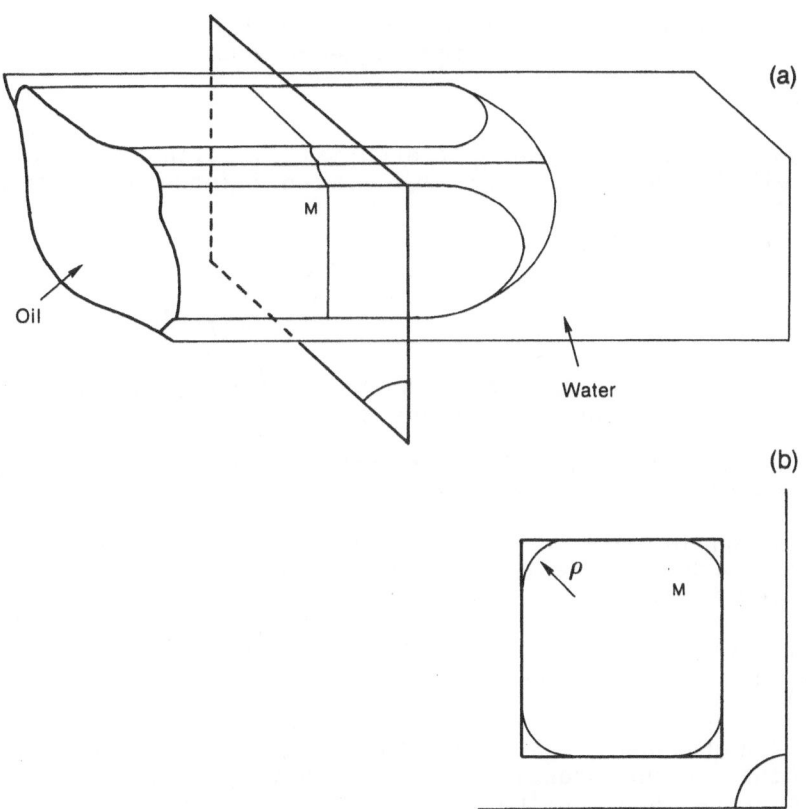

Figure 2. Configuration of the oil-water interface in a square-sectional capillary tubing.

cal equilibrium with the bulk oil phase. Its thickness can be obtained by solving the augmented Young-Laplace equation given by the following expression:

$$\Delta P_c = \sigma \Gamma + \Pi(h) \tag{1}$$

where ΔP_c is the capillary pressure (Pa), h is the film thickness (microns), $\Gamma = 1/R_1 + 1/R_2$ is the mean curvature of the interface, σ is interfacial tension between oil and water (Pa.m) and $\Pi(h)$ represents the disjoining pressure term (Pa).

Disjoining pressure is a macroscopic pressure correction accounting for long-range intermolecular interactions (cf. Deryagin, 1955). For most common solid-liquid interactions, such as the London-van der Waals dispersion forces, the disjoining pressure has the following form:

$$\Pi(h) = A/h^3 \tag{2}$$

239

The adopted value of A = $A_H/6\pi$ = 0.45621x10^{-3} Paμm^3 corresponds to a system consisting of water-quartz-nonane (A_H is the Hamaker constant).

Thickness of the thin film in a corner of the capillary is given by a 2-nd order differential equation obtained from eq. (1) by replacing R_1^{-1} by the well-known formula from analytic geometry:

$$\frac{\Delta P_c - \Pi(h)}{\sigma} - \frac{1}{R_2} = \frac{d^2h/dx^2}{\{1 + (dh/dx)^2\}^{3/2}} \quad (3)$$

Equation (3) has been solved assuming the following boundary conditions (see Figure 3):

a. the oil-water interface is parallel to the porewall in the center of the capillary, i.e. dh/dx = 0,

b. the approximate distance between the film interface and the corner of the tubing , H, is known from experimental investigations.

Thus, one has to resolve a two-point boundary value problem, e.g. by using the method of shooting (cf. Ascher et al., 1988).

The threshold value of the capillary pressure corresponding to the transition from a wetting regime in which the aqueous phase entirely covers porewalls to that in which it appears only in corners of tubings is given by the following expression (cf. Joanny, 1985):

$$\Delta P_c^{crit} = (2S/3)^{3/2}(A/6\pi)^{-1/2} \quad ; \quad S = \sigma_{os} - (\sigma_{ow} + \sigma_{ws}) \quad (4)$$

where S is the spreading coefficient and $\sigma_{\alpha\beta}$ is the interfacial tension between α and β phase ("o" stands for oil, "w" stands for water and "s" is the solid phase).

The final step of the estimation procedure at the pore level is determination of the limiting thin film thickness corresponding to the threshold value of the capillary pressure (eq. 4) using the augmented Young-Laplace equation (3).

ESTIMATION OF WETTABILITY AT THE LEVEL OF A NETWORK OF PORES

As pointed out in the introductory section evaluation of wettability at the level of a network presuposes knowledge of the distribution function of pore sizes. This function can be derived theoretically provided that some assumptions can be made about the structure of the porous medium. One possibility is to assume that the medium is a (deterministic) fractal. A number of recent papers support the hypothesis, that at least some reservoir rocks are indeed fractal (cf. Sen et al., 1981; Katz et al., 1985; Wong, 1987; Hansen et al., 1988).

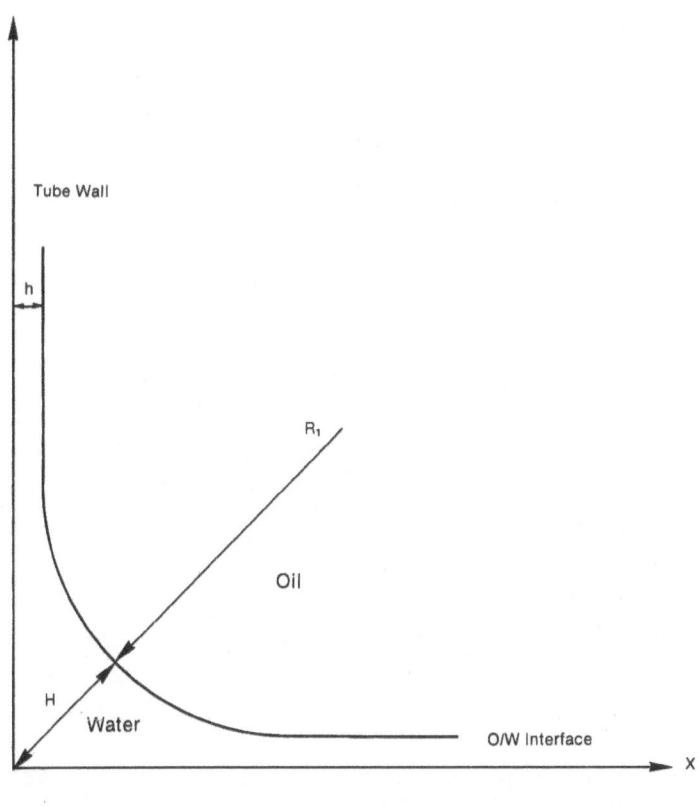

Figure 3. Configuration of the oil-water interface confined
by tube walls and an oil drop constrained in a
square-sectional capillary.

In order to determine the distribution function of pore
sizes consider a fractal porous medium consisting of polygonal grains and square-sectional pores in a certain scale
range, $l_1 \leqslant l \leqslant l_2$ (see Figure 4).
An exact expression for the distribution function can be
found by applying an analog of the transfer matrix method
(cf. Mandelbrot et al., 1985, Mosolov et al., 1987). At each
step of the iterative procedure the porosity, $\emptyset$, of the
fractal porous medium changes as follows:

$$\emptyset_i \longrightarrow (4x + x^2)\emptyset_{i+1} \qquad (5)$$

where $\emptyset_i$ denotes the porosity existing at the i-th step and
$x = 1/3^i$ is the scaling parameter. After N steps of the
iteration the distribution function is given by the coefficient in front of x in the following expression:

$$F_N(x) = (4x + x^2)^N = \sum_{k=0}^{N} \binom{N}{k} 4^{N-k} x^{k+N} \qquad (6)$$

Thus after N iteration steps there remain 4^N largest pores
of size 3^{-N}. In addition, there are N families of smaller

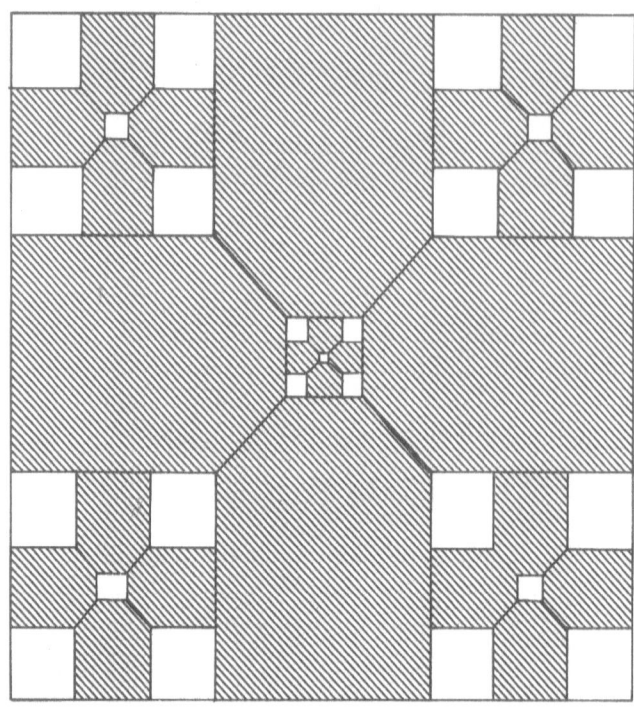

Figure 4. Fractal porous medium with square-sectional pores
and polygonal grains.

pores. Each family contains $\binom{N}{k} 4^{N-k}$ pores of size $3^{-(N+k)}$
($1 \leq k \leq N$).

In the final step of the estimation procedure the
initial water saturation is compared with the critical
saturation defined as the amount of water necessary to fill
all pores smaller than that corresponding to the threshold
value of the capillary pressure, i.e. those pores in which
the wetting films are most likely to appear only in corners
of capillaries. If the initial water saturation is less than
the critical saturation, the reservoir zone under consideration will have been, at least to some extent, depleted
of the wetting films normally covering pore surfaces. The
wetting regime of this type is referred to as the mixed
wettability state (cf. Salathiel, 1973). Needless to say the
shape of the pore size distribution function is a dominating
factor in the evaluation procedure.

CONCLUDING REMARKS

The strategy for the evaluation of wettability described in this paper has several advantanges as compared to
other methods, e.g.:

a. it allows to obtain an in situ estimate of wettability in
   a porous network,

b. it allows to introduce a broad range of wettability pre-
ferences from extremely water-wet to extremely oil-wet.

Much more work is necessary to transform the described
strategy into a useful tool in exploration of hydrocarbon
reservoirs. Among the most important problems requiring
clarification are those of reliable identification of the
dominating solid-liquid interactions and an appropriate
selection of the disjoining pressure term in the augmented
Young-Laplace equation.

REFERENCES

1. Ascher, U.M., Mattheij, R.M., Russell, R.D., 1988, Nu-
    merical Solution of Boundary Value Problems for Ordinary
    Differential Equations, Prentice Hall Inc., Englewood
    Cliffs, New Jersey.
2. Brown, C.E., Fatt, I., 1956, Transactions of the AIME,
    207:262.
3. Chattoraj, D.K., Birdi, K.S., 1984, Adsorption and the
    Gibbs Surface Excess, Plenum Press, New York.
4. De Gennes, P.G., 1985, Rev. Mod. Phys., 57(1):827.
5. Deryagin, B.V., 1955, Colloid Journal of the USSR,
    17:191.
6. Frenkel, J., 1955, Kinetic Theory of Liquids, Dover, New
    York.
7. Hansen, J.P., Skjeltorp, A.T., 1988, Phys. Rev. B,
    38(4):2635.
8. Joanny, J.F., 1985, These, Univ. Paris 6.
9. Katz, A.J., Thompson, A.H., 1985, Phys. Rev. Letters,
    54(12):1325.
10. Mandelbrot, B.B., Gefen, Y., Aharony, A., Peyriere, J.,
    1985, J. Phys. A: Math. Gen., 18:335.
11. Melrose, J.C., 1982, Technical paper no. 10972, Society
    of Petroleum Engineers, Dallas.
12. Mosolov, A.B., Dinariev, O.Yu., 1987, Zh. Tekn. Fiz.,
    57:1679.
13. Salathiel, R.A., 1973 (October),
    J. Petroleum Technology, 1216.
14. Sen, P.N., Scala, C., Cohen, M.H., 1981, Geophysics,
    46(5):781.
15. Winter, A., 1987, in: Migration of Hydrocarbons in Se-
    dimentary Basins, B. Doligez (ed.), Editions Technip,
    Paris.
16. Wong, P., 1987, in: Physics and Chemistry of Porous
    Media, J. Banavar, J. Koplik & K.W. Winkler (eds.),
    American Institute of Physics, New York.

FINGERING INSTABILITY OF A SPREADING DROP

S.M. Troian, X.L. Wu., E. Herbolzheimer and S.A. Safran

Exxon Research and Engineering Co.

Clinton Twp., Rt. 22E, Annandale, New Jersey 08801

Most studies of the dynamics of wetting have focused on the spreading of *pure* non-volatile drops on smooth and *dry* surfaces. For the most part, such spreading processes are well understood, with good agreement between experiment [1,2] and theory [3,4]. In all such studies of spreading on smooth substrates, the macroscopic wetting process, whether spontaneous or forced, proceeds by the movement of a uniform and circular contact line (the boundary where the air, liquid and solid meet). As first discovered by Marmur and Lelah [5], some unusual wetting behavior can result from the addition of surfactant to the spreading liquid. Anionic and non-ionic aqueous surfactant solutions spread on smooth glass by developing fingers, which branch as they grow and leave behind a dendritic-like pattern. The wetting process with surfactant proceeds at a rate which is approximately an order of magnitude faster than the spreading of a pure liquid, although the surface coverage is quite non-uniform. We summarize below some results of an experimental [6] and theoretical study [7] of the effect of adding surfactant to a spreading drop. The interested reader is referred to refs. 6 and 7 for all further details.

Contrary to earlier reports [5], we have confirmed that this novel hydrodynamic instability only occurs for a drop of surfactant solution spreading over a pre-exising film of pure water coating the glass surface. The instability disappears for spreading onto a dry surface. The dependence of the rapid spreading and fingering on the presence of the initial water film (and on the drop surfactant concentration) strongly suggests that the instability is driven by the Marangoni effect [8], which describes flow produced by variations in surface tension. Surface tension gradients, which can be established by gradients in surfactant concentration (as in our experiments) or temperature, cause traction along the air-liquid interface which must be balanced by a shear stress in the fluid. The surface traction induces flow in the interface and adjoining liquid layers in the direction of increasing surface tension.

In our experiments, for example, a 1mM aqueous Aerosol OT (sodium bis(2-ethylhexyl sulfosuccinate) drop has a surface tension of 40 dynes/cm, significantly lower than the surface tension, 73 dynes/cm, of the pre-existing water film. Therefore, upon deposition of the surfactant drop on the pure water film, the Marangoni force draws a layer of liquid out from the drop edge which thins the pre-existing water film as it quickly spreads radially outward. This initial "sheet" of spreading liquid appears to establish a surface tension gradient over a fairly long region ahead of the macroscopic drop. (For low surfactant concentrations, the surface tension is linearly and negatively proportional to the surfactant concentration.) The remainder of the drop then spreads over this thinned film. The velocity of the fingering front is strongly dependent on the initial drop surfactant concentration and weakly dependent on the thickness of the pre-existing water film.

We discuss here only the patterns observed as the thickness of the pre-existing water film is varied and the AOT concentration is held fixed. Fig. 1 shows two typical patterns obtained for spreading of a 1mM drop of aqueous AOT on films of thickness estimated to be (a) $0.1\mu m$ and (b) $1\mu m$. Spreading on the thicker film leads to broader, rounded fingers, while spreading on the thinner film leads to narrow, sharply tipped, more ramified fingers. In all the cases using surfactants we have studied, the fingers undergo tip-splitting, implying there is a preferred finger width in this system. We are investigating whether this width might be set by the competition between the (destabilizing) Marangoni force and the (stabilizing) force of surface tension associated with the curvature of the three-dimensional air-liquid surface.

In addition to tip-splitting, the fingers also undergo shielding and spreading, processes also observed in other fluid flow instabilities [9]. We note, however, that the miscible fluids used (pure water and aqueous surfactant solution with less than 0.1% surfactant by weight) have negligible viscosity difference. In addition, the experiments are performed in an *open cell* geometry with no external pressure gradient forcing movement of the spreading front. These two features rule out the presence of a Saffman-Taylor instability as seen in viscous fingering experiments [10].

To characterize the patterns, we have measured the growth rate of the fingers by defining the radius R(t) of the circular envelope circumscribing the fingers minus the initial radius of the drop. Fig. 2 is a plot of the radius as a function of time. The fingers follow a power law growth in time, whether spreading on a thick or thin film. The proportionality constant is larger for spreading on the thicker water film since viscous dissipation effects are smaller. We have begun to study the fractal dimension of the fingering contours. Preliminary results indicate a perimeter fractal dimension consistent with $D_F \approx 1.7$ for spreading on different thickness films and for all fairly developed patterns.

The Marangoni effect is well understood but most studies have focused on an explanation of the uniform (unperturbed) front flow [11] in the absence of capillary terms or on the instability which gives rise to roll cells when a fluid is heated from above or below [12]. The instability we have modelled occurs at the *moving front* of the spreading drop. Before attempting the stability analysis, the unperturbed (base) flow profile is required of a drop spreading due to a surface tension gradient, including capillary terms. Although the fingering patterns we have seen have features in common with viscous fingers, this system differs from Hele-Shaw flow in two important ways. First, the spreading front is a real free surface and perturbations at the moving edge can cause corresponding changes in the local fluid height. Second, the motion and height of the fluid are strongly coupled to the local concentration distribution of surfactant.

The steady state solution of the two flow equations (in a lubrication approximation) coupling the local fluid height to the local surface tension reveals the presence of three distinct regions in the flow. There exists a region (1) extending across the radius of the macroscopic drop where the capillary term leads to an almost spherical cap shape and a small region (2) near the drop edge in which the droplet shape is smoothly matched to a long region (3) of constant height ahead of the drop. In this third region the surface tension changes from the value of the pure water to the value of the surfactant drop, a change in our experiments of over 30 dynes/cm. Region (1) proves uninteresting since the average flow essentially proceeds as in normal wetting. Region (2) is small in extent and hence accounts for a small fraction of the total change in surface tension. In the stability analysis, this region is collapsed onto an interface separating regions (1) from (3).

In an approximation which ignores height fluctuations in region (3), a simple linear stability analysis about the interface separating regions (1) and (3) indicates that the flow is unstable to perturbations in the position of the interface or in the local surface tension. The stability analysis does not include the

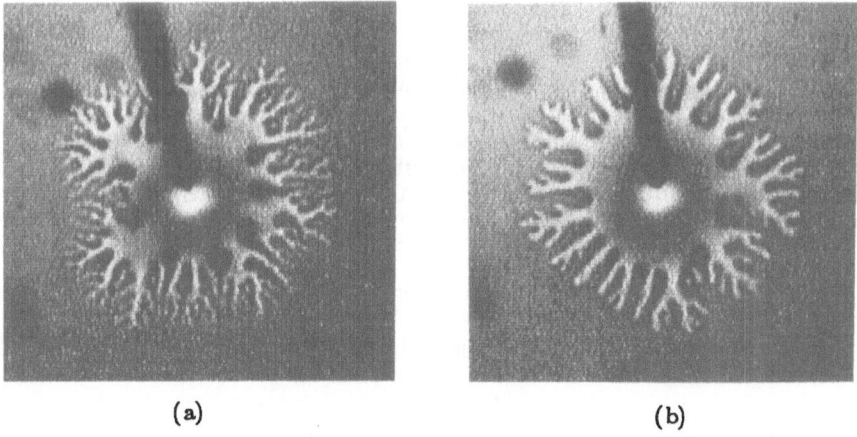

Fig. 1 Fingering of a drop of 1mM aqueous AOT spreading on an (a) thin
($\approx$ 0.1 μm) and (b) thick ($\approx$ 1 μm) water film. The outer radius of the
drops is $\approx$ 0.9 cm. The dark needle corresponds to the syringe tip used for
depositing the drops on the water film.

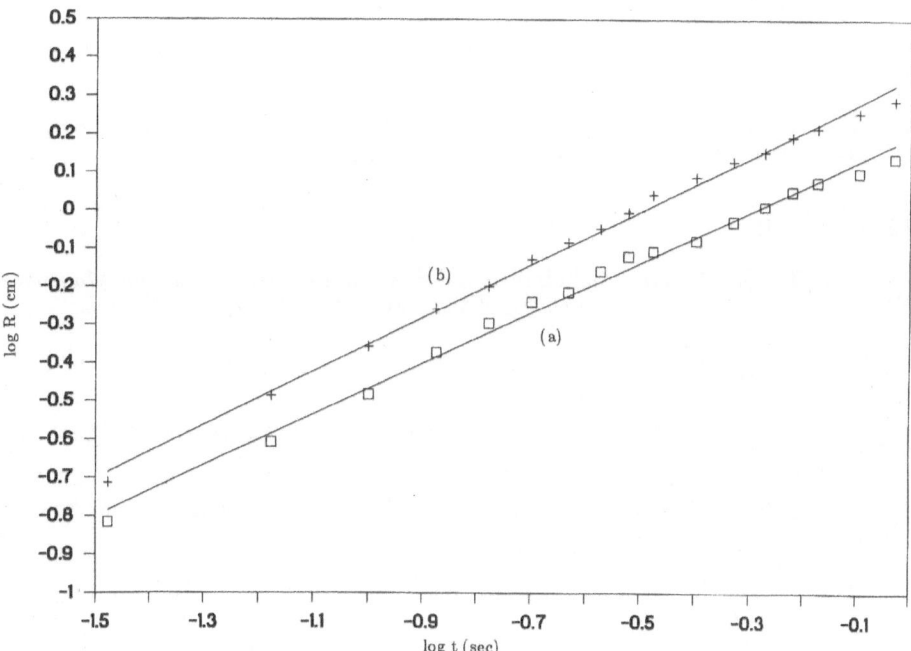

Fig. 2 Log-log plot of the fingering drop radius as a function of time. The
squares and crosses correspond to the spreading of the drops in Fig. 1a and
1b, respectively. The solid line is a linear fit to the experimental data. The
slopes obtained from the fits are (a) 0.66 and (b) 0.70.

presence of a stabilizing mechanism and, therefore, unrealistically predicts that the smallest wavelength disturbances are most unstable. Whether the stabilizing mechanism is the surface tension associated with the curvature of the air-liquid interface needs to be studied in more detail.

The onset of the instability is intimately connected to the presence of the long flat region ahead of the macroscopic drop which sustains a gradient in surface tension. Within the approximations stated above, the local surface tension satisfies Laplace's equation everywhere. In analogy to the Saffman-Taylor instability, a protrusion in the interface experiences a larger surface tension gradient at the tip. Since the Marangoni flow velocity is directly proportional to the gradient in surface tension, the tips move faster than the surrounding environment, and this should lead to fingering at the moving front. This type of instability is not peculiar to surfactant systems and we expect it to occur in many other thin film systems able to sustain surface tension gradients.

References

1. P. Levinson, A.M. Cazabat, M.A. Cohen Stuart, F. Heslot, and S. Nicolet, *The Spreading of Macroscopic Droplets,* Revue Phys. Appl. **23:** 1009 (1988).

2. L. Leger, M. Erman, A.M. Guinet-Picart, D. Ausserre, C. Strazielle, J.J. Benattar, F. Rieutord, J. Daillant, and L. Bosio, *Spreading of non volatile liquids on smooth solid surfaces; the role of long range forces,* Revue Phys. Appl. **23:** 1047 (1988).

3. P.G. de Gennes, *Wetting: Statics and Dynamics,* Rev. Mod. Phys. **57:** 827 (1985).

4. J.F. Joanny, *Dynamics of wetting: Interface profile of a spreading liquid,* J. Theor. Appl. Mech., special issue, 249 (1986).

5. A. Marmur and M. D. Lelah, *The Spreading of Aqueous Surfactant Solutions on Glass,* Chem. Eng. Comm. **13:** 133 (1981).

6. S.M. Troian, X.L. Wu. and S.A. Safran, *Fingering Instability in Thin Wetting Films,* Phys. Rev. Lett. **62:** 1496 (1989).

7. S.M. Troian, E. Herbolzheimer, and S.A. Safran, *Theory of the Marangoni Fingering Instability in Thin Spreading Films,* submitted to Phys. Rev. Lett.

8. V.G. Levich, "Physico-Chemical Hydrodynamics", Prentice-Hall, Englewood Cliffs, N.J. (1962).

9. G.M. Homsy, *Viscous Fingering in Porous Media,* Ann. Rev. Fluid Mech. **19:** 271 (1987).

10. P.G. Saffman, *Viscous Fingering in Hele-Shaw cells,* J. Fluid Mech. **173:** 73 (1986).

11. M.S. Borgas and J.B. Grotberg, *Monolayer flow on a thin film,* J. Fluid Mech. **193:** 151 (1988).

12. J.R.A. Pearson, *On convection cells induced by surface tension,* J. Fluid Mech. **4:** 489 (1958).

RIGID AND FLUCTUATING SURFACES: A SERIES OF SYNCHROTRON X-RAY

SCATTERING STUDIES OF INTERACTING STACKED MEMBRANES

C. R. Safinya

Exxon Research and Engineering Company
Route 22 East
Annandale, NJ   08801

ABSTRACT

In these lectures we discuss fluctuation phenomena encountered in interacting multilayered fluid membranes using synchrotron x-ray scattering as the primary tool. We consider very dilute L_α phases with interlayer separations as large as ≈ 600 Å. While most L_α phases consist of flat membranes with large bending rigidity $k_c \gg k_B T$, with their interlayer interactions determined by detailed microscopic interactions such as hydration and van der Waals, the stability of these phases is due to an effectively long-range interaction arising from the mutual hinderance of fluctuating membranes with a very small rigidity $k_c \approx k_B T$. This regime, which because of its entropic origin exhibits universality, can be accessed from the microscopic regime by thinning and thus lowering the modulus k_c of an initially rigid membrane.

I.   INTRODUCTION

The understanding of the surface properties of both fluid and more ordered membranes has recently attracted much experimental and theoretical attention.[1-13,21-29] To a large degree, this focus of attention stems from the inherent interest in elucidating the statistical physics of two-dimensional random surfaces. More generally, multilayered membranes which are lyotropic liquid crystals, are of high scientific interest because they are prototype models for elucidating the nature of phases and their phase transitions in two-dimensional systems.

The surfaces that we consider are made of surfactants which are amphipathic molecules. They contain a hydrophilic (polar) head group and a hydrophobic (oily) hydrocarbon tail (or tails) shown schematically in Fig. 1. While the polar head "likes" water, the tails will be expelled from it (because of the hydrophobic interaction). For example, surfactants or lipids (e.g. biological surfactants with usually two tails) in water may self-assemble to form a bimolecular sheet which we call a membrane (Fig. 1a). A multilayer system then consists of stacks of alternating membranes and water as we show in Fig. 1a. If the constituent molecules are allowed to diffuse freely inside each two-dimensional bilayer (referred to as a fluid membrane), then the resulting multilayered structure is in the lamellar L_α phase (which is in fact

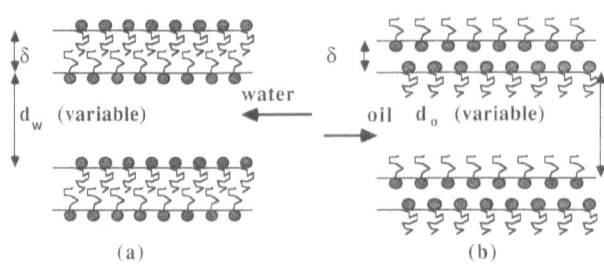

Fig. 1. Top shows schematic of the surfactant SDS. 1a and 1b show two membrane layers separated by either water or oil.

the most biologically relevant state of the membrane). From a structural view point, the L_α phase has the same symmetry as the smectic-A phase of thermotropic liquid crystals. While most biologically active membranes consist of lipids and proteins (e.g. biopolymers which function as receptors, molecular pumps, enzymes, and so on), we will be studying model membranes consisting of one or two types of surfactants. Now suppose one adds a third oily component (say a simple straight chain oil like dodecane) to the binary surfactant-water system. Then the resulting lamellar phase will have an "inverted" membrane structure (consisting of a thin water layer coated with surfactant) as shown in Fig. 1b. In these lectures, when we talk about a "water (oil) dilution", we mean we are following a water (oil) dilution path in the phase diagram where the intermembrane separation $d_w(d_o)$ increases but the membrane thickness δ remains approximately constant (Figs. 1a and b).

In some sense the physical properties of fluid membranes are unique because they have negligible surface tension. Consequently, their free energy is governed by their geometrical shape and its fluctuations. The rigidity k_c associated with the restoring force to layer bending is then the important modulus which in many cases will determine the physical state of the membrane. From a biophysical view point the physical nature of a fluid membrane surface may in some cases have a profound influence on the precise mechanism of membrane-membrane interactions; which influence processes such as cell-cell contact.

In usual biological multilayer membrane systems composed of a single type of lipid, the bending modulus has been measured[6] to be between 20 $k_B T$ and 40 $k_B T$ in the fluid L_α phase. This implies that these rigid interfaces with $k_c \gg k_B T$ are flat and possess macroscopic persistence lengths[7] of order microns so that thermal fluctuations on significantly smaller length scales are not important.[8] The assumption of flat interfaces is also borne out by numerous experiments. For example, in studies[9-11] designed to measure the interaction of two such membranes the free energy is consistent with that between two infinitely flat charge neutral membranes interacting via electrodynamic (van der Waals) and hydration forces (classical molecular regime). In these flat multimembrane systems the attractive van der Waals forces prevent the layers from separating more than 20 Å to 30 Å.

In these lectures, we shall only briefly talk about ordered membranes (with various degrees of positional, orientational bond, and molecular tilt order). We shall also not discuss a class of ordered membranes, which has been the focus of recent elegant theoretial work (see ref. 13, articles by Kantor et. al., Nelson, and Aronovitz and Lubensky), for which at high enough temperature and low rigidity, one may expect a crumpling transition. Rather, our emphasis will be on fluid membranes. This is because from an experimental view point, while most fluid and ordered membranes which have been studied to date are rigid with $k_c \gg k_B T$, there has been a recent important discovery[1-5] of a new class of flexible fluid membranes with low rigidity $k_c \approx k_B T$. We expect then that thermal fluctuations play an important role in our understanding of the statistical physics of these membranes. Furthermore, as we shall describe, when these fluctuating membranes are stacked in multilayers they lead to a new regime of stability for the L_α phase where very large dilutions with intermembrane distances of several hundred (≈ 600 Å) Angstroms,[1-5] and even of a few thousand Angstroms have been reported.[4] These dilute L_α phases are important models for enhancing our current understanding of various physical systems whose descriptions are based on continuum statistical mechanical considerations. We will summarize the principal results of the work done on these phases.

First, we will review recent synchrotron x-ray scattering experiments[2,3,5] which show that the thermodynamic stability of these dilute phases is due to an effectively long range repulsive interaction, first elucidated by Helfrich[12] (which we refer to as undulation forces), arising from the mutual hindrance of fluctuating membranes with a very small bending rigidity $k_c \approx k_B T$. This undulation interaction arises from the difference in entropy between a fluctuating "free" membrane and a "bound" membrane in a multilayer system. The strength of this interaction scales as k_c^{-1}, and as we shall see for flexible membranes ($k_c \approx k_B T$), it completely overwhelms the attractive van der Waals interaction and stabilizes the membranes at large separations. This new regime is distinct from the classical regime in which largely separated membrane sheets are stabilized because of their mutual electrostatic repulsion. Furthermore, because the undulation forces are entropically driven, this interaction is universal with its dependence determined entirely by geometric and elastic parameters such as the layer spacing and the layer rigidity. We shall show that this universal regime consisting of fluctuating (flexible) membranes can be accessed by thinning and thus dramatically lowering the modulus k_c of an initially rigid membrane. Using renormalization group techniques R. Lipowsky and S. Leibler[13] have in fact demonstrated that in the very strong fluctuating regime, this effective long range repulsive force may lead to a complete unbinding of membranes. On a more general level this type of interaction is found in other condensed matter systems. For example, the dominant contribution to the free energy of a polymer confined between walls[14] or that associated with wandering walls of incommensurate phases[15] are similarly entropic in origin.

Second, these dilute L_α phases allow for an obvious demonstration of the Landau-Peierls[16] effect; which is, that the mean square layer displacements of a system of stacked fluid layers diverge logarithmically, $\langle u^2 \rangle \sim \ell n(L/a)$, with sample size L, destroying conventional long-range order (a is of order of the intermolecular distance). For the x-ray structure factor, the consequences are dramatic: Caillé[17] has shown that conventional δ-function Bragg peaks are replaced by singularities characterized by a power-law exponent η_m which is a function of the bulk elastic moduli and the harmonic order m, of the L_α phase. We shall see that the same thermal fluctuation effects which are responsible for the stability of these dilute L_α phases, also lead to an unambiguously large Landau-Peierls effect. This is in contrast to the much smaller effects observed[18] in the smectic-A phase which has a similar structure as the L_α phase. Furthermore the theoretical prediction of the scaling of the structure factor power-law η_m with m^2, which constitutes a severe test of the theory, has only been possible for the L_α phase (in most thermotropic smectic-A phases only the lowest order first harmonic is observable).

Third, dilute L_α phases can be regarded as one-dimensional colloidal suspensions of sheets in analogy to the familiar three-dimensional suspensions of charged polystyrene spheres[19] (e.g. poly-balls). We shall see that the Poisson-Boltzmann equation in one-dimension accurately describes the intermembrane interactions for L_α phases where the dilution is a consequence of long range electrostatic repulsion (rather than undulation forces). This happens when charged sheets are separated by water containing only the counter-ions.

In these lectures we present synchrotron x-ray scattering data for two lipidic and three surfactant systems in their multilayered phases. In the first surfactant system[2] the neutral membrane consists of water layers coated with a mixture of the surfactant sodium-dodecyl-sulfate (SDS) and the cosurfactant (pentanol). The layers are separated by dodecane (i.e. an oil dilution study). In the second and third surfactant systems,[3] the negatively charged membranes consist of a mixture of SDS and pentanol, while the solvent separating the membranes is either pure water or brine ($\simeq 0.5$ mole ℓ^{-1} of NaCl). The lipidic systems consists of a ternary mixture of di-myristoyl-phosphotidyl-choline (DMPC), water, and the cosurfactant pentanol,[5] and the binary DMPC-water system.[11,27]

Before we begin our discussion of the experimental technique and the results (sections III and IV), we consider a phenomenological description of the free energy of the fluid surfactant membranes (interfaces) we shall be discussing, and also of the interactions between membranes in multilayered structures.

II. FREE ENERGY OF FLUID MEMBRANES AND INTERMEMBRANE INTERACTIONS

II.1. Flat Interfaces with $k_c \gg k_B T$

Let us take a water-oil interface in the x-y plane separated by surfactant molecules as shown in Fig. 2. We consider a surface of total area A consisting of N_s surfactants each with an area per polar head of S so ($A = N_s S$). For now we ignore deviations (height fluctuations) of the interface from the x-y plane. Following DeGennes and Taupin[7] we can write down a simple expression for the free energy of the interface:

$$F_i = N_s G(S) + A\gamma_o, \tag{1}$$

where the second term is just the bare interfacial free energy in the absence of surfactants (when $N_s = 0$), where γ_o ($\simeq 50$ dyne/cm) is the bare oil-water surface tension. The first term is the free energy per surfactant which is a function of S and accounts for surfactant-surfactant repulsions (e.g. due to electrostatic repulsions between polar heads). We plot the two terms in Fig. 2 where the solid bold line giving their sum is the free energy per surfactant $f = F_i/N_s$ plotted versus S. Now because there are no external constraints which predetermine the total area A occupied by the surfactants, (which is not the case for monolayer experiments done on a trough), we assume that the surfactants choose there optimum head area S=S* where f is a minimum determined by:

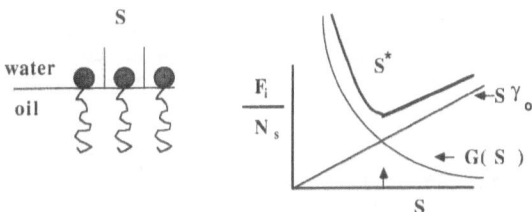

Fig. 2. Left shows surfactants at a water-oil interface with polar head
area S. At the right we plot (solid bold line) the interfacial
free energy per surfactant as a function of S. (The lines
show the separate competing terms discussed in the text that
produce a minimum in $f = F_i/N_s$.)

$$\gamma = dF_i/dA = \gamma_0 - \pi = 0 \qquad\qquad (2)$$

where $\pi = -dG/dS$ is the (Langmuir) surface pressure of the film. Eq.
(2) is known as the Schulman-Montagne[20] condition which fixes the polar
head area at S* where the "effective interfacial tension" γ vanishes.
We will assume that our film is always (approximately) in this state.
We can now expand the interface free energy about this minimum (harmonic
approximation):

$$F(S) = F(S*) + 1/2s(S-S*)^2/S*^2 \ , \qquad\qquad (3)$$

where F is the free energy per unit area and $s = S*^2(d^2F/dS^2)|S=S*$. s
is essentially the (in-plane) inverse compressibility of the film. The
complete elastic free energy of our film is then gotten by considering
deviations of the film from its flat reference plane. This is done by
expanding in the curvature of the film which was first considered by
Helfirch.[21]

II.2. Membrane Elastic Free Energy With Curvature Fluctuations

We consider a liquid membrane (interface) of area A which is almost
parallel to a reference x-y plane as shown in Fig. 3. We describe small
displacements of the membrane from this plane by a height function
h(x,y). For small deviations, the curvatures C_1 and C_2 referring to any
point $(x,y,h(x,y,))$ on the membrane will be given by: $C_1 = 1/R_1 =$

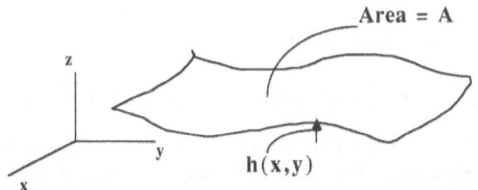

Fig. 3. A fluctuating liquid membrane which is almost parallel to a
reference x-y plane. Each point on the surface is given by
$\vec{r} = (x,y,h(x,y))$.

$\partial^2 h/\partial x^2$, and $C_2 = 1/R_2 \cong \partial^2 h/\partial y^2$. Here, R_1 and R_2 are the principal radii of curvature, and the subscripts 1,2 refer to the x,y axes. Following Helfrich, we write down the elastic free energy per unit area as an expansion in the curvatures keeping all symmetry allowed terms up to second order:

$$F_e = 1/2s(S-S*)^2/S*^2 + 1/2k_c[C_1 + C_2 - C_0]^2 + kC_1C_2 \qquad (4)$$

Here, $C_o = 2/R_o$ where R_o is the natural radius of curvature and describes the tendency of the interface to bend towards either the water ($C_o > 0$) or the oil ($C_o < 0$). k_c and k are the elastic bending moduli of mean and Gaussian curvatures respectively. The first term deals with the in-plane elasticity characterized by the coefficient s (inverse compressibility) of Eq. (3). We will assume that s is always large, and that the system equilibrates at S=S* and satisfies the Schulman criterion for a saturated interface (discussed above). This is a central assumption in our description of the film properties; that is, <u>the interface consists of an incompressible film of surfactant molecules</u> which undergoes curvature fluctuations while keeping its total interfacial area constant. Thus, the first term in Eq. (4) becomes irrelevant. Similarly, because the integral of the third term (Gaussian curvature) over a closed surface is a topological constant (Gauss-Bonnet theorem), it is also irrelevant to our considerations of the elastic free energy integrated over our (approximately) infinite membrane surfaces. Finally, because membranes form bilayers with the same solvent (oil or water) on either side, the natural curvature C_o will be set to zero. F_e then takes on a very simple form:

$$F_e = 1/2k_c(\partial^2 h/\partial x^2 + \partial^2 h/\partial y^2)^2 \qquad (5)$$

We are now in a position to understand under what conditions thermal fluctuations should be important. DeGennes and Taupin[7] introduced the notion of a persistence length ξ_p for a membrane defined as the characteristic length scale of the surface normal-normal correlation function: $\langle \vec{n}(\vec{x}) \cdot \vec{n}(0) \rangle \sim e^{-x/\xi_p}$ ($\vec{n}(\vec{x})$ is the unit vector normal to the membrane surface). Therefore, for distance scales $\ell < \xi_p$ the surface is flat, and for $\ell \gg \xi_p$ the membrane appears "rough" or "crumpled". Using F_e given by Eq. (5), and the equipartition theorem, one finds that[7]:

$$\xi_p = a \exp(2\pi k_c/k_BT) \qquad (6)$$

Where a is a microscopic lower distance cutoff $\cong$ 5Å. Thus, one expects that if $k_c/k_BT \gg 1$, which is what is found in most single lipid or surfactant membranes, (e.g. k_c/k_BT ranges between 20 and 50), the membrane is flat on macroscopic length scales and fluctuations are not important. These are the class of rigid membranes. On the other hand, when $k_c/k_BT \cong 1$ we expect that the membranes will be flexible and layer undulations will be important. We now consider the effect that thermal fluctuations have on the interlayer interactions for multilayers consisting of flexible membranes. We shall return to the role of the persistence length in understanding the ultimate stability of extremely dilute L_α phases in the next section.

II.3. <u>Interacting Stacked Membranes: Undulation Forces</u>

In deriving the undulation forces, we first outline a qualitative but illuminating argument due to Helfrich and Servass.[22] We start by considering height fluctuations for a single membrane introduced in section II.2 (Fig. 3). Using the equipartition theorem and Eq. (5), we find that the mean square height amplitude is given by:

$$\langle h^2 \rangle = 1/4\pi^3 (k_B T/k_c) A , \tag{7}$$

and diverges violently with the size A of the membrane! (this is due to the inherent statistical properties of a low dimensional object such as a surface). We point out that k_c controls the restoring force to the undulation. In a multilayered system, the large height fluctuations of a membrane will be cut off by its neighbors. In fact, we expect that when $(\langle h^2 \rangle)^{\frac{1}{2}} \simeq d$ (where d is the interlayer separation), collisions will occur. Using Eq.(7), we find that patches of membrane of area of order A_p will experience collisions:

$$A_p = 4\pi^3 (k_c/k_B T) d^2 \tag{8}$$

Each collision leads to an increase of the entropic contribution to the free energy by an amount of order $k_B T$ (e.g. because of the steric constraint that membranes cannot cross each other, a collision leads to a loss of configuration for the membrane, and decreases its entropy). We show this schematically in Fig. 4, where a free membrane has more phase space to explore compared to a bound membrane between two walls (its neighbors). Thus, we can expect that the effective (entropic) free energy increase per unit area is of order $F_u \simeq k_B T/A_p = 1/4\pi^3 (k_B T)^2/k_c d^2$. This is the physical origin of the undulation force which was first elucidated in a seminal paper by Helfrich.[12] More accurately, Helfrich derived the undulation interaction for a multilayer system by use of the Landau-DeGennes[22] elastic theory of smectic-A liquid crystals (which is analogous to the L_α phase) with energy density:

$$F/V = \{B(\partial u/\partial z)^2 + K[(\partial^2 u/\partial x^2) \tag{9}$$

$$+ (\partial^2 u/\partial y^2)]^2\}/2,$$

where u(r) is the layer displacement in the z direction normal to the layers, and B and K are the bulk moduli for layer compression (erg/cm^3) and layer curvature (erg/cm).

The undulation interaction per unit area is given by:

$$F_u = \alpha[(k_B T)^2/k_c (d-\delta)^2] , \tag{10}$$

where for multilayers the coefficient is derived to be $\alpha = 0.23$.

It is important to realize that this repulsive undulation interaction is long range, and as we shall see for flexible membranes with low rigidity $k_c \simeq k_B T$, will dominate the attractive long range van der Waals interaction.

We now return to the persistence length and its role in understanding the stability of dilute L_α phases. From Eq.(8), we see that the length $\ell = (A_p)^{\frac{1}{2}}$ is a measure of a typical distance required before a correlated section of a membrane experiences a collision with

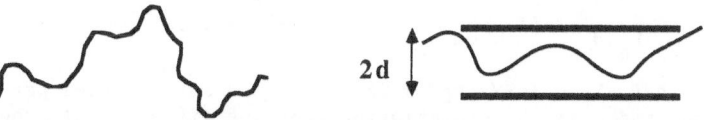

Fig. 4. Schematic of a free membrane compared to one confined between neighbors in a multilayer system. The former has more space to explore and is in a higher entropic state.

its neighbor. As long as this distance satisfies $\ell < \xi_p$, (which is the length scale beyond which the normals of a membrane are no longer correlated) we expect the L_α phase to be stable. For $\ell > \xi_p$, a membrane will begin to perform a two-dimensional random walk before experiencing a collision and one expects the L_α phase to melt into a disordered phase of random sheets. The nature of such a melted phase will be covered by Michael Cates in his lectures on "Random surfactant assemblies and microemulsions". We now consider the more classical interactions between membranes such as those due to electrodynamic and electrostatic forces in the absence of undulations.

II.4. Classical Molecular Interactions Between Flat Membranes: Continuum Limit

The main molecular interactions between two flat membranes separated by a distance d (refer to Fig. 1a), is given by the long range attractive van der Waals (vdW), the electrostatic, and a short range repulsive hydration interaction.

The van der Waals force is given by a simple expression if retardation effects are neglected:[10]

$$F_{vdW} = - \frac{H}{12\pi} \left[\frac{1}{(d-\delta)^2} + \frac{1}{(d+\delta)^2} - \frac{2}{d^2} \right] \qquad (11)$$

this interaction is attractive varying as $\cong \dfrac{1}{(d-\delta)^2}$ for small $d \cong \delta$ with a crossover to $1/d^4$ for large d.

$H = \left(\dfrac{\epsilon_w - \epsilon_m}{\epsilon_w + \epsilon_m}\right)^2 k_B T \sim k_B T$ is the Hamaker constant where ϵ_w and ϵ_m are the dielectric constants for the water and the membrane (oily) regions.

As was demonstrated by Parsegian, Fuller, and Rand,[9] the dominant force governing the interbilayer interactions in rigid ($k_c \gg k_B T$) neutral phospholipids for small separation distances is the hydration force. This strong repulsive potential acts to prevent the approach of phospholipid bilayers embedded in water. The hydration energy per unit area may be represented empirically as:

$$F_{hyd} = A_h e^{-(d-\delta)/\lambda_h}, \qquad (12)$$

where $A_h \simeq 4k_B T/\text{Å}^2$ at room temperature, and λ_h is a microscopic length of order 2Å. This interaction dominates for distances less than 10 Å but is negligible for distances larger than 30 Å.

For charged membranes (refer to Fig. 1a), where the only free ions in solution are the dissociated membrane counter-ions, the electrostatic interaction per unit area can be calculated from the Poisson-Boltzmann equation:[23,3]

$$F_{elec} = \frac{\pi^2 k_B}{L_e 4 d_w} \left[1 - \frac{S}{\alpha L_e d_w} + \left[\frac{S}{\alpha L_e d_w}\right]^2 + \ldots \right], \qquad (13)$$

where $d_w = d-\delta$, $L_e = \pi e^2/\epsilon_w k_B T \simeq 22$Å, S is the surface area per charged polar head, and α is the dissociation constant (if $\alpha = 1$ all the charges are dissociated, otherwise $\alpha < 1$). Thus, the dominant term, due to this long range force varies as $1/d_w$. For sufficiently large additions of ions plus counterions, such as in a NaCl solution, the Debye screening

256

length $\lambda_D = (k_BT\epsilon_w/4\pi \sum_i n_iQ_i^2)^{\frac{1}{2}}$ (n_i = number density of free ions of charge Q_i), is now smaller than the interlayer spacing d_w and the free energy per unit surface decays exponentially[10] with d_w:

$$F_{elec} = E_0 e^{-d_w/\lambda_D}. \tag{14}$$

Here, λ_D (in Å) = 3.04/ c, c is the salt concentration in mole.1^{-1} and, $E_0 = \sigma^2/c$, where σ is the surface charge density. We now turn to a discussion of the experimental technique used to study these dilute L_α phases.

III. EXPERIMENTAL TECHNIQUE: X-RAY STRUCTURE FACTOR AND THE LANDAU-PEIERLS EFFECT

X-ray measurements normally give information on the structure of the system studied. In special cases however, such as the lyotropic L_α and the thermotropic smectic-A phases of liquid crystals or two dimensional solids, thermodynamic properties can be obtained reliably from the analysis of the shape of the structure factor.

As was first pointed out by Peierls and Landau[16], three dimensional structures whose densities are periodic in only one direction such as the lyotropic L_α phase, are marginally stable to thermal fluctuations which destroy the long range order and replace the δ-function bilayer stacking structure factor peaks at $(0,0,q_m=mq_0=m2\pi/d)$ (d=interlayer spacing and m=1,2, ... is the harmonic number) by weaker algebraic singularities. Caillé[17] has derived the scattering cross section for a smectic-A liquid-crystal phase, which has the analogous elastic free energy as that for the L_α phase given by Eq. (9).

The asymptotic form of the structure factor is described by power laws: $I(0,0,q_z) \alpha |q_z - q_m|^{-2+\eta_m}$ and $I(q_\rho,0,q_m) \sim q_\rho^{-4+2\eta_m}$. q_ρ and q_z are components of the wave vector parallel and normal to the layers and:

$$\eta_m = m^2 q_0^2 k_B T/(8\pi(BK)^{0.5}), \tag{15}$$

where η_m is the exponent which describes the algebraic decay of layer correlations. The moduli B (for layer compression) and K = k_c/d (for layer curvature) were introduced earlier in section II.3. While the asymptotic forms for I(q) are simple, to quantitatively analyze the data we fit to the exact expression for the structure factor:

$$I(q) \sim \int dz \int dp \; S(z,\rho) e^{-R^2\pi/L^2} [(\sin qR)/qR] e^{-iq_m z}. \tag{16}$$

Here, $S(z,p) \sim (1/\rho)^{2\eta_m} \exp\{-\eta_m[2\gamma+E_1(\rho^2/4\lambda z)]\}$ is the layer-layer correlation function,[17] where $R^2=z^2+\rho^2$, γ is Euler's constant, $E_1(x)$ is the exponential integral function, and $\lambda = \sqrt{(K/B)}$. In Eq. (16), the exponential term incorporates a finite size effect because of the observed finite lamellar domain sizes typically between ~2000 and ~10,000 Å (L^3 is the domain volume). Also because the samples consist of randomly oriented domains, we perform an exact powder average over all solid angles in reciprocal space. The precise steps leading to Eq. (16) have been discussed elsewhere.[2,3] The analysis consists of simultaneous fits of Eq. (16) (convoluted with the resolution function), to either two or three harmonics (depending on whether the third harmonic is observable), where the scaling of η_m (=$m^2\eta_1$) is incorporated. The powder averaging results in the asymptotic power-law behavior for I(q) ~ | q - q_m |$^{-p}$ where, the exponent p is approximately equal to 1 - η. To unambiguously study the lineshape of the structure factor one must use a specialized spectrometer. In fact, the experimental set-up must have a very high resolving power capable of determining length scales on the order of a micron to probe the interesting long wavelength behavior of the system. In addition, the tail of the resolution function must decrease as a function of the wave vector q faster than $1/q^2$ which is the falloff expected for the structure factor of a Landau-Peierls system with a small value of η. These two requirements are fulfilled when using the high resolution spectrometer[2,3] shown in Fig. 5. The experiments were carried out at the National Synchrotron Light Source on the Exxon Beam line X-10 A and at the Stanford Synchrotron Radiation Laboratory on beam line VI-2. The monochromator (Fig. 5) consists of a double bounce Si(1,1,1) crystal. The analyzer is a triple bounce Si(1,1,1) channel cut crystal. The configuration yields a very sharp in-plane Gaussian resolution function with very weak tail scattering with Half Width at Half Maximum (HMHM) 8 x 10^{-5} Å^{-1}. This then enables us to resolve length scales as large as 1.5 μm. The channel-cut crystals cut down the tails of the resolution function; indeed it is known[24,18] that after n reflections from a perfect crystal the resolution is given by the resolution of a single reflection to the nth power. In our case, since the tail intensity fall off is limited by the monochromator double bounce we have checked that the tails decrease approximately as $1/q^4$. The out-of-plane resolution was set by narrow slits which yielded an out-of-plane Guassian resolution function with HWHM = $1/1000$ Å^{-1}. The samples were contained in sealed quartz capillaries with diameters of 1 and 2 mm, which yielded randomly oriented lamellar domains.

IV. EXPERIMENTAL RESULTS

IV.I. Dilute L_α Phase Stabilized by Undulation Forces: A Dodecane
Dilution Study

The first system where the existence of the undulation forces was quantitatively established was in the lamellar L_α phase of the quaternary mixture of sodium dodecyl sulfate (SDS, the surfactant), pentanol (cosurfactant), water, and dodecane as a function of dodecane dilution[2]. Figure 6 shows a cut of the phase diagram discovered and mapped out by Roux and Bellocq[25] represented on a standard triangular phase diagram in the plane with a constant water/SDS weight ratio equal to 1.55. This

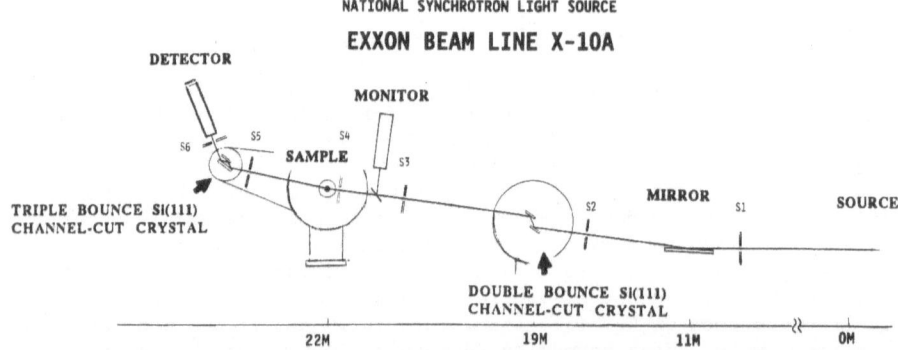

Fig. 5. Schematic drawing of the high resolution X-ray set-up used in this study. The crucial high resolution elements are the double-bounce Si(1,1,1) monochromator crystal and the triple-bounce Si(1,1,1) analyzer crystal as discussed in the text.

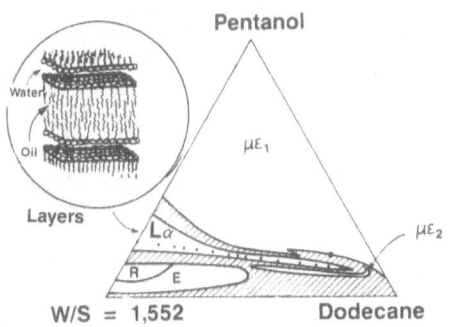

Fig. 6. A cut of the phase diagram of the quaternary mixture of SDS, pentanol, water and dodecane shown in the plane with a constant water/SDS weight ratio=1.55 (Ref. 25). The dots in the lamellar phase labeled L_α correspond to the mixtures studied.

cut contains five one-phase regions: the microemulsion phases $\mu\epsilon_1$ and $\mu\epsilon_2$ and the liquid crystalline hexagonal (E), rectangular (R), and lamellar (L_α) phases.

The L_α phase consists of water layers (inverted membrane) embedded in oil (refer to Fig. 1b) (Fig. 6). This system affords distinct advantages for studies of intermembrane interactions. First, the water layers are charge neutral so that there are no intermembrane electro-static interactions. Second, and very significantly for this plane of the phase diagram, we are able to dilute over an unusually large oil range between 0 and 80 wt.%, which corresponds to intermembrane separa-tions between ~2.0 and larger than 50.0 nm. Third, studies[1] of the curvature elasticity in similar dilute lyotropic L_α phases had indicated unusually small values of $k_c \approx k_BT$. These unique properties of this L_α phase thus allows for a comprehensive study of the long-range van der Waals and fluctuation induced undulation interactions.

In the dodecane dilution series, detailed studies have been carried out for 10 distinct mixtures: x=0, 0.07, 0.13, 0.18, 0.29, 0.35, 0.47, 0.54, and 0.62. Here x is the percent dodecane by weight of the mix-tures. We show in Fig. 7 typical scattering profiles for longitudinal scans through the first harmonic of the structure factor for x between 0 and 0.54 in the L_α phase where the total layer spacing $d = 2\pi/q_0$ in-creases from 3.82 nm to 11.5 nm. In the mixtures studied, the dilution corresponds to a path where water layers with approximately constant thickness $\delta \sim 18$ Å are pushed apart with d_0/δ varying from ~1 to ~7.

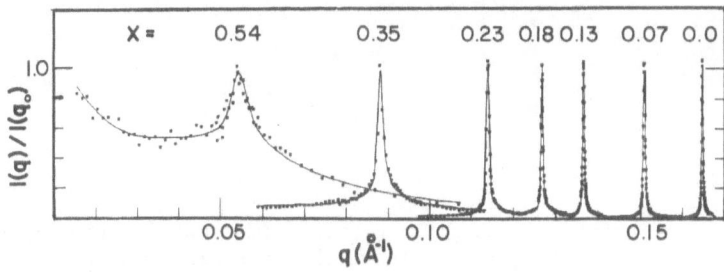

Fig. 7. Longitudinal profiles of the first harmonic for seven different mixtures along the dodecane dilution path. The percentage dodecane by weight of the mixtures (x) is indicated above each profile. All peak intensities are normalized to unity. The solid lines are fits by the Caillé power-law line shape Eq.(16).

($d_0=d-\delta$ is the oil thickness between water layers). A striking
feature of the profiles in Fig. 7 is the tail scattering which becomes
dramatically more pronounced as d increases. This effect is further
elucidated in Fig. 8(c), where we plot on a logarithmic intensity scale
versus $q-q_0$ the scattering for three mixtures, x=0.07, 0.23, and 0.35.
The significant difference in the profiles over the entire dilution
range is now immediately clear. It is qualitatively clear that $\eta_1(d)$,
which characterizes, the asymptotic scattering profile and which is a
measure of the ratio of the tail to peak intensity scattering, is
increasing as d increases. For a quantitative analysis the data are fit
to the exact expression for the structure factor Eq.(16).

We plot in Fig. 8(a) on a log-log scale the intensity versus $q-q_0$
for x=0.23, where the solid line is a result of the fit yielding $\eta_1=0.25$
and L=8640 Å. Two features in the scattering profile and the theoretical cross section are immediately apparent. While at large $q-q_0 \geq 2\pi/L$
the scattering exhibits power-law behavior $S(q) \sim |q - q_0|^{-P}$ with
$P \sim 1-\eta$, at $q \approx q_0$, the finite size effects round off the observed profile
with characteristic width $\sim 1/L$. The data also confirm the scaling of η_m
with m^2. We show in Fig. 8(b) the profiles and fits for the first and
second harmonics for x=0.07 on a normalized logarithmic intensity scale.
We find $\eta_1(q_0)=0.14\pm0.02$, $\lambda=8.59\pm1$ for the first harmonic and
$\eta_2(2q_0)=0.57\pm0.02$, $\lambda=8.13\pm2$ for the second harmonic.

A further, more subtle aspect of the data is evident in Fig. 8(c),
where the profiles around $q=q_0$ are slightly asymmetric with the high q
scattering more intense than the low q. The parameter $\lambda = (K/B)^{\frac{1}{2}}$, which
enters the real-space correlation function $S(\vec{R})$, is a measure of the
degree of anisotropy present in $S(q_z,q_\perp)^{17}$ of an oriented sample. The

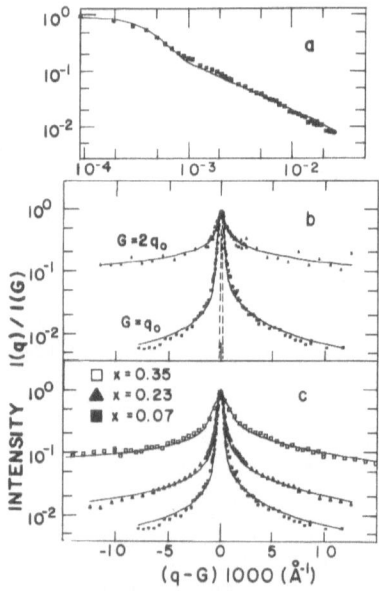

Fig. 8. (a) Profile of the first harmonic (G = q_0) for the mixture
x=0.23 on a log-log scale which shows finite size rounding at
small q-G followed by power-law behavior at larger q-G. (b)
Profile of the first and second harmonics for the mixture x=0.07
on a logarithmic intensity scale. (c) Profile of the first
harmonic (G=q_0) for three mixtures on a logarithmic intensity
scale. All peak intensities are normalized to unity. The solid
lines are fits by the Caillé power-law line shape [Eq.(16)].

powder average of this anisotropic scattering cross section results in the observed I(q) asymmetric profile. For $\lambda > 2\pi/q_0$ the asymmetry is negligible and I(q) is not very sensitive to the precise value of λ.

The value for k_c ($=K.d$) is obtained from the measured values of λ ($=\sqrt{[K/B]}$) and $\eta(\alpha 1/\sqrt{KB})$. k_c varies smoothly between 0.5 and 2 k_BT. This value is significantly lower than that measured for most lipidic membrane systems and is precisely why fluctuations on short wavelength < d and long wavelength >> d scales become important for these dilute membranes. Therefore, since the Helfrich mechanism scales as k_c^{-1}, we expect this interaction F_u(Eq. (10)), which ranges between $0.12k_BT/d^2$ and $0.46\ k_BT/d^2$, to dominate the van der Waals interaction $F_{vdW} \simeq -0.03k_BT/d^2$ (Eq. (11)).

From the free energy per unit surface, we readily calculate the layer compressional modulus B and η as a function of d. Indeed, B is related to the second derivative of the free energy per unit volume:

$$B = d^2 \left| \frac{\partial^2 \frac{F}{V}}{\partial d^2} \right|_n \qquad (17)$$

where n is the number of layers per unit length and F is the free energy. We perform the double derivative at constant number of layers assuming that the thickness of surfactant layers δ remains constant. In this manner we obtain an expression for B as a function of d using Eq.(10). The value of η_1^{und} predicted by the Helfrich theory is then obtained using Eq.(15) together with $K = k_c/d$:

$$\eta_1^{und} = 1.33 \cdot (1-\delta/d)^2 \ . \qquad (18)$$

We plot in Fig. 9 η_1(d) resulting from fits to the profile at the first harmonic as a function of d. The solid line, which agrees well with the experimental data, is a plot of the predicted value for η_1^{und}(d) (Eq.(18)) of the Helfrich theory. Here, we have taken the effective water thickness $\delta=29$ Å to include the known excluded volume effects [25] of the surfactant tails in the oil. This then provides compelling evidence that in this SDS multimembrane system swollen by dodecane the intermembrane interactions are dominated by the Helfrich mechanism of entropically driven undulation forces.

In contrast to the oil dilution system just discussed, to date, most previous work on dilute L_α phases has been on water dilution systems consisting of charged surfactants where the dilution is achieved through repulsive electrostatic interactions.[26] We now present

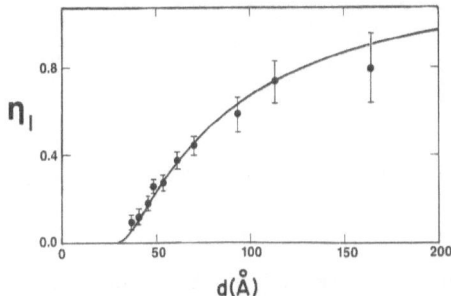

Fig. 9   Power-law exponent η_1^{und} as a function of the intermembrane distance for mixtures along the dodecane dilution path. The solid line is the prediction of the model of Helfrich of entropically driven undulation interactions Eq.(18).

data on a charged system which shows a qualitatively different type of behavior than the dodecane dilution system.

IV.2. Charge Stabilized Dilute L_α Phases

We now turn to consider two charged systems where the membrane consists of a mixture of SDS and pentanol. The first system corresponds to a water dilution[3] (refer to Fig. 1a) where the only ions in the solvent are the dissociated membrane counter-ions. In this case the electrostatic interactions are long range and significantly larger than all other forces such as hydration, van der Waals, and undulations. Using equations (13) and (17), one can readily derive the exponent η_1^{elec} (d):[3]

$$\eta_1^{elec}(d) = (k_BTL_e/2gk_c)^{0.5}[(1-\delta/d)^{1.5}/d^{0.5}], \qquad (19)$$

where $g = [1-3(D/d_w) + 6(D/d_w)^2 + \ldots]$. Here, $d_w = d-\delta$, $D = S/L_e$ ($L_e \approx$ 22Å was defined in section II.4), and all surface charges are assumed to be dissociated. We show in Fig. 10 a series of scattering profiles for longitudinal scans through the first harmonic along the water dilution path for the weight fraction of water (Φ_w) increasing from 0.47 to 0.77.[3] (This corresponds to interlayer separations 30Å < d < 90Å). The scattering profiles are to be compared to those along the dodecane (oil) dilution line (Fig. 7, section IV.1) where undulation forces dominate. Qualitatively, we see a striking difference. While in the oil dilution the tail scattering increases dramatically as d increases, for the water dilution (where electrostatic forces dominate) the profiles are quite similar as d increases. Figure 11 shows the results of the comparison between the experimental and theoretical values of η_1^{elec} (Eq. (19)) for the water dilution with long range electrostatic forces (open circles for the experimental data and dashed curve for the theoretical prediction). k_c varies smoothly from 2 to 0.7 k_BT, and S and $\delta=20$Å correspond to the value calculated from the known composition of alcohol and surfactant. The good agreement indicates that a simple model of electrostatic interactions with no adjustable parameters is sufficient to interpret the experimental data. This behavior has also been veri- fied for the SDS-hexanol water dilution system up to larger layer spacings d=120Å.[5]

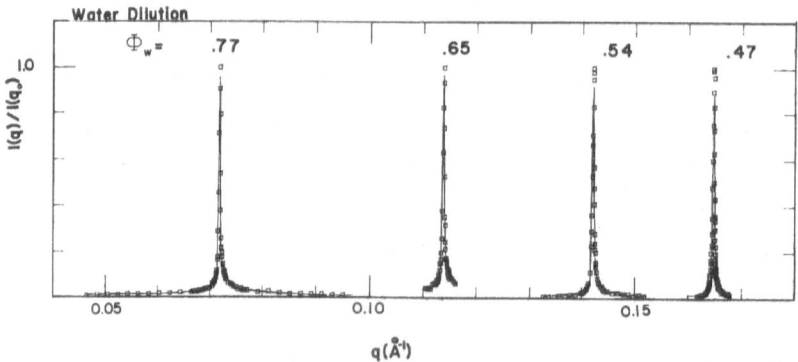

Fig. 10. Longitudinal profiles of the first harmonic of four different mixtures along the water dilution path. The percentage water by weight of the mixture Φ_w is indicated above each profile. The solid lines are fits by the full Caillé expression given by Eq.(18). All peak intensities are normalized to unity.

We now consider a charged system consisting of the same membrane (SDS and pentanol), but where the solvent separating the layers is brine (0.5 mole/ℓ of NaCl). In this case, the addition of free ions (NaCl) to the solvent yields a small Debeye screening length $\lambda_d \simeq 6\text{Å} \ll d$, and electrostatic forces should be negligible. We expect again that undulation forces will dominate since $k_c \simeq k_BT$. The solid line in Fig. 11 is the prediction of η_1^{und} with $\delta=20\text{Å}$. The data (open squares) are the results of η_1 for the first harmonic in the brine dilution series. Quite clearly, when only short range electrostatic forces are present for flexible membranes, undulation forces dominate. We stress the large difference in behavior in $\eta_1(d)$ over the dilution range between the electrostatically stabilized dilute membranes (open circle; Fig. 11) and those stabilized by entropically induced undulation forces (open square; Fig. 11): while in the former case η_1 changes by less than 30%, it varies by about an order of magnitude in the latter systems.

IV.3 Rigid Membranes and "Bound" L_α Phases

In the previous two sections, we have established the existence of a novel dilute (almost unbound) regime for membranes whose thermodynamic stability is due to entropically induced undulation interactions (e.g. the oil and brine dilutions) rather than classical electrostatic repulsions (e.g. the water dilution of the previous section). The strength of this interaction is controlled by the bending rigidity k_c of a single membrane. For flexible membranes with $k_c \simeq k_BT$ this repulsive interaction completely overwhelms the van der Waals attraction. In this section we shall consider the L_α phase consisting of rigid ($k_c \gg k_BT$) membranes where the undulation interaction is "turned off" and plays no role. Most biological membranes are believed to be rigid.

We consider the biological multimembrane system of di-myristoyl-phosphotidyl-choline (DMPC) and water. The binary DMPC-water system

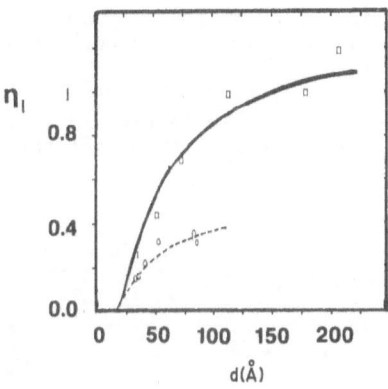

Fig. 11. Variation of the exponent η as a function of the repeat distance d for the water dilution (open circles) and for the brine dilution (open squares). The solid line corresponds to the prediction of the pure undulation interaction (Eq.18), the dashed line is the solution of the Poisson-Boltzman equation (Eq.19). In each case, the values of all parameters are determined experimentally. $\delta=20\text{Å}$ for both cases. For the water dilution: k_c ranges from 2 k_BT to 0.07 k_BT, S ranges from 80 to 190Å^2.

consists of uncharged lipid molecules stacked in double layers separated by water (refer to Fig. 1a). Its temperature versus concentration phase diagram is shown in Fig. 12.[27] The $L_{\beta'}$ is an ordered lamellar phase region (whose precise structural nature we shall return to consider), in which the hydrocarbon chains are frozen in the all-trans conformation and are ordered within the layers. The $P_{\beta'}$, also known as the "rippled phase," has mostly frozen chains and greater in-plane order than the L_α, but its most distinguishing feature is a long-wavelength (100-200Å) in-plane modulation of the layers.

In the L_α phase the rigidity k_c has been measured to be about 25 $k_B T$.[6] As one increases the amount of water in the L_α phase, d increases and reaches a limiting value d^* (d_w = 25Å, δ = 35Å, d^* = 60Å). At this point the excess water phase-separates from the L_α phase indicating a minimum in the interaction energy as a function of the d-spacing. This minimum comes from the competition between the long range van der Waals attraction and the short range repulsive hydration interaction. We show schematically the shape of this intermembrane potential in Fig. 12. The behavior of this system is characteristic of a "bound" L_α phase, with the phase separation (at 40 wt% water) being a consequence of the competition between the interlayer interactions. This is in contrast to the dilute L_α phases[2,3,5] that we considered earlier (sections IV.1 and IV.2), where those "almost unbound" systems ultimately phase separated because of the existence of other thermodynamically stable nearby phases with a different structure and a lower free energy. For example, the oil dilution L_α phase (Fig. 6 section IV.1) can be diluted up to about 80% dodecane by weight, beyond which it phase separates with a phase of multiply connected random bilayer sheets.[28] Cates will discuss this in more detail in his lectures.

The structure factor for these rigid and bound L_α phases is quite distinct from the dilute L_α phases. We shall discuss x-ray studies obtained on well aligned freely suspended films of DMPC at controlled temperature and relative humidity.[11] By varying the humidity, which is directly related to the chemical potential of water [$\Delta\mu$ = RTln(RH/100), where R is the gas constant], one continuously varies the amount of water in the layers. (The relative humidity RH = 100(P/P$_0$), P = partial water vapor pressure over the film, P$_0$ = vapor pressure over pure water.)

The T,RH phase diagram is shown in Fig. 13 (left). When compared to Fig. 12(left) we see that the $L_{\beta'}$ is in fact three distinct phases.

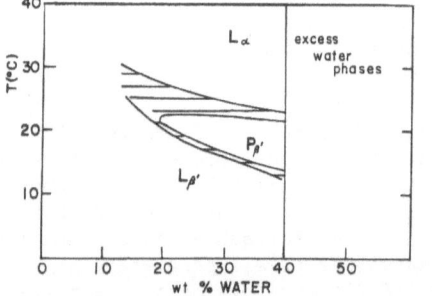 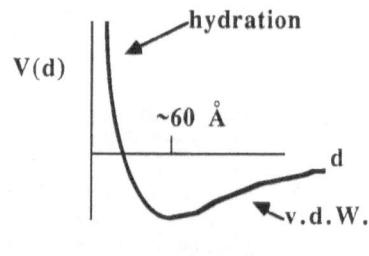

Fig. 12. Left: the temperature-concentration phase diagram for the DMPC-water-binary system. Right: intermembrane potential due to hydration and van der Waals.

The phases are distinguished by the direction of the molecular tilt with respect to the local two-dimensional distorted hexagonal lattice. In $L_{\beta F}$, the tilt is between nearest neighbors; in $L_{\beta I}$ it is toward a nearest neighbor; and in $L_{\beta L}$, it varies continuously between the two, with second-order transitions on both ends. These distinct $L_{\beta'}$ phases were discovered[11] only because of the ability to prepare well aligned samples.

In the L_α phase, we compare scattering data along q_z from the first two harmonics in films of the ternary system SDS, pentanol, and water (the same water dilution of section IV.2) in a sealed cell, to data from the sixth harmonic of the DMPC-water system (Fig. 13 (right)).

As discussed previously (section III) we expect the structure factor to have the asymptotic form: $I(0,0,q_z) \sim |q_z - q_m|^{\eta_m - 2}$ with η_m given by Eq.(15) (remember that for the unoriented L_α systems (sections IV.1 and IV.2) the powder averaging results in the asymptotic behavior $\sim |q_z - q_m|^{\eta_m - P}$ with $P \approx 1$.) For the SDS system we see that the first harmonic (open squares Fig. 13 (right)) and the second harmonic (closed squares) obey the expected m^2 scaling and give $\eta_1 = 0.23 \pm 0.03$ consistent with the unoriented results of section IV.2 (see Fig. 11). In the DMPC system, eight harmonics of the structure factor are observed[27]; since η_8 ($= 64\eta_1$) has to be less than 2 (or the scattering has no singularity), this immediately suggests that thermal fluctuations are weak. From the data of Fig. 13 (right) for the DMPC system (solid diamond) an upperbound is found for $\eta_1 = \eta_6/36 < 0.003$. Smith et. al.[27] have shown that this small value is consistent with the large rigidity of the DMPC membrane and the hydration forces which dominate the interlayer interactions.[9]

IV.4. Universality in Interacting L_α Phases: Cross-Over From Rigid to Flexible Membranes

In these lectures, we have considered two limiting states of the L_α phase in systems where the electrostatic repulsion is negligible: one rigid ($k_c \gg k_B T$) and "bound", the other flexible ($k_c \approx k_B T$) and dilute ("almost unbound"). In the rigid limit the interlayer interactions are

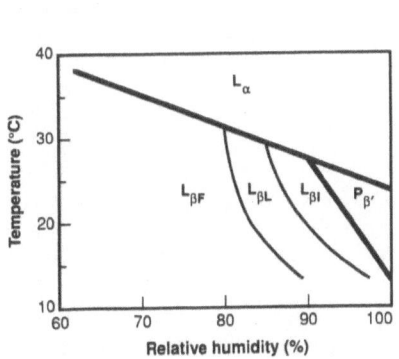
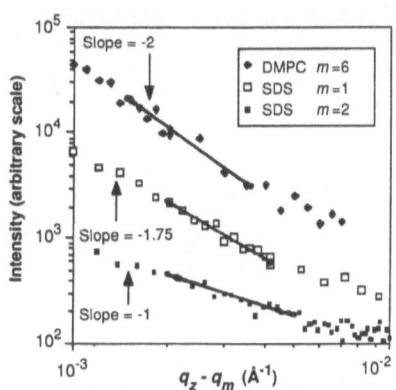

Fig. 13. Left: The temperature-humidity phase diagram of DMPC. Because both axes are related to thermodynamic potentials, there are no two-phase regions. Note that the phase previously known as $L_{\beta'}$ is, in fact, three distinct phases, $L_{\beta F}$, $L_{\beta L}$, and $L_{\beta I}$. Right: Log-log plot of the scattering intensity from the tails of the first and second harmonics of the ternary mixture [SDS (30% by weight) pentanol (23%), water (47%)] and the sixth harmonic of DMPC-water (90% RH, 31°C).

due to the classical van der Waals and hydration potential. In the flexible limit undulations dominate. We are intrested in knowing how to control the rigidity k_c which takes us between these two limits. On theoretical ground, Lipowsky and Leibler[13] have shown that by tuning and decreasing the Hamaker constant of the vdW interaction, one may obtain a complete unbinding of membranes.

The main difference between the L_α phases of these two limiting systems lies in the nature of the bilayer film, which for the dilute systems[1-5] usually contains cosurfactant in addition to surfactant molecules. From a geometric view point, the addition of cosurfactant (e.g. pentanol) molecules (with a cosurfactant/surfactant ratio of about 3/1) leads effectively to a significantly thinner membrane ($\delta \approx 20 Å$) than the pure systems which are rigid. (For DMPC $\delta \approx 35 Å$).

Harmonic spring models and scaling arguments[29] lead to simple predictions of a power law dependence of the bending rigidity on the thickness of the membrane: $k_c \sim \delta^p$ with p between 2 and 3.

The thickness dependence of $k_c(\delta)$ can be studied by varying the cosurfactant chain length.[5] Fig. 14 (left), shows a plot of k_c as a function of the membrane thickness for seven homologous SDS-alcohol-water systems (alcohol—from pentanol to dodecanol). In each case, the cosurfactant/surfactant/water weight ratio was 0.21/0.15/0.64 and the principal purpose of changing the cosurfactant tail length is to change the membrane (cosurfactant plus surfactant) thickness. We see, while k_c/k_BT is between 1 and 3 for the first three samples (C_5OH, C_6OH, C_7OH) the data show an abrupt increase around $\delta \approx 23 Å$ towards a significantly larger value for the rigidity for the latter four samples ($C_8OH, C_9OH, C_{10}OH, C_{12}OH$).[30] The solid line in Fig. 14 (left) follows a δ^3 scaling behavior which when extrapolated to $\delta \approx 35 Å$ (corresponding to the thickness for the pure DMPC multimembrane in water), gives a value of about 20 k_BT consistent with typical values obtained by other techniques.[6] While the behavior appears to be discontinuous (shown as dashed line) rather than power law, the data show an unambiguous trend in the correlation of k_c with the thickness of the membrane. Thus, the primary effect of replacing surfactants with shorter chain cosurfactants in a mixed system is the thinning of the interface which in turn leads to a reduction of the rigidity modulus k_c.

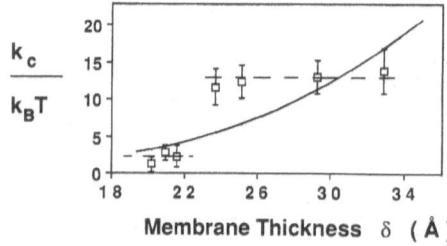 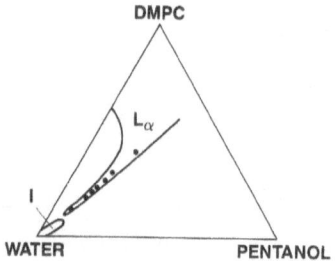

Fig. 14. Left: k_c as a function of the membrane thickness for the SDS-cosurfactant-water mixtures for seven varying cosurfactant chain lengths. The solid line follows a δ^3 law which may be expected from elasticity theory.[29] The dashed line indicates discontinuous behavior. Right: Schematic phase diagram of the ternary DMPC-pentanol-water system studied along the water dilution line (mixtures studied are indicated as dots).

This effect of the thinning of a rigid membrane has been studied in the DMPC-water system.[5] Fig. 14 (right) shows the region of stability for the L_α phase at T = 25°C for the ternary DMPC-pentanol-water system. Along the DMPC-water line, the L_α phase is stable up to a maximum dilution of 40% water which corresponds to a maximum interlayer separation (d-δ) of 25Å. This corresponds to the spacing for which the vdW attraction equals the hydration repulsion. What is striking is that once each DMPC surfactant has been replaced with several cosurfactants (between two and four) one is able to separate the membranes to very large spacings with d > 200Å (corresponding to a multimembrane system with more than 80% water by weight). One expects that this dilute L_α region (samples studies are shown as dots in Fig. 14), corresponds to the regime where undulations dominate (remember that the membrane is charge neutral). Because the power-law exponent η_1^{und} (Eq. (18)) has a simple universal functional form, it is instructive to plot it versus $(1-\delta/d)^2$ for the flexible membrane systems discussed in these lectures (where electrostatic repulsions are negligible). Fig. 15 plots η_1 versus $(1-\delta/d)^2$ for the DMPC-pentanol-water system as solid squares, together with the SDS-pentanol membranes diluted respectively with (i) dodecane (open squares) and (ii) brine (open circles). The theoretical prediction for η_1 is drawn as a solid line. The <u>universal behavior</u> is now clear and is in remarkable agreement with the prediction of the Helfrich theory for multimembranes where undulation forces dominate. Thus, the embedding of cosurfactant molecules in a rigid membrane system, thins the bilayer, and leads to a large reduction in the bending rigidity for the fluid bilayer. This effect leads to a cross-over from a microscopically driven, to a fluctuation induced universal regime for intermembrane interactions.

V. CONCLUSION

In these lectures we have seen that the lyotropic lamellar L_α phase exhibits in a pronounced manner, the Landau-Peierls effect for dilute stacked membrane systems with low rigidity $k_c \simeq k_BT$. The x-ray structure factor was seen to exhibit power-law behavior with the exponent $\eta(d)$ describing the algebraic decay of layer correlations. In turn, $\eta(d)$ is directly related to the intermembrane interactions.

The synchrotron x-ray work reviewed here has focused on the precise nature of competing interactions in both neutral and charged layered surfactant systems. The basic stabilizing force between rigid ($k_c \gg k_BT$) membrane sheets is known to originate in usual cases from a balance

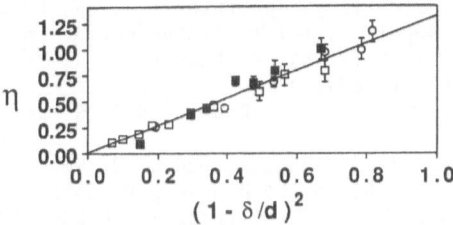

Fig. 15. Power-law exponent η as a function of $(1-\delta/d)^2$ for three dilution systems. The solid squares are for the DMPC-pentanol-water system while the open circles and squares are for the SDS-pentanol membranes along brine and oil dilution lines respectively. The solid line is the prediction of the Helfrich theory of undulation interactions.

between van der Waals (electrodynamic), electrostatic, and hydration forces.

We have demonstrated the existence of a novel universal regime for floppy membranes with small rigidity moduli $k_c \simeq k_B T$, where the interactions originate from entropically induced undulation forces and depend only on geometric and elastic parameters such as the layer spacing and the layer rigidity. The data were seen to be quantitatively consistent with the theoretical work of Helfrich on undulation forces. This repulsive interaction, which is due to the constraint that a meandering membrane cannot cross its two neighbors, is the two-dimensional analogue of interactions associated with wandering one-dimension walls of incommensurate phases close to the commensurate phase transition.

Finally, we have seen that thinning a rigid membrane, by replacing lipid surfactants with shorter chain cosurfactants, leads to a significant reduction of the rigidity modulus k_c. In this case then repulsive undulation forces completely overwhelm the van der Waals attraction and one crosses over from the classical microscopic regime for "bound membranes" to the floppy "almost unbound" regime for fluctuating dilute membranes.

VI. ACKNOWLEDGMENTS

The experimental work described here has been carried out in close collaboration with D. Roux, G. S. Smith, E. B. Sirota and N. A. Clark. S. K. Sinha, P. Dimon and A. M. Bellocq also made important experimental contributions during the early part of the work. The author has benefited from numerous stimulating discussions with P. A. Pincus, S. A. Safran, and S. Milner. The experiments reported here were carried out on the Exxon beam lines at the National Synchrotron Light Source, Brookhaven National Laboratory, and the Stanford Synchrotron Radiation Laboratory. Both facilities are supported by the U.S. Department of Energy. A part of this work was supported by a joint Industry/ University NSF Grant No. DMR-8307157.

VII. REFERENCES

1. J. M. Dimeglio, M. Dvolaitsky, and C. Taupin, 89:871 (1985); J. M. Dimeglio, M. Dvolaitsky, L. Leger, and C. Taupin, Phys. Rev. Lett., 54:1686 (1985).
2. C. R. Safinya, D. Roux, G. S. Smith, S. K. Sinha, P. Dimon, N. A. Clark, and A. M. Bellocq, Phys. Rev. Lett., 57:2718 (1986).
3. D. Roux and C. R. Safinya, J. Phys. (Paris), 49:307 (1988).
4. F. Larche, J. Appel, G. Porte, P. Bassereau, and J. Marignan, Phys. Rev. Lett., 56:1700 (1986); P. Bassereau, J. Marignan, G. Porte, J. Physique, 48:673 (1987).
5. C. R. Safinya, E. Sirota, D. Roux, and G. S. Smith, Phys. Rev. Lett., 62:1134 (1989).
6. M. B. Schneider, J. T. Jenkins, and W. W. Webb, J. Physique, 45:1457 (1984); I. Bivas, P. Hanusse, P. Bothorel, J. Lalanne and O. Aguerre-Charriol, J. Phys. (Paris), 46:855 (1987).
7. P. G. de Gennes and C. Taupin, J. Phys. Chem., 86:2294 (1982); S. A. Safran, D. Roux, M. E. Cates, D. Andelman, Phys. Rev. Lett. 57:491 (1986).
8. W. Helfrich, J. Phys. (Paris), 46:1263 (1985); L. Peliti and S. Leibler, Phys. Rev. Lett., 54:1960 (1985).
9. A. Parsegian, N. Fuller, and R. P. Rand, Proc. Natl. Acad. Sci., 76:2750 (1979); R. P. Rand, Ann. Rev. Biophys. Bioeng., 10:277 (1981).

10. J. N. Israelachvili, "Intermolecular and Surface Forces," Academic Press, Orlando (1985); J. Mahanty and B. W. Ninham, "Dispersion Forces," London (1976).

11. G. S. Smith, E. B. Sirota, C. R. Safinya, and N. A. Clark, Phys. Rev. Lett., 60:813 (1988); E. B. Sirota, G. S. Smith, C. R. Safinya, R. J. Plano, and N. A. Clark, Science, 242:1406 (1988).

12. W. Helfrich, Z. Naturforsch, 33a:305 (1978).

13. R. Lipowsky and S. Leibler, Phys. Rev. Lett. 56:2561 (1986); Y. Kantor, M. Kardar, and D. R. Nelson, Phys. Rev. Lett., 57:791 (1986); D. R. Nelson and L. Peliti, J. Phys. (Paris), 48:1085 (1987); J. A. Aronovitz and T.C. Lubensky, Phys. Rev. Lett., 60:2634 (1988); For a broader discussion see D. R. Nelson, "Statistical Mechanics of Membranes and Surfaces," proceeding of the Jerusalem Winter School, edited by Nelson, Piran, and Weinberg (1987).

14. P. G. DeGennes, "Scaling Concepts in Polymer Physics," Cornell Univ. Press (1979).

15. See for example, J. Villain and P. Bak, J. de Physique (France), 42:657 (1981); S. G. J. Mochrie, A. R. Kortan, R. J. Birgeneau, P. M. Horn, Z. Phys. B., 62:79 (1985).

16. L. D. Landau, in Collected Papers of L. S. Landua, edited by D. Ter. Haar, Gordon and Breach, New York (1965), p. 209; R. E. Peierls, Helv. Phys. Acta. 7, Suppl., 81 (1934).

17. A. Caillé, C. R. Acad. Sci. Ser., B274:891 (1972).

18. J. Als-Nielsen, J. D. Litster, R. J. Birgeneau, M. Kaplan, C. R. Safinya, A. Lindegaard-Andersen, and S. Mathiesen, Phys. Rev. B, 22:312 (1980).

19. P. Pieranski, Contemp. Phys., 24:25 (1983); N. A. Clark et al., J. Phys. Colloq. C3, 43:137 (1985).

20. J. H. Schulman and J. B. Montagne, Ann. N.Y. Acad. Sci., 92:366 (1961).

21. W. Helfrich, Z. Naturforsch, 28c:693 (1973); W. Helfrich and R. M. Servuss, IL Nuovo Cimento, 3D:137 (1984).

22. P. G. DeGennes, "The Physics of Liquid Crystals," Clarendon, Oxford (1974).

23. A. C. Cowley, N. L. Fuller, R. P. Rand, V. A. Parsegian, Biochem. 17:3163 (1978); S. Leibler, R. Lipowsky, Phys. Rev. B., 35:7004 (1984).

24. V. Bonse and M. Hart, "Small Angle X-ray Scattering," edited by H. Brumberger, NY, Gordon & Breach).

25. D. Roux and A. M. Bellocq, "Physics of Amphiphiles," edited by V. DeGiorgio and M. Corti North-Holland, Amsterdam, (1985).

26. P. Ekwald, "Advances in Liquid Crystals," Ed. G. H. Brown, Academic Press (1975).

27. G. S. Smith, C. R. Safinya, D. Roux, and N. A. Clark, Mol. Cryst. Liq. Cryst., 144:235 (1987).

28. M. E. Cates, D. Roux, D. Andelman, S. T. Milner, and S. A. Safran, Europhysics Lett. 5:8:733 (1988); also for a more general discussion of structure and phase equilibria of surfactants in solution see D. Andelman, M. E. Cates, D. Roux and S. A. Safran, J. Chem. Phys., 87:12:7229 (1987).

29. Elastic models lead to p=3 if the stress is uniformly distributed through the thickness of a bent sheet; e.g. L. D. Landau and E. M. Lifshitz, "Theory of Elasticity," Pergamon, New York (1970); S. T. Milner and T. A. Witten, M. E. Cates, Europhysics, Lett. 5, 413 (1988); I. Szleifer, D. Kramer, A. Ben-Shaul, D. Roux, and W. M. Gelbart, Phys. Rev. Lett., 60:1966 (1988).

30. For this SDS-alcohol series which consists of negatively charged membranes separated by water the electrostatic interactions are long range (Eq. (13)) and significantly larger than all other

forces such as hydration (Eq. (12)), van der Waals (Eq. (11)), and undulations (Eq. (10)). Consequently, the power-law exponent η is dominated by electrostatic forces and gives a direct measurement of k_c (Eq. (19)).

SPONTANEOUS AND INDUCED ADHESION OF FLUID MEMBRANES

Wolfgang Helfrich

Fachbereich Physik
Freie Universität Berlin
Arnimallee 14, D-1000 Berlin 33, FRG

ABSTRACT

Soft membranes have recently been shown to possess many novel properties due to their out-of-plane fluctuations. In particular, spontaneous adhesion including a temperature-driven unbinding transition and adhesion induced by lateral tension have been observed with electrically neutral biological model membranes. We review existing theories of adhesion in
terms of direct forces and undulation forces. An outline of the experimental results is also given. Some contradictions between theory and experiment suggest an unknown submicroscopic roughness of these membranes. It
should far exceed the contribution of undulations, i.e. of the out-of-
plane fluctuations controlled by bending rigidity and lateral tension.

1.   INTRODUCTION

The "hard" membranes of engineers have been known for a long time.
Here we are interested in "soft" membranes, either fluid or solid, but
always affected in their properties by thermal fluctuations. In contrast
to biological membranes with their enormous compositional and structural
complexity, the soft membranes we have in mind are simple and may consist
of a single species of molecules, thus being suitable objects for doing
physics. Fluid monolayers of amphiphilic molecules are, perhaps, the best-
known example. They form micelles in water or inverse micelles in oil of
spherical or cylindrical shape.[1] These can be disconnected or connected,
disordered or ordered. If the medium contains both water and oil the
amphiphilic monolayers tend to form coated drops of one material in the
other. The resulting microemulsions are, in special cases, bicontinuous
so that infinite systems of oil and water interpenetrate.[2] Amphiphilic
monolayers of alternating polarity arranged in multilayer systems with
intercalated water or oil represent lyotropic smectic liquid crystals.
Of particular interest are normal bilayers, i.e. paired monolayers embedded in water. If exposed to an excess of water, stable bilayers may
form single-walled vesicles and other unilamellar structures including a
bicontinuous water/water system.[3] Single bilayers of amphiphilic molecules
in water are in some respects very similar to biological model membranes
if made of lipid molecules extracted from the latter.

Tethered membranes are another type of soft membranes. They may be obtained by polymerizing fluid membranes so that a continuous network of chemically bonded molecules is formed. An idealized representative is a hexagonal crystal with indestructible but flexible and stretchable bonds. Such systems cannot melt and keep their topological order when being deformed. Regular solid bilayers and monolayers may also be included among soft membranes. Solid lecithin bilayers were found to be easily deformable without breaking in micromechanical experiments.[4] Regular solid membranes can, of course, melt and are obtained by freezing from the fluid state.

Soft membranes attracted the attention of physicists in recent years because of novel properties related to their nonplanarity. Typically, thermal energies are sufficent to produce marked deviations from the planar state. In the case of fluid membranes, these fluctuations depend primarily on their bending elasticity. The energy of bending per unit area, g_c, is usually expressed by a quadratic form in the principal curvatures, c_1 and c_2 which are splays in the language of liquid crystals[5]

$$g_c = \frac{1}{2}\varkappa(c_1 + c_2 - c_0)^2 + \bar{\varkappa} c_1 c_2$$

Here $\varkappa$ is the bending rigidity and $\bar{\varkappa}$ the elastic modulus associated with Gaussian curvature $c_1 c_2$. A linear term arises from the spontaneous curvature c_0 of asymmetric layers. In the case of fluid membranes, solely this elasticity controls the equilibrium shapes of vesicles and determines whether flat layers, closed spheres, or periodic bicontinuous structures are the configuration of minimal elastic energy.[6] The out-of-plane fluctuations of fluid membranes are called undulations if they are governed by the bending rigidity (there might be additional mechanisms; see below). They give rise to a number of effects, such as an "absorption" of membrane area,[7] repulsive undulation forces between parallel membranes,[8] a crumpling of the single membrane when it is larger than the persistence length of orientation,[9] and before crumpling a decrease of the effective bending rigidity with membrane size[10]. Crumpling may occur at the droplet size of microemulsions or at more than astronomical dimensions, the critical size depending exponentially on the bare bending rigidity.

There is no doubt that tethered membranes are essentially flat at low temperatures. They have been predicted[11] to undergo a crumpling transition when being heated past a critical temperature, but very recent computer simulations seem to indicate that complete crumpling is prevented by the excluded-volume effect associated with overhangs.[12] The melting of regular solid membranes also differs from known theory when they are free to escape into the third dimension. It has been argued that the energy of forming a single dislocation may be finite so that in a strict sense there would be no solid state at all at nonzero temperatures.[13]

II.   INTRODUCTION TO ADHESION OF SOFT MEMBRANES

It is a frequent task in colloid science to prevent the collapse of solutions of particles in the face of van der Waals attraction. Electrostatic repulsion by surface charges and steric repulsion by grafted polymers frequently serve this purpose. In the case of soft membranes, the repulsion resulting from out-of-plane fluctuations is in principle capable of overcoming van der Waals attraction. This has been shown theoretically for fluid and solid membranes. The pure steric interaction energy drops as $1/\bar{z}^2$ for the former if due to undulations[8] and approximately as the third power of $1/\bar{z}$ for the latter,[14] $\bar{z}$ being the mean spacing.

In the following, we will consider the question of adhesion versus separation which has been studied experimentally for some electrically neutral biological model membranes. These relatively stiff lipid bilayers turned out to separate[15-17], in contrast to what is generally believed. One of the materials has recently been found to undergo a reversible unbinding transition in salt solution.[18] While spontaneous adhesion is restricted to this particular material, all of them display mutual adhesion induced by ultralow lateral tensions. The emphasis here is on collecting some theoretical predictions, both for spontaneous and induced adhesion, all of which are made on the assumption that intermembrane repulsion is due to undulations. The theoretical results may be applicable in the future to other systems. The biological model membranes satisfy some predicted scaling laws, but are much more sensitive to lateral tension than the theory permits. This seems to indicate that on a submicroscopic scale these bilayers are much rougher than is to be expected from their undulations.

III. INTERACTIONS OF FLUID BILAYERS IN WATER

Soft membranes interact with each other "directly" through force fields and "indirectly" through out-of-plane fluctuations. The three direct interactions[19] are best written down for two (or more) parallel membranes of uniform spacing z. We begin with van der Waals interaction, expressing its energy per unit area by

$$g_{vdw} = - \frac{H}{12 \pi z^2} \qquad (1)$$

or

$$g_{vdw} = - \frac{H}{12 \pi} \left[\frac{1}{z^2} - \frac{2}{(z + b)^2} + \frac{1}{(z + 2b)^2} \right] \qquad (2)$$

where H is the Hamaker constant. Eq. (1) is the so-called half-space approximation, while the finite membrane thickness b is taken into account in Eq. (2). The latter formula presupposes additivity of molecular interactions. However, while the variation with $1/z^4$ at large spacings is correct, the very large dielectric constant of water was predicted to increase the range of approximate validity (within a factor of two) of the half-space approximation to three or four times the membrane thickness.[20] The scaling laws of electrostatic repulsion may be expressed, for fixed surface charge, by[15]

$$g_{es} \sim \ln \frac{z_o}{z} , \frac{1}{z} , \text{ or } e^{-z/\lambda_D} \qquad (3)$$

where the length z_o is an integration constant and λ_D is the Debye screening length. The first law represents an ideal gas of ions, the middle one may intervene at larger spacings, and the last one cuts off the interaction at spacings of the order of λ_D. The third direct interaction, again repulsive, results from the hydration of the membranes and is usually written as

$$g_{hyd} \sim e^{-z/\lambda_h} \qquad (4)$$

where λ_h is a characteristic length of the order of 2 Å.

The indirect interaction of fluid amphiphilic bilayers is generally assumed to arise from their undulations, i.e. from out-of-plane fluctuations controlled by the bending rigidity. Pure steric interaction is characterized by a hard core potential for the direct interaction. Its energy per unit area in a stack of membranes was calculated to be[8]

$$g_{und} = \frac{3 \pi^2}{128} \frac{(kT)^2}{\varkappa \bar{z}^2} \qquad (5)$$

where $\bar{z}$ is the mean spacing of adjacent membranes. The numerical factor of this formula has been confirmed within ± 20 percent by Safinya, Roux, and coworkers[21] in experiments with membranes of bending rigidities so low ($\varkappa \approx kT$) that van der Waals attraction is negligible. In the case of biological model membranes ($\varkappa$ = 3 kT to 50 kT) indirect as well as direct interactions have to be taken into account. The superposition of both is hardly ever exact but may be qualitatively correct, provided $g_{dir} \sim (-1/z^n)$ with n < 2. We will immediately return to this point which was made by Lipowsky and Leibler.[22]

IV. THEORY OF THE UNBINDING TRANSITION

The simplest criterion of whether to expect spontaneous adhesion or separation may be obtained from the comparison of undulatory repulsion and van der Waals attraction in the half-space approximation since both obey the same inverse square law with respect to spacing.[8] From (1) and (5) and putting $z = \bar{z}$, we obtain

$$\frac{H_c}{12 \pi} = \frac{3\pi^2}{128} \frac{(kT)^2}{\varkappa}$$

(6)

for the "critical" Hamaker constant on the borderline. Actually, H_c is likely to be somewhat smaller because z is locally less than $\bar{z}$ and, in addition, its distribution is distorted by the direct interaction. We recall that the superposition underlying the comparison is only marginally applicable (see also below).

A theory of the unbinding transition was put forward, for a pair of undulating membranes, by Lipowsky and Leibler[22] who did a renormalization group calculation for the interaction potential. Starting from a direct potential composed of (2) and (4), they found in numerical computations a critical Hamaker constant nearly identical to that derived from (6). They also predicted the transition to be continuous, the mean spacing varying as

$$\bar{z} \sim (H - H_c)^{-1}$$

(7)

just below the critical point. The superposition of direct and undulatory interactions would lead to a first-order transition. Its failure is linked to the fact that at large spacings the direct interaction is given by (2), thus varying as $1/z^4$. A difficulty with the renormalization group method is a strong dependence of H_c on the step size of renormalization.[23]

Noting that the flow equations of renormalization are the same for a fluid membrane in 3d space as for a linear interface in 2d space, Lipowsky[24] pointed out that for a pair of membranes the question of binding versus unbinding can also be handled in terms of a Schrödinger equation. One recovers (7), but the numerical, i.e. nonuniversal, factors may be as inaccurate as in the renormalization group calculation on which the analogy is based.

Since the critical exponent of unbinding is unity and the distribution of membrane spacings can be described by a squared "wave function", one may hope to find a very straightforward access to the unbinding transition. We propose now such an approach, calling it "reflection model".[25] Considering first a single membrane interacting with a box-like potential well (see Fig. 1), we may write for its total energy per unit area the following sum of undulatory and direct interaction energies.

$$g_{total} = \alpha \frac{(kT)^2}{\varkappa} \frac{1}{w^2} - \beta \frac{Ud}{w}$$

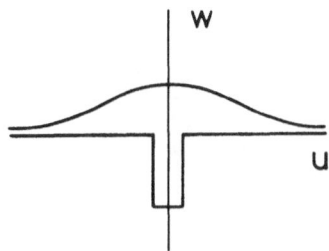

a. Distribution function w(u) of membrane bound by potential well.

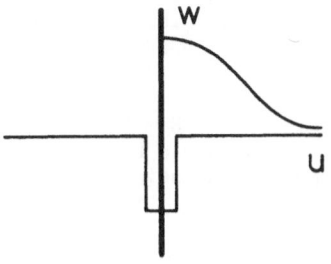

b. Distribution function w(u) of membrane bound by potential well with a rigid wall at u=0.

c. Reflection of membrane by the rigid wall and associated bending of membrane.

Fig. 1.

Here α and β are factors near unity, U and d are the depth and the width, respectively, of the potential well, and w ($\gg$ d) is the width of a Gaussian distribution function for the distance u of the membrane from the origin. Minimization of g_{total} leads to

$$w \sim \frac{(kT)^2}{\varkappa\, Ud} \tag{8}$$

According to this equation, the membrane is bound by any attractive potential regardless of its strength. Next, we place a hard wall in the center of the potential well (see Fig. 1), forcing the membrane to stay on one side of it. With the simplifying assumption that the membrane bends back like being reflected within the remaining potential well (see Fig. 1), we may write

$$g_{dir} = - \beta \frac{(U - B)d}{w}$$

Here B is the bending energy of reflection per unit area where the membrane is in the potential well. Note that B is not entirely constant but varies as φ^3/d where $\varphi^2 \sim \ln(w/w_0)$ is the mean square deflection angle due to undulations.[26] Replacing finally U in (8) by U - B we obtain

$$w \sim (U - B)^{-1}$$

which suggests a continuous unbinding transition at U_c = B, again with the same critical exponent. The reflection model seems suitable to deal with logarithmic corrections of the unbinding transition. We think that the increase of B with w makes the transition very weakly first order, but refrain here from a further discussion of this possibility. The problem of a membrane in front of a wall can be easily mapped on that of two interacting membranes.

Accurate predictions of the critical value of the Hamaker constant (or some other parameter instead of it) appear impossible by any of the four methods. Such a situation calls for computer simulation, first results of which have already been published.[27]

V.   THEORY OF ADHESION INDUCED BY LATERAL TENSION

Lateral tensions in soft membranes are essentially positive since these systems buckle so easily. Their action can only diminish the strength of out-of-plane fluctuations, thus weakening steric interaction. This may induce mutual adhesion where in the absence of stress the membranes would separate. In the following we deal with the effect of lateral tension on undulation forces and with the resulting equilibrium spacing. For simplicity the direct interaction is represented by the half-space approximation (1) of van der Waals attraction with $H < H_c$. In a first treatment, we will employ superposition although we are only on the borderline of its applicability.

The two forces between membranes may then be expressed by the negative van der Waals pressure

$$P_{vdw} = - \frac{H}{6 \pi z^3} \tag{9}$$

obtained from (1) and the positive undulation pressure

$$P_{und} = \frac{3\pi^2}{64} \frac{(kT)^2}{\varkappa \bar{z}^3} \frac{S(0)}{S(\sigma)} \tag{10}$$

obtained from (5) but with the factor $S(0)/S(\sigma)$ to account for the effect of a lateral tension σ. $S(\sigma)$ is the plaquette size in the sense of an independent-membrane-piece approximation.[25] We use the ratio of the plaquette sizes without and with tension rather than calculate P_{und} ab initio, which is also possible but gives a poor numerical factor. The mean square amplitude of undulations of a spherical or quadratic membrane piece of area S was calculated[26] to be

$$\langle u^2 \rangle = \frac{kT}{4\pi\sigma} \ln \left(1 + \frac{\sigma S}{\varkappa \pi^2} \right) \tag{11}$$

On the other hand, the mean square amplitude of an undulating membrane at zero tension between rigid plates was shown to approximately obey

$$\langle u^2 \rangle = \frac{d}{6}$$

where 2d is the separation of the plates. Putting $\bar{z} = d$, applying the last two formulae without any adjustments to pairs as well as stacks of membranes, and eliminating $\langle u^2 \rangle$ between them, we obtain the plaquette size

$$S(\sigma) = \frac{\pi^2 \varkappa}{\sigma}\left[\exp\left(\frac{2\pi\sigma\bar{z}^2}{3kT}\right) - 1\right].\tag{12}$$

It is easy to see that the right-hand sides of both (11) and (12) tend to finite values independent of σ when σ goes to zero. For $2\pi\sigma\bar{z}^2/3kT \ll 1$ we may expand the exponential in (12), arriving at

$$\frac{S(0)}{S(\sigma)} = 1 - \frac{\pi\sigma\bar{z}^2}{3kT}$$

to first order in $\bar{z}^2$. If this is inserted into (10), the equilibrium condition $P_{vdw} + P_{und} = 0$ becomes

$$\bar{z}^2_{eq} = \frac{H_c - H}{H_c}\,\frac{3}{\pi}\,\frac{kT}{\sigma}\tag{13}$$

where use has been made of (6). For fixed H and T we have the scaling law

$$\sigma \sim 1/\bar{z}^2_{eq}.\tag{14}$$

Next to the equilibrium mean spacing we are most interested in the energy of adhesion. Inspection shows that the energy of bringing membranes under equal tensions together obeys the scaling law

$$\int\limits_{\bar{z}_{eq}}^{\infty}\left[P_{vdw} + P_{und}\,(\sigma)\right]\,d\bar{z} \sim \sigma\tag{15}$$

whenever $S(0)/S(\sigma)$ depends on σ through the combination $\sigma\bar{z}^2$. It may be noted that for $\sigma\bar{z}^2 \ll kT$ the negative exponent in the exponential governing the decay of P_{und} is expected to be proportional to $\sigma 1/2\bar{z}$ rather than its square.[28,29]. However, (15) need not be identical to the adhesion energy as the tension delivers work to the membrane if the membrane area increases. Another quantity relevant in this context is the energy per unit area of stretching a free membrane,[26]

$$g_f = \frac{kT}{8\pi\varkappa}\,\sigma + \frac{1}{2\varkappa_s}\,\sigma^2\tag{16}$$

It is also proportional to σ at low tensions where the flattening of undulations is the predominant effect. The regular, quadratic term is negligible for tensions below ca. 0.1 dyn cm^{-1} as the stretching modulus $\varkappa_s$ is of the order of 200 dyn cm^{-1} for biological model membranes.[4] In dealing with induced adhesion, we have to distinguish between the tensions σ_f and σ_b in a given membrane where it is free and bound, respectively. They are linked according to

$$\sigma_b = \sigma_f \cos\psi\tag{17}$$

through the contact angle ψ made by the free and adhering parts of the membrane. The adhesion energy per unit area may be generally written as

$$g_a = g_f(\sigma_f) - g_b(\sigma_b)\tag{18}$$

where $g_b\,(\sigma_b)$ is the energy of simultaneously stretching parallel membranes which all the time are in mutual equilibrium so that $\bar{z} = \bar{z}_{eq}$. Note that $g_b\,(\sigma_b)$ must be positive in the absence of spontaneous adhesion. Eq. (18) refers to a stack of membranes; its right-hand side has to be multiplied by 2 in the case of symmetric adhesion, i.e. a pair of membranes under equal tensions. From (1), (5), (15) and the linear part of (16) we may infer $g_b \sim \sigma_b$. The proportionalities of g_f and g_b to their respective tensions imply in turn

$$g_a \sim \sigma_f \qquad (19)$$

and a tension-independent contact angle. However, a further equation is needed to obtain the latter. We may adopt the Young equation from the theory of interfaces,

$$g_a = (1 - \cos \psi)\sigma_f \qquad (20)$$

where again a factor 2 has to be inserted on the right-hand side in the case of symmetric pair adhesion. Eqs. (18) and (20) can be solved for g_a/σ_f and ψ, both constant, if the factor of proportionality in $g_b \sim \sigma_b$ is known.

Unfortunately, careful inspection of the derivation[26] of (16) shows that the stretching energy should be independent of $(H_c - H)/H_c$, i.e. $g_b/\sigma_b = g_f/\sigma_f$, at least if the lower cutoff for the undulation wave vector can be disregarded. (This is because the area absorbed by undulations depends always logarithmically on σ.) As a result (18) and (20) would be identical apart from a factor of proportionality, which rules out their solution. Therefore, our model has to be enlarged or modified if it is to explain induced adhesion.

The scaling law (14) has been obtained previously without recourse to the superposition of direct and indirect forces by using two characteristic lengths.[16] One is the lateral correlation length[22]

$$\xi_\parallel = (\frac{\kappa}{kT})^{1/2} \bar{z} \qquad (21)$$

of sterically interacting membranes, the other the orientational correlation length

$$\xi\sigma = (\frac{\kappa}{\sigma})^{1/2} \qquad (22)$$

of a membrane under tension. It can be argued on general grounds that, for $H < H_c$, the total intermembrane pressure should have the functional form

$$P_{tot} = \frac{3\pi^2}{64} \frac{(kT)^2}{\kappa \bar{z}^3} F(\frac{H_c \kappa}{(kT)^2}, \frac{\xi_\parallel^2}{\xi_\sigma^2}) \qquad (23)$$

where the second independent variable in the function $F(\leq 1)$, $\xi_\parallel^2/\xi_\sigma^2 = \sigma\bar{z}^2/kT$, does not depend on κ. The equilibrium condition $\partial P_{und}/\partial \bar{z} = 0$ leads immediately to $\sigma \sim 1/\bar{z}_{eq}^2$. Unlike this scaling law, eq. (13) for the mean spacing as a function of tension and Hamaker constant hinges on the superposition of direct and indirect forces. The use of superposition is problematical since we are dealing with the marginal case $g_{dir} \sim (-1/z^2)$. In fact, primitive estimates based on the wave function analogy indicate that the collapse of tension-free membranes takes place at a final P_{tot} which is less than four times weaker than P_{und}. (At the point of collapse the wave function starts out as $z^{1/2}$ at $z = 0$, which excludes a divergence of its square at zero spacing. Therefore, the membrane is, at worst, "reflected" at $z = 0$, which implies $P_{tot} \leq (1/4) P_{und}$.) This casts doubt on the general validity of (13), suggesting a lower bound for $\bar{z}_{eq}$ as a function of $(H_c - H)/H_c$ which is roughly rendered by

$$\bar{z}_{eq}^2 > \frac{3}{4\pi} \frac{kT}{\sigma} \qquad (24)$$

278

Our predictions concerning induced adhesion rest on the assumption that
the direct interaction is satisfactorily described by van der Waals
attraction in its half-space approximation (1). This restricts the range
of mean spacings where they may be expected to be valid to, perhaps,
5 nm < $\bar{z}$ < 15 nm (see above). Moreover, we have considered only the
stiff-membrane limit, which means that the difference between real membrane area and projected area is neglected except in calculating the
energy of stretching. The relative decrease of projected area due to undulations was shown previously to be[26]

$$\frac{\Delta A}{A} = \frac{kT}{8\pi\kappa} \ln \frac{q_{max}^2}{q_{min}^2} \tag{25}$$

Here $q_{max}^2 = \pi^2/a^2$, a being a molecular length, while $q_{min}^2 = \pi^2/A$ or σ/κ,
whichever is larger, A being the membrane size. The numerical factor in
(25) is $1.6 \cdot 10^{-3}$ and $1.6 \cdot 10^{-2}$ for $\kappa = 1 \cdot 10^{-12}$ erg and $1 \cdot 10^{-13}$ erg,
respectively. The ratios of projected to real area depend only logarith-
mically on tension and can be easily incorporated in (18) and (20) which
for small contact angles are relatively sensitive to them.[16] They differ
from unity by no more than a few percent in the case of the larger of the
two rigidities. Because the ratio depends on whether the membrane is free
or bound this complement can, in principle, make (18) and (20) solvable.

VI.   OUTLINE OF EXPERIMENTAL RESULTS

As stated at the beginning, some electrically neutral biological
model membranes have been investigated for spontaneous and induced adhesion. The materials were swollen in water and the resulting structures
involving single membranes were observed in optical phase contrast microscopy.

A first unbinding transition has very recently been found with
DGDG (= digalactosyldiacylglyceride) in 0.1 M NaCl solution.[18] The
material is a natural product extracted from vegetable membranes. The
fluid bilayers separate in water at room temperature. In salt solution,
they separate only at elevated temperatures and undergo a well defined,
reversible binding transition when being cooled. No hysteresis was observed, in accordance with a continuous phase transition. The occurrence
of the transition is compatible with the criterion (6) and that of
Lipowsky and Leibler on the basis of available data, $\kappa = (0.12 - 0.21) \cdot 10^{-12}$ erg and H = $(3.1 - 7.5) \cdot 10^{-14}$ erg.

Adhesion induced by lateral tension has been studied extensively
for egg lecithin (= egg phosphatidylcholine) membranes in pure water and
salt solutions.[16] The material parameters of the bilayers are
$\kappa = (0.8 - 2) \cdot 10^{12}$ erg and H = $(1.3 - 6) \cdot 10^{-14}$ erg. A first series of
experiments utilized the rare adhesive contacts seen after the "disorderly"
swelling in water of very small quantities of the material. The tension
was calculated from a rounding of the adhering membrane next to the contact
area. Contact rounding is governed by the orientational correlation
length (22), thus being optically resolvable only at ultralow tensions.
In the range of measured tensions, ca. 10^{-6} to 10^{-4} dyn cm^{-1}, the contact
angles were, independently of σ_f, ca. 40° for symmetric adhesion and ca.
70° for adhesion of a single membrane to a stack. The constancy of the
contact angles agrees with the above predictions. Adhesion energies
$g_a \approx (1/2) \sigma_f$ are obrtained, if the contact angles are inserted in the
Young equation (20). They are in sharp conflict with the upper bound

$$g_a \leq \frac{kT}{8\pi\varkappa} \, \sigma_f$$

derived from (18) which, with $\varkappa = 1 \cdot 10^{-12}$ erg and $kT = 10^{-14}$ erg, becomes $g_a \leq 1.6 \cdot 10^{-3} \, \sigma_f$. The contradiction is almost equally sharp in the case of DGDG with contact angles of induced adhesion of ca. 60° for symmetric adhesion and ca. 85° for adhesion to a stack.[18] A preliminary study of a few cephalins (= phosphatidylethanolamines) showed bending rigidities and contact angles very similar to those of egg lecithin.[30]

In a second series of experiments,[15,17] slightly hydrated egg lecithin was squeezed flat to a final cell thickness of ca. 20 µm and then exposed to free water which filled the remainder of the sample cell. The procedure resulted in well-ordered multilayer systems, the membranes being mostly parallel to the glass slides but forming concentric semi-circles at the border to water. The "orderly" swelling of these systems resulted in compact myelin cylinders growing from the border into the free water. The lipid density, monitored by the intrinsic fluorescence of egg lecithin in the planar regions, decreased in the course of days or weeks. Most of the time, it was ca. two times lower in the curved regions, i.e. border and myelin cylinders, than in the planar regions. Induced adhesion[17] was achieved in a controlled and reversible way be cooling the samples after swelling at elevated temperatures. It manifested itself by a "phase separation" in the curved regions into single membranes running between water pockets and, on the other hand, domains of adhesion. The lateral tension was self-limiting, staying just below the strength necessary to induce adhesion in the planar regions. The experimental data appear to reproduce $\sigma \sim 1/\bar{z}_{eq}^2$, i.e. the scaling law (14), the tensions ranging from $3 \cdot 10^{-5}$ to $2 \cdot 10^{-3}$ dyn cm^{-1}. The spacings are surprisingly small, e.g. $\bar{z} = 10$ nm at $\sigma_f = 1 \cdot 10^{-4}$ dyn cm^{-1}. Insertion of these values into (13) gives $(H_c - H)/H_c \approx 10^{-3}$. This suggests a very close competition of undulatory and van der Waals attraction which is unlikely for a number of reasons. First, if the competitition were indeed so close, very small changes of the experimental parameters should produce a binding transition which was never observed with egg lecithin membranes. Second, the repulsive forces between the membranes in the absence of tension which could be approximately determinded were found to be of the order of pure undulation forces. Third, the wavefunction analogy, if it is reliable, rules out a weakening of the repulsion by more than a factor 4 as compared to pure undulatory interaction.

A submicroscopic roughness of electrically neutral biological model membranes, hitherto unknown, has been postulated to explain the large adhesion energies and small mean spacings. The membrane area absorbed by this roughness would have to be quite large, reaching perhaps 50 % of the projected area at vanishing lateral tension, which by far exceeds the area absorbed by undulations. A theoretical model of the conjectured roughness has been proposed[31] and experiments are under way to check the hypothesis.[32]

VII. CONCLUSION

Spontaneous and tension-induced adhesion of fluid membranes, including the unbinding transition, is a very new field of physics, both experimentally and theoretically. We have tried to give a reveiw of present activities, emphasizing theoretical concepts. The latter are all based on the assumption that the undulations controlled by the bending rigidity are the only out-of-plane fluctuations of these membranes. The theoretical models were developed with a view to electrically neutral biological model membranes. They fail in part when applied to these systems.

However, unless they are wrong, the failure means that biological model membranes are more complex than expected.

ACKNOWLEDGEMENT

I am grateful to R. Lipowsky for discussions.

REFERENCES

1. S. A. Safran, L. A. Turkevich, and P. Pincus, J. Physique Lettres, 45:L-69 (1984).
2. M. E. Cates, D. Andelman, S. A. Safran, and D. Roux, Langmuir, 4:802 (1988).
3. M. E. Cates, D. Róux, D. Andelman, S.T. Milner,and S. A. Safran, Europhys. Lett., 5:793 (1987).
4. E. Evans and D. Needham, J. Chem. Phys., 91:4219 (1987).
5. W. Helfrich, Z. Naturforsch., 28c:693 (1973)
6. W. Helfrich and W. Harbich, in "Physics of Amphiphilic Layers", J. Meunier, D.Langevin, and N. Boccara, eds., Springer Proceedings in Physics, Vol. 21 (1987).
7. W. Helfrich, Z. Naturforsch., 30c:841 (1973).
8. W. Helfrich, Z. Naturforsch., 33a:305 (1978).
9. P. G. de Gennes and C. Taupin, J. Phys. Chem., 86:2294 (1982).
10. W. Helfrich, J. Physique, 48:285 (1987); and references cited therein.
11. Y. Kantor, M. Kardar, and D. R. Nelson, Phys. Rev. Lett., 57:791 (1986).
12. F. F. Abraham, W. E. Rudge, and M. Plischke, Phys. Rev. Lett., 62:1757 (1989)
13. D. R. Nelson and L. Peliti, J. Physique, 48:1085 (1987).
14. S. Leibler and A. C. Maggs, to be published.
15. W. Harbich and W. Helfrich, Chem. Phys. Lipids, 36:39 (1984).
16. R. M. Servuss and W. Helfrich, J. Phys. France, 50:809 (1989).
17. W. Harbich and W. Helfrich,to be published.
18. M. Mutz and W. Helfrich, Phys. Rev. Lett., 62:2881 (1989).
19. J. N. Israelachvili, "Intermolecular and Surface Forces", Academic Press, Orlando (1985).
20. J. Mahanty and B. W. Ninham, "Dispersion Forces", Academic Press London (1976).
21. C. R. Safinya, D. Roux, G. S. Smith, S. K. Sinha, P. Dimon, N. A. Clark, and A. M. Bellocq, Phys. Rev. Lett., 57:2718 (1986).
22. R. Lipowsky and S. Leibler, Phys. Rev. Lett., 56:2541 (1986); 59:1983(E) (1987).
23. R. Lipowsky, private communication.
24. R. Lipowsky, Europhys. Lett., 7: 255 (1988).
25. W. Helfrich, unpublished.
26. W. Helfrich and R. M. Servuss, Il Nuovo Cimento, 30:137 (1984).
27. R. Lipowsky and B. Zielinska, Phys. Rev. Lett 62:1572 (1989).
28. D. Sornette, Europhys. Lett., 2:715 (1986).
29. W. Helfrich, unpublished.
30. M. Mutz and W. Helfrich, in preparation
31. W. Helfrich, Liquid Crystals, (in press).
32. B. Klösgen and W. Helfrich, in preparation.

IMPURITY-MODULATED INTERFACE FORMATION
IN LIPID BILAYER MEMBRANES
NEAR THE CHAIN-MELTING PHASE TRANSITION

John Hjort Ipsen and **Ole G. Mouritsen**
Department of Structural Properties of Materials
The Technical University of Denmark, Building 307
DK-2800 Lyngby, Denmark

Leonor Cruzeiro-Hansson
Department of Crystallography, Birkbeck College
University of London, Malet Street
London, WC1E 7HX, England

I. INTRODUCTION

The gel-to-fluid chain-melting transition in pseudo-two-dimensional lipid bilayer membranes induces formation of lipid domains of gel-like lipids in the fluid phase and and fluid-like lipids in the gel phase. The average domain size and in particular the average length of the one-dimensional interfaces between lipid domains and bulk have a dramatic temperature dependence with anomalies at the transition temperature. These anomalies are related to similar anomalies in response functions. The interfacial area may be modulated by intrinsic impurities which are interfacially active molecules such as cholesterol [1,2]. The properties of the interfacial area provide a means for understanding aspects of the functioning of certain biological membrane processes like the passive permeability of small ions and the activity of some membrane enzymes.

II. MODEL

For the description of the chain-melting phase transition of pure lipid bilayer membranes the microscopic model of Pink and collaborators has been adopted. This model takes into account the acyl-chain conformational statistics and the van der Waals interaction between various conformers in a detailed way, while the excluded volume effect is accounted for by assigning each lipid chain to a site in a triangular lattice. The acyl chain conformations are represented by ten single chain states $"\alpha"$, each described by a cross-sectional area A_α, an internal energy E_α and an internal degeneracy D_α. The second membrane component is assumed to be a stiff, hydrophobically smooth molecule with no internal degrees of freedom and a cross-sectional area A_C. The model parameters will be chosen so that the mixture display the properties of the DPPC-cholesterol bilayer system for small concentrations of cholesterol [3]. The impurity will hereafter be called $"$cholesterol$"$.

The Hamiltonian takes the form

$$
\begin{aligned}
\mathcal{H} = {} & \sum_i \sum_{\alpha=1}^{10} (E_\alpha + \Pi A_\alpha)\mathcal{L}_{\alpha,i} + \Pi A_C \sum_i \mathcal{L}_{C,i} \\
& - \frac{J_0}{2} \sum_{i,j} \sum_{\alpha,\beta=1}^{10} I_\alpha I_\beta \mathcal{L}_{\alpha,i}\mathcal{L}_{\beta,j} \\
& - \frac{J_0}{2} \sum_{i,j} \sum_{\alpha=1}^{10} I_\alpha I_C (\mathcal{L}_{\alpha,i}\mathcal{L}_{C,j} + \mathcal{L}_{C,i}\mathcal{L}_{\alpha,i}) - \frac{J_0}{2} \sum_{i,j} I_C^2 \mathcal{L}_{C,i}\mathcal{L}_{C,j}
\end{aligned}
\tag{1}
$$

J_0 is the strength of the van der Waals interactions. $\mathcal{L}_{\alpha,i}$ and $\mathcal{L}_{C,i}$ are occupation variables of lipid molecules (in the various chain states) and cholesterol molecules. The interaction strengths are given by the shape-dependent nematic parameters I_α and I_C [4]. Π is an internal pressure to assure bilayer stability.

III. METHOD OF CALCULATION

Monte-Carlo simulations of the equilibrium properties of the model are carried for a triangular lattice of 100×100 sites. The equilibrium is provided by a combination of Glauber (single-chain excitations) and Kawasaki dynamics (conserving the "cholesterol" content) [5]. Monte Carlo simulations allow for an accurate determination of thermal quantites like the isothermal compressibility:

$$
\chi(T) = -\left(\frac{\partial A}{\partial \Pi}\right) = (k_B T)^{-1}(< A^2 > - < A >^2)
\tag{2}
$$

or the average cluster-size:

$$
\bar{\ell}(T, x_C) = \sum_l l n_l^\alpha(T, x_C) / \sum_l n_l^\alpha(T, x_C)
\tag{3}
$$

where $\alpha = f$ in the gel phase and $\alpha = g$ in the fluid phase. n_l^α is the cluster-size distribution function where l is the cluster size.

IV. RESULTS

Monte Carlo simulations have been carried out for three pure lipid bilayer systems composed of lipids with different acyl chain lenghts: DMPC(14 carbons), DPPC(16 carbons) and DSPC(18 carbons). The Monte Carlo data are basis for calculations of isothermal compressibility and the average cluster sizes. The compressibilities display sharp peaks at the transition temperatures T_m and pronounced wings around T_m indicating strong lateral density fluctuations in the transition region. The lateral density flutuations are increasing away from the T_m for decreasing acyl-chain length. The strong lateral density fluctuation are accompanied by the formation of lipid domains, cf. Fig. 1. The average cluster-size distribution for simulations of DMPC, DPPC and DSPC, cf. Fig. 2, display the same characteristic features as the isothermal compressibility. The sizes of the domains are increasing for decreasing acyl-chain length. This is in accordance calorimetric measurements [6], which indicate that the main transition is accompaied with formation of clusters of increasing sizes for decreasing acyl-chain length.

The parameter I_C in the model of the lipid-cholesterol mixture Eq. (1) is fixed by relating a mean-field calculation of the phase diagram of the lipid-cholesterol mixture to the experimentally observed phase behaviour [3]. The phase diagram is characterized by a very narrow coexistence region terminating in a critical end-point at $x_C = 0.48$ and a modest freezing-point depression. Monte-Carlo data for the average cross-sectional area shows that the chain-melting transition is broadened by the precence of cholesterol. Furthermore cholesterol has a condensing effect on the bilayer system at high temperatures. Calculations of the isothermal lateral compressibility by Monte Carlo

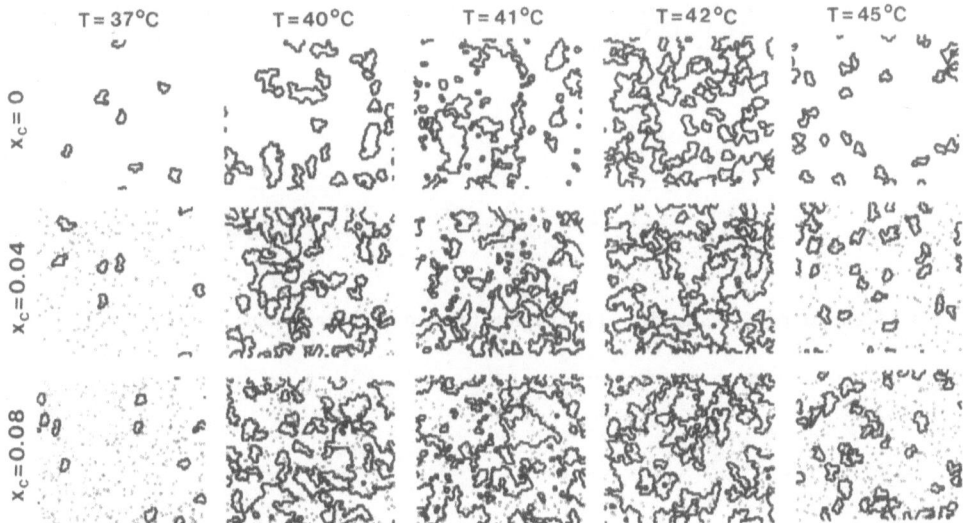

Fig. 1. Snapshots of membrane microconfigurations for varying temperature and cholesterol content. The lipid-domain interfaces are marked with black symbols and cholesterol molecules with open circles.

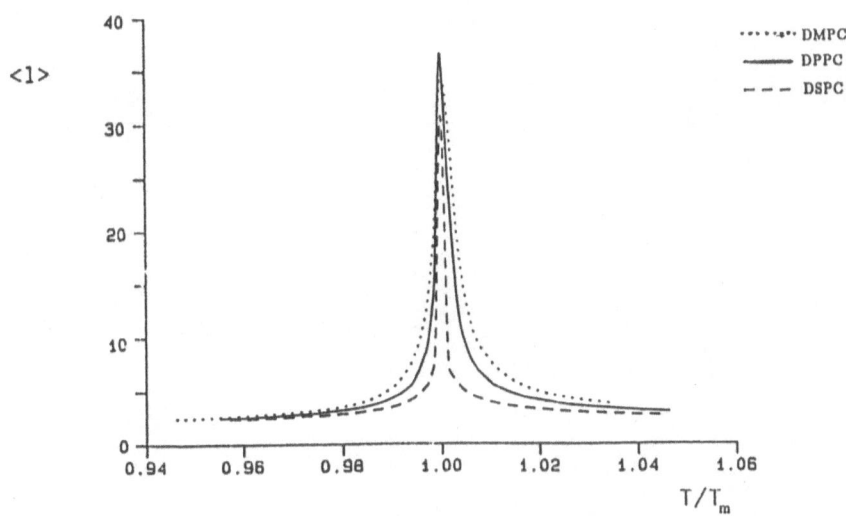

Fig. 2. Average cluster-size distribution for simulations of DMPC, DPPC and DSPC.

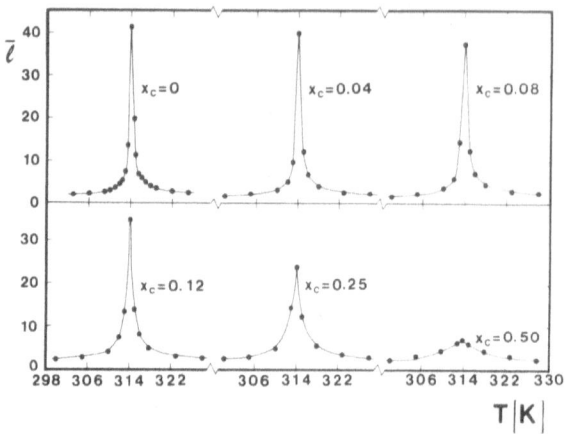

Fig. 3. The average cluster size, $\bar{\ell}(T, x_C)$, for different cholesterol concentrations, x_C.

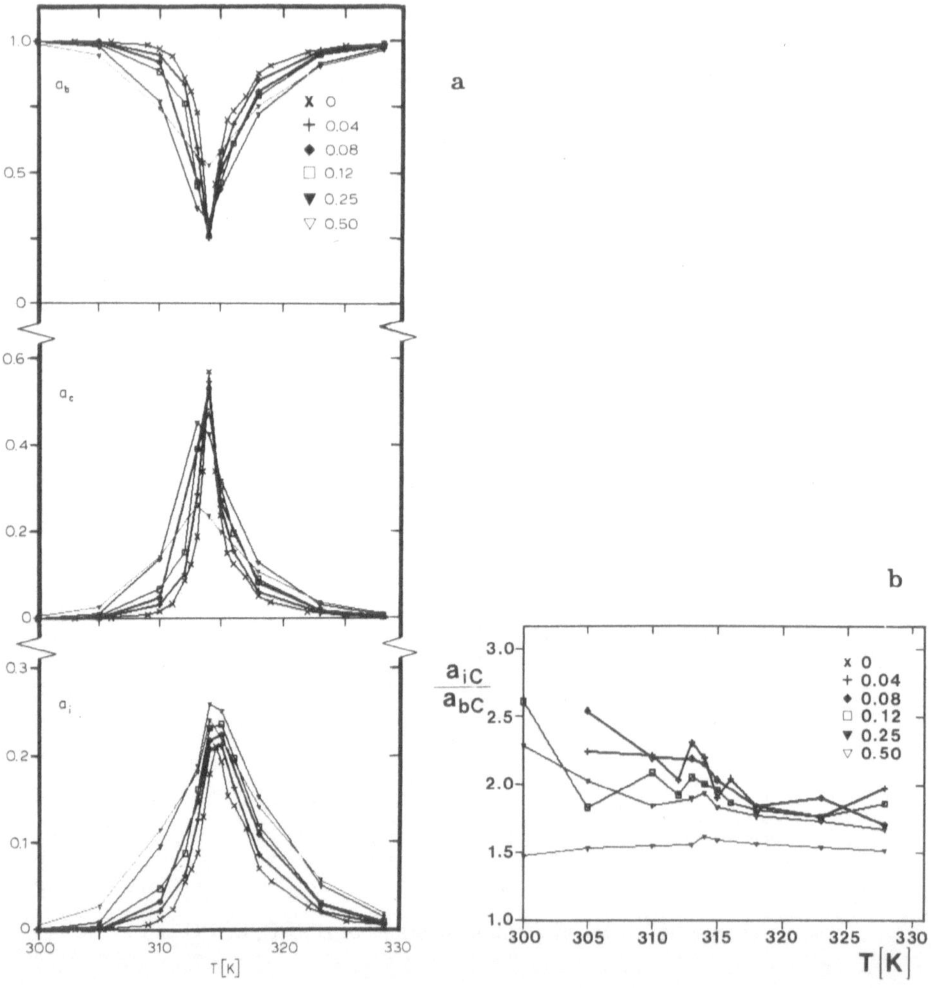

Fig. 4 (a): The variation with temperature and cholesterol content of fractional bulk, cluster and interfacial areas. **(b):** The ratio of the cholesterol concentration in the interface to that of the bulk.

techniques show that the peak at the transition temperature is reduced as x_C is increased. Concominantly, the compressibility broadens, i.e., there is an enhancement of the fluctuations away from the transition! The average cluster size, $\bar{\ell}(T, x_C)$, cf. Fig. 3 shows a sharp peak that loses its intensity in the centre as x_C is increased. Concominantly, there is an increase of $\bar{\ell}$ away from the transition temperature. This suggest in analogy with the result for the lateral compressibility that cholesterol enhances the thermal fluctuations and accompanying cluster formation phenomena in the wings of the transition. The microscopic phenomena wich accompany the strong lateral density fluctuations near the chain-melting transition are illustrated by a series of snapshots of the microconfigurations for different temperatures and cholesterol contents in Fig. 1. Only the lipid-domain interfaces and cholesterol molecules are marked in the figure. Cholesterol clearly induces larger and more ramified clusters. This is quantified in Fig. 4a where the variation with temperature and cholesterol content of the three areas in which the membrane can be divided: the bulk area a_b, the cluster area a_c and the interfacial area a_i. a_i has a maximum slightly above the main transition temperature and it generally increases as x_C increases. The figure clearly illustate that cholesterol enhance the lipid-domain interfaces away from the transition temperature.

Fig. 1 furthermore indicates a clustering of the clusters themselves and the resulting superclusters appear to be 'glued' together by locally elevated levels of cholesterol. Even more striking is the qualitative observation that the distribution of the cholesterol molecules in the plane of the membrane is not random. Indeed, cholesterol molecules tend to accumulate in the lipid-domain interfaces. This is quantified in Fig. 4b which shows the ratio of the cholesterol concentration at the interfaces to that of the bulk. This figure shows that there is more than twice the amount of cholesterol at the lipid-domain interfaces than in the bulk. Cholesterol is clearly not randomly distributed in the membrane plane, but tend to accumulate in lipid-domain interfaces.

V. CONCLUSION

The formation of lipid domains and associated interfaces in lipid membranes can be modulated by system parameters like acyl-chain length and impurity content. The origin of the lipid-domain formation in lipid bilayer membranes is pseudo-critical flutuations acompanying the chain-melting transition. For decreasing acyl-chain length the transition is driven closer to a critical point giving rise to the enhanced fluctuations. Cholesterol molecules prefer to be situated in lipid-domain interfaces, where they cause a lowering of the interfacial energy and the formation of larger and more ramified clusters away from the transition.

These findings may be important for the understanding of the regulation a number of biological membrane processes, which is believed to be controlled by interfacial properties. Some enzymatic processes associated with membranes are suggested to be related to lipid-domain interfaces. A striking example is that of pancreatic phospholipase for which it is known that hydrolysis only occurs in the transition region [7]. This observation has been linked to the lateral density fluctuations and the formation of a particular interfacial inviroment which supports the active conformation of the enzyme. The addition of small amounts of cholate, which is similar to cholesterol, to a lipid membrane containing phospholipase strongly increases the activity of the enzyme at temperatures away from the main transition temperature. This is consistant with the interpretation that cholate increases the interfacial area. Another important membrane process which can be related to lateral density fluctuations is the passive permeability of small ions. It has recently been shown [8] that the temperature variation of the permeability can be well described by assuming that lipid-domain interfacial lipids give rise to higher ion leakeage probability than lipids in the bulk or the clusters. Predictions of temperature dependence of the passive permeability for lipid bilayers with small concentrations of cholesterol are presented in Ref. 1.

ACKNOWLEDGEMENTS

This work was supported by the Danish Natural Science Research Council under grants Nos. 5.21.99.72 and 11-7498.

REFERENCES

1. L. Cruzeiro-Hansson, J. H. Ipsen, and O. G. Mouritsen, Biochim. Biophys. Acta **979**, 166 (1989).
2. N. Gheriani-Gruszka, S. Almog, R. L. Biltonen, and D. Lichtenberg, J. Biol. Chem. **263**, 11808 (1988).
3. T. N. Estep, D. B. Mountcastle, R. L. Biltonen, and T. E. Thompson, Biochemistry **17**, 1984 (1978).
4. A. Caillé, D. A. Pink, F. de Verteuil, and M. J. Zuckermann, Can. J. Phys. **58**, 581 (1980).
5. O. G. Mouritsen, *Computer Studies of Phase Transitions and Critical Phenomena* (Springer-Verlag, New York, 1984).
6. E. Freire and R. Biltonen, Biochim. Biophys. Acta **514**, 54 (1978).
7. M. Menashe, G. Romero, R. L. Biltonen, and D. Lichtenberg, J. Biol. Chem. **261**, 5328 (1986).
8. L. Cruzeiro-Hansson and O. G. Mouritsen, Biochim. Biophys. Acta **944**, 63 (1988).

TEMPERATURE-DEPENDENT GROWTH OF FRACTAL
AND COMPACT DOMAINS

Hans C. Fogedby
Institute of Physics, Aarhus University
DK-8000 Aarhus C, Denmark

Erik Schwartz Sørensen
Department of Physics
University of California at Santa Cruz
Santa Cruz, CA 95064, USA

Ole G. Mouritsen
Department of Structural Properties of Materials
The Technical University of Denmark, Building 307
DK-2800 Lyngby, Denmark

I. INTRODUCTION

Recent studies[1,2] of physical systems exhibiting fractal growth patterns have made it clear that the fractal morphology is usually a consequence of a non-equilibrium condition which is introduced either as an intrinsic irreversibility, an effective zero-temperature condition, or simply as a constraint in observation time. By lifting the non-equilibrium condition, one will observe restructuring and compact growth[3-6] of the system which eventually, in thermodynamic equilibrium, will have a compact morphology. A proper study of the crossover from non-equilibrium fractal growth to equilibrium compact growth requires a temperature as well as a relaxation mechanism. The way to meet these requirements is to introduce a Hamiltonian which governs a spontaneous thermally driven ordering process and which controls the full scenario from nucleation, fractal growth and coarsening of ordered domains, to equilibrium compact growth of the domains.

II. THE MODEL

We have performed a computer-simulation study[7,8] of a a microscopic two-dimensional interaction model which permits an investigation of the temperature-dependent growth of fractal and compact domains and the crossover from non-equilibrium fractal growth to equilibrium compact growth. The model, which is the site-diluted two-state model first suggested by Doniach,[9] is formulated in terms of Ising spin variables, $\sigma_i = \pm 1$, where $\sigma_i = +1$ refers to the fluid state and $\sigma_i = -1$ refers to the solid state. The two states carry different internal entropies, $S_{+1} \gg S_{-1} = 0$, and internal energies, where we have introduced the occupation variables, $\eta_i = 0, 1$, and where the second summation is only over nearest neighbors of the square lattice. For a suitable set of

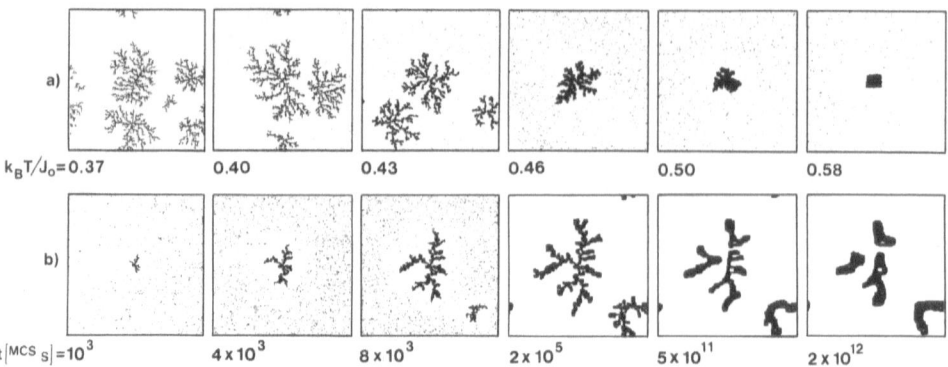

Fig. 1. (a) Typical configurations in the early-time regime, $t = 5000$ MCS/S, for different temperatures T. Both fluid and solid particles are denoted by black dots. The connected structures are solid. (b) Typical configurations at different times, t, in the late-time regime at a temperature $k_B T/J = 0.43$.

$E_{+1} > E_{-1} = 0$. The Hamiltonian is given by

$$\mathcal{H} = \sum_i (E_{\sigma_i} - T S_{\sigma_i}) \sigma_i \eta_i - J \sum_{<i,j>} (1 - \sigma_i)(1 - \sigma_j) \eta_i \eta_j, \tag{1}$$

model parameters this model has a stiff equilibrium first-order phase transition from a low-temperature phase characterized by particles predominantly in the $\sigma_i = -1$-state (the solid phase) to a high-temperature phase characterized by particles predominantly in the $\sigma_i = +1$-phase (the fluid phase).

III. COMPUTATIONAL TECHNIQUES

The dynamical process associated with the interaction model is devised as a combination of two types of transitions: (a) Kawasaki-type nearest-neighbor pair exchange of fluid particles and vacancies, and (b) Glauber-type single-site fluid-solid conversion. The dynamics of the solidification process has two different time regimes: (i) early-time nucleation, growth and total solidification, and (ii) late-time restructuring and compact growth. In the early-time regime we have employed conventional Metropolis Monte Carlo sampling. Since the late-time regime involves breaking and reformation of bonds which in the present model is much slower than the events dominating the early-time regime, we have in this regime used the continuum-time method of Bortz et al.[10] The time parameter, t, is in units of MCS/S (Monte Carlo steps per site).

The calculations are carried out on lattices with 256×256 sites subject to periodic boundary conditions. The Hamiltonian parameters are $E_{+1}/J = 2.42$ and $S_{+1}/k_B = 20.72$ and the concentration of particles is $\langle \eta_i \rangle = 0.1$. Other choices of these parameters lead to results similar to those reported below. The particle concentration has to be low, however, to capture an early-time fractal solidification.[11,12]

IV. RESULTS

In Fig. 1a are shown typical early-time configurations of the system at $t = 5000$ MCS/S for an increasing series of temperatures. From the visual appearance of the solid regions it is clear that the temperature has a significant influence on the morphology. As the growth temperature is increased there is distinct crossover from tenuous, fractal structures to bulk domains. As clearly seen on Fig. 1a, this crossover takes place concurrently with a decrease in the high local vacancy concentration in the active growth zone. Hence the crossover may also be considered a transition from vacancy-controlled unstable growth to stable growth. It is noted that at low temperatures there is a high

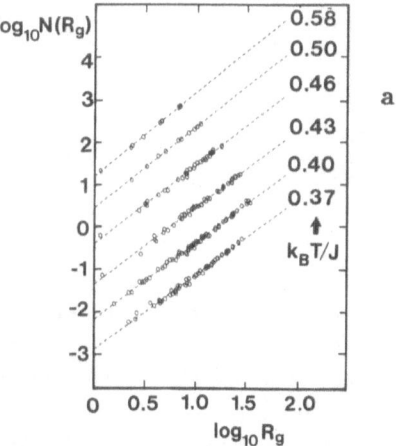

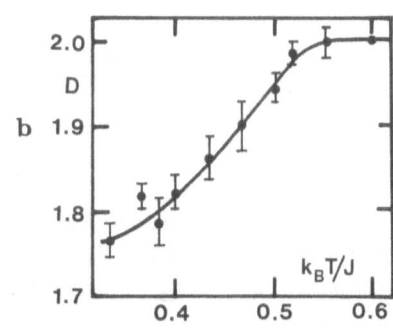

Fig. 2. (a) Double-logarithmic plot of particle content, $N(R_g)$, of solidified clusters vs radius of gyration, R_g, at time $t = 5000$ MCS/S. Results are shown for different temperatures, $k_B T/J$, cf. Fig. 1. For the sake of clarity, the ordinate has been displaced by arbitrary amounts for curves with $k_B T/J < 0.58$. The dashed lines indicate linear fits to the data according to Eq. (2). (b) Resulting fractal dimension D, as a function of temperature.

probability that several nucleation centres are formed within the observation time. It is also observed that the equilibrium shape of the solid domains are strongly influenced by the underlying square lattice structure which is expected since the phase transition is strongly first order. In this early-time regime, restructuring is not important for the morphology. This is observed in two different ways. Firstly, a double-logarithmic plot, Fig. 2a, of the number of solidified particles, $N(R_g)$, in a cluster vs its radius of gyration, R_g, shows the same linear relationship

$$\log N(R_g) \sim D \log R_g \qquad (2)$$

for all clusters. Since larger radii of gyration for a given cluster correspond to a longer time lapse since its nucleation, one would not expect this to be the case if a significant restructuring took place during the growth process. Secondly, the time evolution of the growth process up to $t = 5000$ MCS/S directly shows that only single-bonded solid particles detach due to thermal fluctuations. This corresponds effectively to a lowering of the sticking probability. The fractal dimension, D, of the solid clusters as a function of temperature is obtained for each temperature by averaging over 10-70 runs, depending on the temperature. The results given in Fig. 2b show that there is a smooth crossover from $D \simeq 1.76 \pm 0.02$ at $k_B T/J = 0.37$ to $D \simeq 2.00 \pm 0.01$ at $k_B T/J = 0.58$. As expected, the fractal dimension at low temperatures in the early-time regime is close to that of standard DLA.[2]

In Fig. 1b are shown typical late-time configurations as a function of time for $k_B T/J = 0.43$. As time lapses, all of the material solidifies, the tenuous structures start breaking up, restructuring occurs via melting, and the resulting fragments eventually compactify. The final stage, which is extremely slow, will eventually lead to formation of a single solid domain. The overall crossover from fractal morphology to compact solid domains occurs earlier in time the higher the temperature is.

The results presented here are obtained at high dilution. For less diluted systems, the value of the effective early-time fractal exponent increases.[11] In that case, the clusters will be fractal only over length scales up to a certain correlation length (which depends on the degree of dilution) beyond which there is a crossover to teneous clusters with a uniform density ($D = 2$).[12]

a 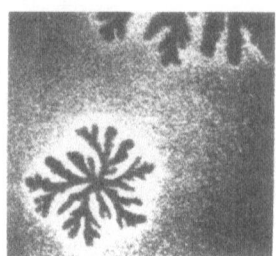 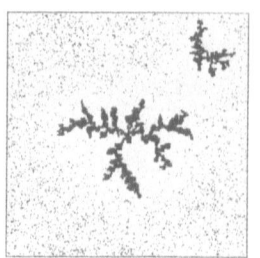 b

Fig. 3. Fractal growth patterns in lipid monolayers. (a): Fluorescence microscopic picture of a dimyristoyl phosphatidylethanolamine monolayer doped with a fluorescent dye impurity. The monolayer has been rapidly compressed through a fluid-solid phase transition. The dark fractal patterns are solid domains being formed in the fluorescing fluid phase. The radius is about 50 μm. [From Ref. 13 by courtesy of Helmuth Möhwald.] (b): Fractal domains of solid as obtained from Monte Carlo simulation of the growth model of the present paper. The dark dots in the white background denote lipid chains in the fluid conformational state.

V. COMPARISON WITH EXPERIMENT

The pressure-induced transition from a fluid to a solid phase in phospholipid monolayers spread on an air/water interface is associated with aggregation processes leading to smooth solid domain shapes as well as fractal or dendritic morphologies,[13] depending on the experimental circumstances. In these monolayers, it is the diffusion of a dye impurity only miscible in the fluid phase which is responsible for the tenuous solid domains.[11,13] The two-dimensional model presented in the present work not only identifies and clarifies the non-equilibrium fractal-forming mechanism underlying the experimental observations but also describes qualitatively the experimentally observed crossover[13] from non-equilibrium fractal growth to equilibrium compact solid domains at late times.

In Figs. 3a and b is shown that the simulated finite-time, finite-temperature pattern af solidified domains is very similar to that observed experimentally in dimyristoyl phosphatidylethanolamine monolayers on the air/water interface.[13]

ACKNOWLEDGEMENTS

This work was supported by the Danish Natural Science Research Council under grants Nos. 5.21.99.72 and 11-6836 and via a fellowship to one of us (ESS).

REFERENCES

1. *On Growth and Form. Fractal and Non-Fractal Patterns in Physics*, edited by H. E. Stanley and N. Ostrowsky (Martinus Nijhoff Publ., Boston, 1986).
2. T. A. Witten and M. E. Cates, Science **232**, 1607 (1986); L. M. Sander, Nature **322**, 789 (1986).
3. C. Aubert and D. S. Cannell, Phys. Rev. Lett. **56**, 738 (1986). 4. P. Dimon, S. K. Sinha, D. A. Weitz, C. R. Safinya, G. S. Smith, W. A. Varady, and H. M. Lindsay, Phys. Rev. Lett. **57**, 595 (1986).
5. W. Y. Shih, I. A. Aksay, and R. Kikuchi, Phys. Rev. A **36**, 5015 (1987).
6. P. M. Mors, R. Botet, and R. Jullien, J. Phys. A: Math. Gen. **20**, L975 (1987).
7. E. S. Sørensen, H. C. Fogedby, and O. G. Mouritsen, Phys. Rev. Lett. **61**, 2770 (1988); Phys. Rev A **39**, 2194 (1989).
8. J. Naudts reported on the Geilo-1987 meeting a related study of fractal growth in dilute Potts models.
9. S. Doniach, J. Chem. Phys. **68**, 4912 (1978). 10. A. B. Bortz, M. H. Kalos, and J. L. Lebowitz, J. Comp. Phys. **17**, 10 (1975).
11. H. C. Fogedby, E. Schwartz Sørensen, and O. G. Mouritsen, J. Chem. Phys. **87**, 6706 (1987).
12. P. Meakin and J. M. Deutz, J. Chem. Phys. **80**, 2115 (1983); R. F. Voss, Phys. Rev. B **30**, 334 (1984).
13. A. Miller, W. Knoll, and H. Möhwald, Phys. Rev. Lett. **56**, 2633 (1986); A. Miller and H. Möhwald, J. Chem. Phys. **86**, 4258 (1987).

STRUCTURAL PROPERTIES OF A LECITHIN-CHOLESTEROL

SYSTEM: RIPPLE STRUCTURE AND PHASE DIAGRAM

Kell Mortensen

Physics Department
Risø National Laboratory
Roskilde, Denmark

Walter Pfeiffer and Erich Sackmann

Biophysics Group
Technical University of Munich
Munich, F.R.G.

Wolfgang Knoll

Max Planck Institute for Polymer Research
Mainz, F.R.G.

INTRODUCTION

Cholesterol is an amphipathic lipid molecule which occurs as a major component of cellular membranes and of lipoproteins. In plasma membranes, the ratio of cholesterol to the other plasma membrane lipids, phospholipids etc. is about 0.7-0.8 mol/mol, whereas in intracellular membranes it is typically 0.1-0.2 mol/mol. Cholesterol serves as a precursor for various acids and hormones, however, it has also major influence on the membrane function and structure itself. Numerous studies have been performed during the last decades in order to learn about the role of cholesterol[1,2]. But still there is a severe difference of opinion on basic issues such as the location of the cholesterol in the cell membranes and how cholesterol affects various structural features[3,4].

In this paper we report a neutron scattering study performed on a model membrane system incorporated with various amount of cholesterol. The experimental data gives detailed insight into the phospholipid-cholesterol phase diagram, and shows how the observations depends on the thermal history of the sample[5]. The data show also, however, that much more structural studies must be performed in order to get a complete understanding of the effect of cholesterol on membrane structures.

EXPERIMENTAL

The model membrane used for the study was based on dimyristoylphosphatidylcholine with perdeuterated fatty acid chains

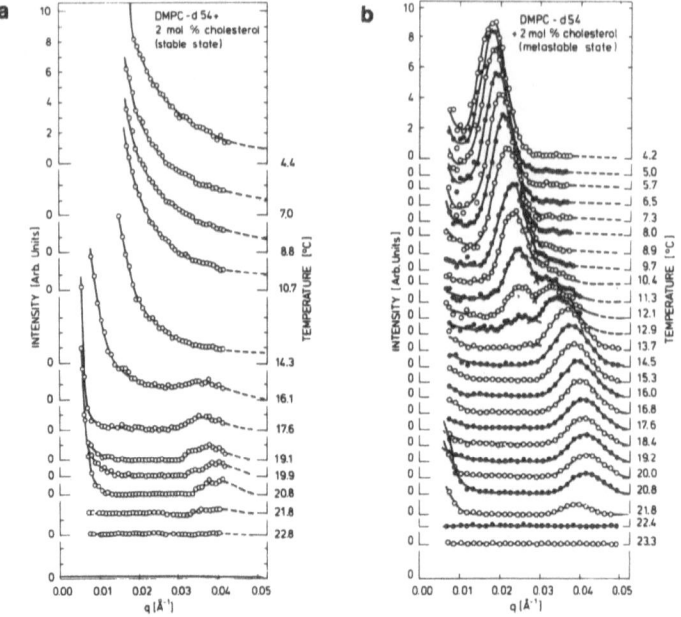

Fig. 1. Neutron scattering results of DMPC-d54 with 2 mol pct. cholesterol.

(DMPC-d54). The membranes were prepared to include 17 pct. D_2O, which is just below water saturation. The bilayers form then well defined lamellar structures well suited for diffraction studies. The neutron scattering was performed at the Risø Small Angle Neutron Scattering facility.

EXPERIMENTAL RESULTS

The neutron scattering data exhibits a variety of structural characteristics, which gives evidence of disordered concentration fluctuations as well as crystalline features. Studies of the dependence on temperature and on cholesterol content of these properties gives details of the phase diagram. In Fig. 1 is shown the scattering data of DMPC-d54 with 2 mol pct. cholesterol. Fig. 1a (marked stable) represents data as observed after long time storage at 5°C, whereas 1b (marked meta-stable) represents data of a freshly made sample or a sample after being annealed to 30°C. It clearly appears that the low temperature phase (T < 15°C) of the annealed (or freshly made) sample is not in a stable state, but will undergo a transition into another state. The time for this phase change is of the order of months. For time periods up to some weeks, however, the meta-stable properties (Fig. 1.b) are perfectly reproducible and independent of whether the sequence of measurements are made during heating or cooling. The scattering of the stable phase is characterized by major small angle scattering, which appears to have fractal characteristics with the associated dimension equal 2.8.

The scattering peaks evident from the figure are associated with the rippled structure observed in both the intermediate P_β structure and the low temperature gel state L_β.

Samples with other concentrations of cholesterol shows equally characteristic behavior as a function of time and temperature. The low temperature stable state, denoted L_c, is observed for cholesterol concen-

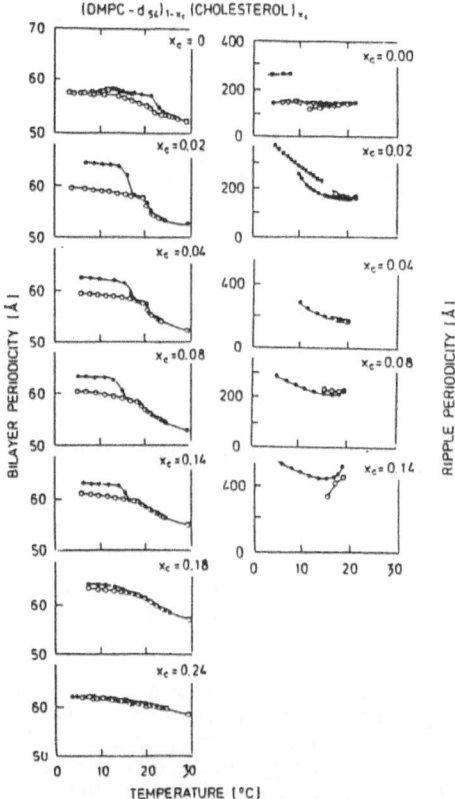

Fig. 2.   Bilayer and ripple periodicity as obtained by neutron scattering. Open symbols represents data of the stable state, whereas closed symbols represents data of the meta-stable states.

trations up to 18 pct. The sample with 24 pct. cholesterol shows no evidence of this state at all, but seems on the contrary to remain in the fluid L_α state even at 5°C. The intensity of the characteristic small angle scattering of the low temperature L_c phase appears to be linearly dependent on the content of cholesterol, thus indicating that the scattering reflects aggregation of cholesterol at low temperatures. Pristine DMPC-d54 exhibits also a stable low temperature phase, which is reached after some time only, namely a highly crystalline state. However, while DMPC-d54 reach the L_c state after some hours, the samples with cholesterol require months to stabilize.

Bragg reflections associated with the ripple structure are observed for samples with cholesterol content up to 14 pct. The sample with 18 pct. cholesterol shows no Bragg reflections in this region, but major small angle scattering may reflect some disordered, irregular ripples. In Fig. 2 is shown the periodicity of the observed ripples, along with the bilayer periodicity observed at larger q-values. It appears that while the ripple periodicity of pristine DMPC only shows little temperature dependence, those with cholesterol shows all major and characteristic T-behavior, thus giving evidence of active influence of cholesterol in the ripple formation. As the

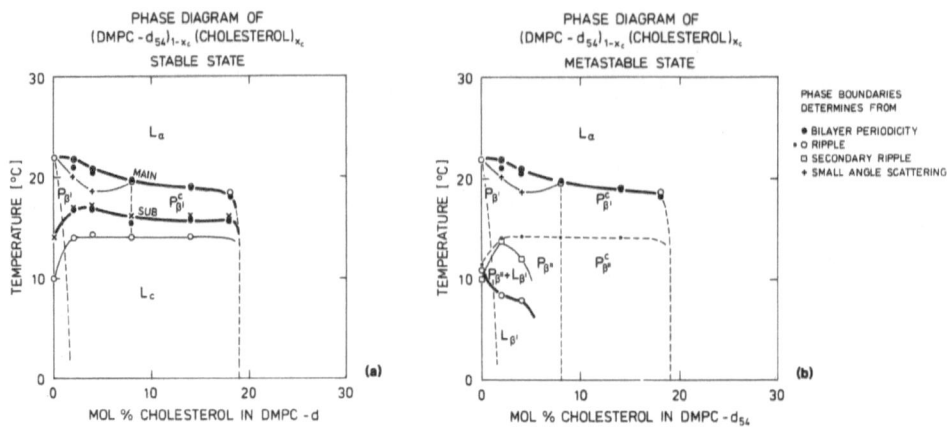

Fig. 3.   Phase diagram of DMPC-d54 cholesterol mixtures as obtained
from neutron scattering.

ripple periodicity moreover to some extent scales with the cholesterol
content, we suggest that within the L_β and P_β gel states, cholesterol phase
separates microscopically within the defect structures of the ripples[5].

The left column of Fig. 2 shows the bilayer periodicity. Changes in
this value gives clear evidence of the various phase transitions. It is
especially striking that the low temperature L_C phase is associated with such
a drastic increase in periodicity going from pristine DMPC-d54 to that with 2
pct. cholesterol. Already in the L_C phase of DMPC-d54, the acyl chains are
namely supposed to be perpendicular to the membrane plane.

In Fig. 3 is shown the phase diagram of DMPC-d54-cholesterol, as
obtained from the discussion above.

REFERENCES

1.   R. A. Demel and B. Kruyff, Biochim. Biophys. Acta 457:109 (1976).
2.   F. T. Presti, "The role of cholesterol in regulating membrane fluidity",
     in: Membrane fluidity in Biology, Vol. 4, Academic Press 1985.
3.   B. R. Copeland and H. M. McConnel, Biochim. Biophys. Acta 599:95
     (1980).
4.   A. Hicks, M. Dinda, and M. A. Singer, Biochim. Biophys. Acta 903:177
     (1987).
5.   K. Mortensen, W. Pfeiffer, E. Sackmann and W. Knoll, Biochim.
     Biophys. Acta 945:221 (1988).

POLYMER BLENDS IN SOLUTION

Jean-François Joanny   Ludwik Leibler

E.N.S.   46Allée d'Italie   E.S.P.C.I. 10 rue Vauquelin
69364 Lyon Cedex 07 France   75005 Paris France

In recent years, studies of solutions of polymer blends and of copolymers have aroused a substantial theoretical and experimental interest. This is motivated by both numerous applications and more fundamental issues concerning the usefulness of the scaling and universality concepts to describe the thermodynamic properties and the phase transitions in these systems. In this lecture, chain interactions in dilute and semidilute solutions are reviewed and it is discussed how and when the interactions between chemically different monomers lead to a macroscopic phase separation in the case of ternary polymer A-polymer B- solvent systems and to a mesophase formation in diblock-copolymer solutions. The important conclusion is that due to both the overall monomer concentration fluctuations (excluded volume effects) and the composition fluctuations, the classical Flory theory often fails. This requires the use of the renormalization method and of scaling concepts to give a correct description of the phase diagrams and the critical phenomena observed in these complex systems. We give only here a brief outline, a complete review has been published elsewhere[1].

I - Polymer blends in solution

The interaction between flexible polymer chains of different chemical nature A and B is measured   by their second virial coefficient G_{AB}. We have calculated G_{AB} [2] using Descloizeaux direct renormalization method[3] in a solvent which is either a good or a Θ solvent for the chains when the A-B interaction is repulsive.

When both polymers are in a Θ solvent, the excluded volume effects are absent between chains of same chemical nature, however chains of different chemical nature and equal radius interact as hard spheres, their mutual virial coefficient is proportional to their volume. The proportionality constant is universal and has been calculated as an expansion in powers of $\varepsilon = 4-d$ where d is the space dimension.

In a common good solvent, chains of equal radius behave as hard spheres due to the excluded volume effects. In the asymptotic limit of very long chains, different polymers cannot tell each other apart : a polymer A cannot distinguish between a polymer A and a polymer B. Interactions between unlike monomers manifest themselves only through corrections to the asymptotic hard sphere behavior. These corrections to scaling vanish in the limit of infinitely long chains and decay as a power law of the molecular mass of the polymer. This accounts well[4] for recent systematic studies of the effective virial coefficient between monomers $\Delta B = B_{AB} - 1/2(B_{AA} + B_{BB})$ for both compatible and incompatible systems which show that ΔB decays as a power law of molecular weight. The measured exponent seems to be in excellent agreement with the calculated value (~ 0.45). In a selective solvent, say good for A and Θ for B, the mutual interaction is also a hard sphere interaction for chains of same radius.

The very fact that flexible chains interact as hard spheres in dilute solutions is related to their fractal[5] nature. Two fractal objects exclude each other if the sum of their fractal dimensions is larger than the space dimension d. This is actually the case for flexible polymers in good and Θ solvents. The fractal character of the polymers is even more relevant when considering the interactions between chains of different radii. The excluded volume for the small chain is not that of a sphere of radius equal to that of the long chain[6], it is equal to the fractal volume of the long chain measured with a length scale equal to the radius of the smaller chain.

A very different behavior is expected for blends of rigid rods and flexible chains. Chains and rods interpenetrate almost freely and their mutual virial coefficient is proportional to their number of possible contacts i.e. to the product of the molecular masses of the two polymers as in the classical theories. Logarithmic corrections are expected if the flexible chain is in a Θ solvent[7].

In a more concentrated solution the interactions between unlike polymers provoke a demixing phase transition. In a common Θ solvent, the critical concentration scales as the overlap concentration c* (in the symmetric case). In a common good solvent the critical concentration c_k lies well in the semidilute regime and is governed by interactions related to corrections to the scaling behavior; a direct study of these interactions in the semidilute regime leads to $c_k \sim M^{-b}$ with an exponent b equal to 0.62. This result is in good agreement with recent experimental results.

The critical behavior has been studied in details for polymer blends in a common good solvent. The composition fluctuations play an important role and the critical behavior is not of the mean field type except in the limit of extremely long chains. Still, the critical behavior of these systems differs from that of usual ternary mixtures and the singularities are characterized by unrenormalized Ising exponents. The theoretical study of the shape of the coexistence curve, the correlation length, the osmotic compressibility and the interfacial tension between the two phases leads to a universal molecular weight dependence of the critical amplitudes. This is well confirmed by MonteCarlo simulations[8]. Far from the critical point excluded volume effects influence the interface structure[9]. The interfacial tensions are low and the solution has unusual emulsifying properties.

II - Diblock copolymers in solution

Mesophase formation in diblock copolymer solutions has been studied when both blocks are in a common good solvent or in a highly selective solvent (good solvent for one block and poor solvent for the other block).

In a common good solvent[10], in a semidilute solution, the interactions between unlike chemical species may provoke phase separation into a solvent rich disordered (homogeneous) phase and a solvent poor ordered phase (mesophase). The two-phase regions are extremely narrow and in a good approximation the phase diagram of copolymer solutions has a topology similar to that of a pure molten copolymer[11]: the symmetry of the mesophases (spherical, rod-like, lamellar) essentially depends on the copolymer chain architecture. Excluded volume effects and nonclassical corrections arising from composition fluctuations must be taken into account : they lead to a very peculiar molecular weight dependence of the segregation phenomena.

In a highly selective solvent[12] intermolecular organization occurs even in the dilute regime. Micelle formation has been sudied using a Flory-type model which gives scaling laws for the critical micelle concentration and the micellar sizes. The critical micelle concentration decays exponentially with molecular weight and occurs thus in an extremely dilute solution that is hardly measurable experimentally. In more concentrated solutions micelles order and mesophases are formed.

Aknowledgements : Part of the work presented has been done in collaboration with D.BROSETA, C.M.MARQUES, R.C.BALL and G.FREDRICKSON.

References

[1] J.F.JOANNY, L.LEIBLER "Phase transitions in polymer solutions" Proceedings of the ACS meeting in honor of P.G.deGennes (Toronto 1988) L.LEE ed. (Plenum in press)

[2] J.F.JOANNY, L.LEIBLER, R.BALL J.Chem. Phys. $\underline{81}$,4640 (1984)

[3] J.DESCLOIZEAUX, G. JANNINK Les polymères en solution : leur modélisation et leur structure (Editions de Physique Paris1987)

[4] D.BROSETA, L.LEIBLER, J.F.JOANNY Macromolecules $\underline{20}$, 1935, 1987

[5] B.MANDELBROJT The fractal geometry of nature (Freeman San-Francisco 1982)

[6] T.WITTEN, J.PRENTIS J.Chem. Phys. $\underline{77}$,4247 (1982)

[7] J.F.JOANNY J.de Phys. $\underline{49}$, 1981(1988)

[8] A.SARIBAN, K.BINDER J.Chem. Phys. $\underline{86}$,4247 (1982)

[9] D.BROSETA, L.LEIBLER, L.OULD KADDOUR, C.STRAZIELLE J.Chem. Phys. $\underline{87}$,7248 (1982)

[10] G.FREDRICKSON, L.LEIBLER to be published

[11] L.LEIBLER Macromolecules $\underline{13}$, 1602, (1980)

[12] C.MARQUES, J.F.JOANNY, L.LEIBLER Macromolecules $\underline{21}$, 1051, (1988)

SEGREGATION OF POLYMER BLENDS

IN SMALL PORES

Elie Raphaël

Matière Condensée
Collège de France
F-75231 Paris Cedex 05

INTRODUCTION

Many studies have been devoted to the subject of polymer blends, mainly as a result of their major role in the processing of new high-performance materials.

The thermodynamic properties of an A + B polymer mixture in the fluid state are usually described by the Flory-Huggins model :[1] the chains are inscribed on a lattice, all sites being filled either by a monomer A (probability ϕ) or by a monomer B (probability $1 - \phi$). The free energy per site is then given by :

$$F/kT = N_A^{-1} \phi \, \text{Log} \, \phi + N_B^{-1}(1 - \phi) \, \text{Log}(1 - \phi) + \chi\phi(1 - \phi) \quad (1)$$

(where N_A and N_B are the degrees of polymerization of A and B respectively). The first two terms describe the translational entropy of the chains. The last term corresponds to the energy of interaction. The Flory parameter χ is generally positive and favors segregation.

The (mean-field) Flory-Huggins theory is expected to be qualitatively correct, provided that both N_A and N_B are large.[2] For instance, in the symmetric case ($N_A = N_B = N$), the critical value of χ is given by

$$\chi_c = 2/N \quad (2)$$

For $\chi < \chi_c$, the system is entirely miscible, while for $\chi > \chi_c$ the system separates into two phases for a certain range of the relative concentration ϕ .

Our aim here is to understand -by the use of scaling laws- how the critical value (2) is modified when the blend is *confined* in a cylindrical tube of diameter D.

Let us first recall the main results obtained by Brochard and de Gennes for a monodisperse polymer melt of chemically identical chains confined in a tube.[3] When the diameter D of the tube is large, we are dealing with a three dimensional system : the chains are ideal spherical coils of size $R_0 = aN^{1/2}$ (a being a monomer size). For $aN^{1/4} \ll D \ll aN^{1/2}$, each chain is confined in two directions but still spans an unperturbed length R_0 in the direction parallel to the tube axis. For $D \ll aN^{1/4}$, the chains are *spatially segregated* : each chain occupies a given length $R \sim Na^3D^{-2}$ of the tube and the chains lie in sequence one after the other.

SEGREGATION IN A TUBE

We now consider the case of an A + B molten polymer blend confined in a tube of diameter D. For sake of simplicity, we restrict ourselves to the symmetric case $N_A = N_B = N$. We take the wall to be repulsive and assume that the interaction between a monomer and the wall is the same for the two polymer species. If $D \gg aN^{1/2}$, the system is three dimensional and χ_c is given by $\chi_c \sim 1/N$ (Eq.(2)). On the other hand, if $D \ll aN^{1/4}$, we know (see section 1) that the chains lie in sequence one after the other (Fig. 1). The free energy per site is then given by

$$F/kT = N^{-1}\phi \, Log\phi + N^{-1}(1 - \phi) \, Log(1 - \phi) + 2N^{-1}\phi \, (1 - \phi)\varepsilon_{AB}/kT \quad (3)$$

where ε_{AB} represents the energy associated to the boundary between an A-chain and a B-chain. If we denote by L the thickness of this boundary, ε_{AB} can be written as

$$\varepsilon_{AB} \sim LD^2a^{-3}kT\chi \quad (4)$$

where LD^2 represents the volume of the interfacial region. For a *non confined* system, the interfacial thickness was predicted to vary as $\sim a\chi^{-1/2}$.[4] This result can be qualitatively derived[5] by considering a sharp A - B boundary (Fig. 2). A portion (n monomers) of the A-chain will enter the B-phase if $n.(kT \chi) \lesssim kT$. Such a portion extending over a distance $\sim an^{1/2}$, we indeed recover the expression $a\chi^{-1/2}$.

If we assume that the interfacial thickness is not modified by the confinement, i.e. :

$$L \sim a\chi^{-1/2}, \quad (5)$$

Eqs.(3), (4) and (5) lead to a critical value of the Flory parameter[6]

$$\chi_c \sim (D/a)^{-4} \quad (1 < D/a \ll N^{1/4}) \quad (6)$$

Since for $D = aN^{1/4}$ we recover the 3d value $\chi_c \sim 1/N$, we conclude that χ_c departs from its 3d value only when D becomes smaller than $aN^{1/4}$ (see Fig. 3). For $D \ll aN^{1/4}$, χ_c increases, thus *increasing* the blend miscibility.

DISCUSSION

1. It can be shown that the approximation $L \sim a\chi^{-1/2}$ breaks down for $D < a\chi^{-1/4}$. However, the results derived in the previous section do remain valid.[7]

2. It is important to notice that the chains may have difficulty to re-arrange themselves : to pass through a B-chain, an A-chain must get over a potential barrier which becomes large with regard to kT for $D \ll aN^{1/4}$. In that case, equilibration times might thus become very long.

3. In the case of a polymer blend confined in a *slab*, the critical value of χ is expected to be given by the usual three dimensional expression (Eq.(2)), whatever the distance between the two walls.[7] To observe an increase of the blend miscibility, one must therefore confine the system at least along two directions (e.g. in a tube).

4. We have here studied the segregation of two, chemically different, polymers of the same length in a confined geometry. The somewhat related

problem of a confined mixture of long and short -chemically identical-
chains will be discussed elsewhere. [8]

ACKNOWLEDGMENTS

I am greatly indebted to P.G. de Gennes who suggested me this work
and provided me with constant advice. I also thank A. Silberberg and
D. Andelman for stimulating discussions and useful comments.

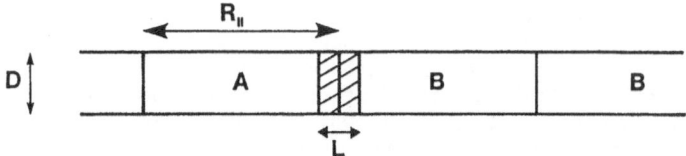

Figure 1. A+B polymer blend confined in a tube of
diameter D in the regime $1 \leq D/a << N1/4$.
Each chain occupies a given length $R_{\|}$ of
the tube and the chains lie in sequence
one after the other. Two adjacent chains
A and B overlap in a region of thickness
L (shaded area).

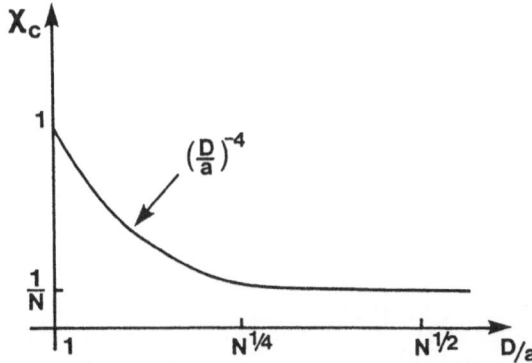

Figure 2. Variation of the critical value of the Flory
parameter with the diameter D of the tube.

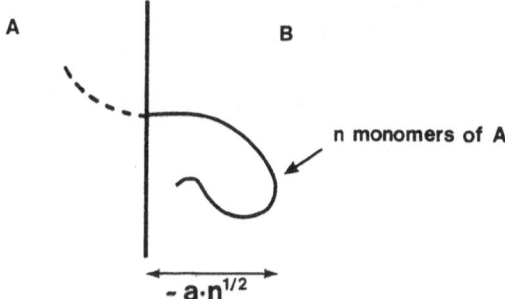

Figure 3. An interface between A and B polymers. A portion
of an A-chain (n monomers) may enter the B phase
if $n \cdot \chi \leq 1$. Such a portion of the chain extends
over a distance $an^{1/2}$ and the overall thickness
of the interface is given by $a \cdot \chi^{-1/2}$.

303

REFERENCES

1. P. J. Flory, "Principles of Polymer Chemistry", Cornell U.P., Ithaca, N.Y., Chap. XII (1971).
2. P. G. de Gennes, J. Phys. Lett. France 38:441 (1977).
3. F. Brochard and P. G. de Gennes, J. Phys. Lett. France 40:399 (1979).
4. E. Helfand and Y. Tagami, J. Polym. Sci. B29:741 (1971).
5. E. Helfand, Acc. Chem. Res. 8:295 (1975).
6. Since we are now dealing with a one dimensional system and short range couplings we cannot strictly define a critical point and a critical value of ε_{AB}. However, there is at least a crossover-point between an uncorrelated mixture and a situation with long correlated "trains" of identical chains. The crossover point corresponds to $\varepsilon_{AB} \sim kT$, which leads to Eq.(6).
7. E. Raphaël, J. Phys. France 50:803 (1989).
8. F. Brochard and E. Raphaël, submitted to Macromolecules (1989).

ADSORPTION OF RANDOM COPOLYMERS

Carlos Marques

E.N.S.L.

46, Alée d'Italie 69364 Lyon Cedex 07 France

Adsorption of polymers has a broad spectrum of industrial applications as a tool to control the stabilization of colloidal suspensions[1]. The problem is to overcome the aggregation tendency of solid particles, due to the their van der Waals attraction and one idea is to sterically protect the colloid with polymers[2]. The steric protection is usually achieved by adsorbing or grafting polymers onto the colloidal particle. If the colloid is much bigger than the polymer one can model the colloidal particle by a flat surface. Adsorption from a good solvent, which swells the chains, leads to a fluffy layer with a thickness of order of R_G, the radius of gyration of the chain in the solvent[3,4]. Grafting the chains by one end at high densities actually forces the chains in a highly stretched conformation and the layer size becomes proportional to N, the polymerization index of the chains[5,6].

The adsorbed layer is well described by a self-similar picture[4] generated by a geometrical constraint on the screening length ξ of the semi-dilute solution near the wall: the only distance in the problem being the the distance z from the wall one must have $\xi = z$ which leads to a concentration profile

$$\phi(z) \simeq \left(\frac{a}{z}\right)^{\frac{4}{3}} \tag{1}$$

The grafted layer can also be viewed as a semi-dilute solution[5] but the correlation length ξ is now equal to the distance D between the anchoring points. If a blob of size D has g monomers, one chain has N/g blobs, the layer thickness is then

$$L = \frac{N}{g}D = Na\sigma^{\frac{1}{3}} \tag{2}$$

where $\sigma = (a/D)^2$ is the density of anchoring points and a the size of a monomer.

In addition to homopolymers a large variety of other architectures can be used to achieve colloidal stabilization. We investigate here the interfacial behavior of diblock and random copolymers. Diblock copolymers adsorb[7,8] in a bi-layer structure which depends on several physico-chemical parameters like the surface affinity, the solvent quality or the mutual incompatibility of the blocks. We discuss next the adsorption from a selective and a non-selective solvent.

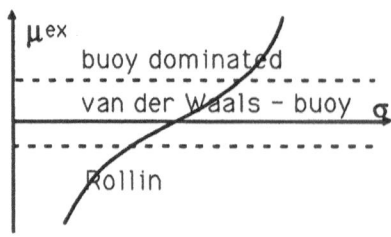

Figure 1. Variation of the chemical potential μ_{ex} as a function of the surface density σ

1. Selective solvent.[7] In this case the solvent is poor for one of the blocks and good for the other. An isolated chain in the bulk would have thus a collapsed block — let's say of N_A monomers of type A (the anchor) — and a swollen block of N_B monomers of type B (the buoy). The collapsed block sticks on the wall in order to avoid contact with the solvent. It forms, in the limit of an infinite selectivity of the solvent, a molten layer of A monomers. This layer anchors the swollen blocks that point towards the solvent building-up a grafted layer.

Diblock copolymers are known to form mesophases[9,10] in selective solvents: micelles, lamellae, worm-like micelles ... These aggregation effects are relevant for the interfacial behavior of the copolymers: the bulk solution acts as a reservoir for the adsorbed layer and imposes the chemical potential μ_{ex}. To study the adsorbed layer we write the surface grand canonical free energy of the layer as

$$\Omega = F_A + F_B - \mu_{ex}\sigma + \Pi_{ex}(L_A + L_B) \qquad (3)$$

where F_A and F_B are the contributions of the anchor A and of the buoy B layers, L_A and L_B their respective thicknesses and Π_{ex} the sum of the external pressures acting on the layer: osmotic, mechanical... The anchor free energy has three different contributions measured in units of $k_B T$:

$$F_A = -S + \frac{3}{2}\frac{L_A^2 \sigma}{N_A} + \frac{A}{6\pi L_A^2} \qquad (4)$$

S is the spreading power that takes in account the various surface tensions[11]. The second term on the right-hand side of equation (4) is the contribution of the stretching energy of the chains in the molten layer[12]. The third term accounts for the direct interactions between the wall-anchor and anchor-solvent interfaces[13], which we approach by a van der Waals energy. A is there the Hamacker constant which is positive in this case: the interaction tends to pull the surfaces apart. In a melt there is no confinement energy of the chains[14].

The buoy is in a grafted layer configuration and has a surface free energy[5]:

$$F_B = N_B \sigma^{\frac{11}{6}} \qquad (5)$$

Since the A blocks form a molten layer

$$N_A a = \frac{L_A}{\sigma} \qquad (6)$$

L_B is also a function of σ given by Eq.(2) and the minimization of the grand canonical potential leads to the equilibrium curve $\sigma = \sigma(\mu_{ex})$ scketched in figure 1. Different regimes should be distinguished for the configuration of the adsorbed layer according to the value of the chemical potential:

(a) *Rollin regime* When the solution is very dilute the chemical potential is strongly negative. At equilibrium the chemical potential term balances the van der Waals energy

$$L_A \sim a \left(\frac{\mu_{ex}}{N_A} \right)^{-1/3} \quad ; \quad L_B \sim \frac{N_B}{N_A^{2/9}} \mu_{ex}^{-1/9} \tag{7}$$

(b) *van der Waals-Buoy regime* In the concentration range where micelles form in the bulk, the chemical potential is[7]

$$\mu_{ex} \simeq N_A^{\frac{2}{5}} \gamma_{AS}^{\frac{3}{5}} \tag{8}$$

For a small asymmetry ($\beta = (R_B/R_A) \simeq 1$) the chemical potential is positive and drives the chain towards the layer. But the insertion of extra chains in the layer increases the stretching energy. The balance between these two terms leads to the equilibrium sizes

$$L_A \sim a N_A^{\frac{12}{25}} \beta^{-2} \sim R_A \quad ; \quad L_B \sim R_B N_A^{\frac{4}{25}} \simeq R_{micelles} \tag{9}$$

(c) *Buoy dominated regime* If micelles are present in the bulk and the asymmetry of the copolymer is large ($\beta \gg 1$) the anchoring layer has a thickness smaller than the radius of the chains in the melt and the chains in the buoy are almost extended:

$$L_A \sim a N_A^{\frac{6}{23}} \beta^{-\frac{10}{23}} \ll R_A \quad ; \quad L_B \sim N_B^{\frac{21}{23}} N_A^{-\frac{12}{23}} \gg R_{micelles} \tag{10}$$

2. Non-selective solvent.[8] In this case the solvent is good for both blocks and one of the blocks adsorbs preferentially on the surface, the other one is strongly repelled by the surface. The layer can be modelled by a self-similar continuous layer anchoring an external grafted layer. As above we define an asymmetry ratio $\beta = (R_B/R_A)$. The free energy has the same form as in the previous case — Eq.(3) — but the contribution from the anchor has a functional dependence on the concentration profile $\phi_A(z)$. Thus one needs to minimize the grand canonical potential with respect to $\phi_A(z)$ and σ, obeying the internal constraint of connection of blocks A and B

$$\sigma = \frac{1}{N_A} \int_0^\infty dz \, \phi_A(z) \tag{10}$$

The profile of the A monomers is given by Eq.(1) but it extends only over a distance L_A smaller than R_G which depends on β. For asymmetry ratios smaller than $N_A^{\frac{1}{2}}$ the characteristics of the layer are

$$\sigma = \frac{1}{N_A}; \quad L_A \sim \frac{N_A^{\frac{1}{2}}}{\beta} a; \quad L_B \sim N_B N_A^{-\frac{1}{3}} \tag{11}$$

For larger asymmetries the thickness of the adsorbed layer becomes of order of a monomer size and the surface density is β dependent

$$\sigma \sim \beta^{-2} \tag{12}$$

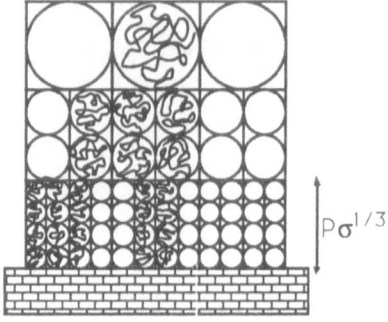

Figure 2. Blob picture for a random chain adsorbed from a dilute solution. Near the wall the chains are stretched. On the outer side the chains build a self-similar layer.

When the asymmetry is very large ($\beta \gg N_A^{3/4}$) the anchor breaks into small individual pancakes but the chains in the buoy remain in a brush configuration.

3. Random copolymers.[15] The copolymer considered in this paragraph is built up from a long repulsive chain of monomers where a fraction f of stickers A has been introduced randomly. Random copolymers often adsorb in an intermediate configuration between the grafted layer made by block copolymers and the adsorbed layer of homopolymers.

We first study the adsorption of an isolated chain : for an idealized (gaussian) isolated chain the adsorption profile extends much further in the solvent than the equivalent adsorbing chain of A monomers. The diagrammatic expansion of a perturbation serie can be resummed to study in detail the behavior of the chain. More qualitatively it is possible to give a simple Flory argument balancing the loss of entropy due to confinement with the adsorption energy gained by contact with the surface. In the case of a penetrable interface this reads

$$\left.\frac{F}{T}\right|_{chain} = -\frac{\delta}{P}\varphi M + \left(\frac{a}{D}\right)^2 M \tag{13}$$

where P is the average number of monomers B between two adjacent stickers A. In the first term, on the right hand side, $\varphi = a/D$ is the fraction of monomers in contact with the surface, a fraction $1/P$ of those (the monomers A) being able to gain an energy $-\delta$. The second term is the loss of entropy of a gaussian chain of M monomers confined in a size D. Minimizing this energy with respect to D leads to

$$D \sim \frac{aP}{\delta} \tag{14}$$

The adsorbed random copolymer thus has loops extending f^{-1} times further in the solvent than a simple homopolymer. For a solid wall there is a depletion layer close to the wall, decreasing the number of B monomers in contact with the interface by an additional P factor. This leads to a confinement length

$$D \sim \frac{aP^2}{\delta} \tag{14}$$

with loops extending even further in the solvent.

In the more realistic case where the chains are adsorbed from a solution of finite concentration we propose a blob model — figure 2 — to describe the layer. If the density of attached stickers is higher then $P^{-\frac{6}{5}}$ the chains in the vicinity of the surface are stretched over a distance of order Pa. For larger distances the layer crosses over smoothly from a grafted to an adsorbed configuration. The outer part of the layer has thus a self-similar profile.

REFERENCES

1. Napper,D.H *Polymeric Stabilization of Colloidal Dispersions* Academic Press London, 1983.
2. Pincus, P.A. Proceedings of "XVII Reunion de Fisica Estadistica", Oaxtepec, Mexico (1988).
3. Auvray, L.; Cotton, J.P. *Macromolecules* **1987**, *20*, 202.
4. de Gennes, P.G. *Macromolecules* **1981**, *14*, 1637.
5. Alexander, S. *J. Phys. (Paris)* **1977**, *38*, 983.
6. Milner, S.; Witten, T.; Cates, M. *Europhysics Lett.* **1988**, *5*, 413.
7. Marques, C.M.; Joanny, J.F.; Leibler, L. *Macromolecules* **1988**, *21*, 1051.
8. Marques, C.M.; Joanny, J.F. *Macromolecules*, in press.
9. Joanny, J.F.; Leibler, L. to be published in A.C.S. Proceedings 1988 , Toronto.
10. Thomas, E.L.; Alward, D.B; Kinning, D.J.; Martin, D.C., Handlin, D.L. Fetters, L.J. *Macromolecules* **1986**, *19*, 2197.
11. Cooper, W; Nuttal, W. *Journal of Agricultural Science* **1915**, *7*, 219.
12. de Gennes, P.G. *Scaling Concepts in Polymer Physics*, Cornell University Press, (1978).
13. Rowlinson, J.S.; Widom, B. *Molecular Theory of Capillarity*, Clarendon Press, Oxford (1982).
14. de Gennes, P.G. *C. R. Acad. Sc. Paris* **1980**, *290*, 509.
15. Marques, C.M.; Joanny, J.F. *Macromolecules*, in press.

NOVEL FACTORS INFLUENCING MICROPHASE SEPARATION

IN SB/SBS BLOCK COPOLYMERS

Richard J. Spontak and Michael C. Williams

Center for Advanced Materials, Materials and Chemical Sciences Division, Lawrence Berkeley Laboratory, and Department of Chemical Engineering, University of California Berkeley, California, 94720, USA

INTRODUCTION

Microphase separation in poly(styrene-butadiene) diblock (SB) and triblock (SBS) copolymers is induced by thermodynamic incompatibility between the blocks and is responsible for the formation of microstructures, or domains, which endow the copolymers with thermo-mechanical properties unlike those of either parent homopolymer or those of a random copolymer of equal composition. Microstructural elements, appearing as either co-continuous lamellae or disperse domains of one phase in a continuous matrix of the other, are on the size scale of the domain-forming block. Previous theoretical investigations[1-6] have shown that the microphase-separation transition (MST) and the resultant microstructure depend on temperature and material characteristics such as the polymer composition, molecular weight (M), and molecular architecture (e.g., SB or SBS). In the present work, we demonstrate the significance of two other factors influencing the MST in the solid state: (1) the molecular-weight distribution (MWD), or polydispersity, and (2) a "tapered" block junction. Predictions are made for system energetics and microstructural features in the strong-interaction limit with extended versions of the Leary-Henderson-Williams (LHW) thermodynamic theory.[3-5]

THERMODYNAMIC THEORY AND RESULTS

As a block copolymer undergoes microphase separation, it seeks to reach a state of equilibrium, wherein the free energy between the microphase-separated copolymer and its homogeneous analog (Δg) is minimized:

$$\Delta g \rightarrow \Delta g_{min} = (\Delta h - T\Delta s)_{min} \tag{1}$$

Since T is set at 298 K here, the sign of Δg_{min} (hereafter referred to as Δg_m) reveals the thermodynamically-favored state at ambient temperature, with $\Delta g_m < 0$ indicating that the microphase-separated state is preferred. Estimates of the coexistence (binodal) conditions are obtained when $\Delta g_m = 0$, and comparisons[5] of these conditions with experimental data show good agreement. The enthalpic contribution (Δh), the details of which are provided elsewhere,[3-5] is based on regular-solution theory and resembles the interaction term of the generalized Flory-Huggins free-energy expression. This expression accounts for complete demixing of the domain and matrix cores, while still retaining the energetics associated with a residually-mixed interphase existing between the microphases. The entropy (Δs) also consists of contributions from the demixed cores and the interphase region, with the individual terms derived from both random-flight chain statistics and the probability functions and elastic perturbations resulting from block confinement.[3,4] Since Δs depends on morphology, as well as on molecular architecture, only the lamellar morphology will be considered here.

Effect of Polydispersity

The general version of the LHW theory assumes that the copolymer molecules in a particular system possess the same molecular weight and are, consequently, monodisperse.[3,4] However, evidence suggests that polydispersity (i.e., a distribution of molecular weights) does affect both the microstructural dimensions[7] and mechanical properties[8] of block copolymers. Extension of the LHW theory to account for polydispersity[9] requires a realistic distribution function, W(M), to imitate a given MWD, for which the Schultz[10] function is chosen. Setting the initial location of the distribution at zero results in simplified expressions for both the weight- and number-average molecular weights (M_w and M_n, respectively) and straightforward calculation of the polydispersity index, p ($=M_w/M_n$).[9] The free energy describing a polydisperse system (Δg_m^*) is given by

$$\Delta g_m^* = \int_0^\infty \Delta g_m(M) \, W(M) \, dM \bigg/ \int_0^\infty W(M) \, dM \qquad (2)$$

where $\Delta g_m(M)$ is provided elsewhere[9] (for $\phi_S=0.50$). It is clear from Fig. 1 that $\Delta g_m^*(p)$ decreases with increasing p. In fact, an increase in p alone can result in thermodynamically-favored microphase separation (at

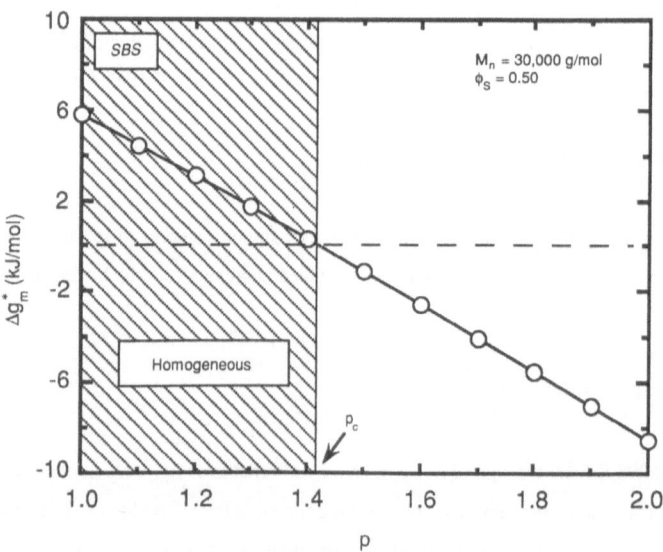

Fig. 1. Predicted $\Delta g_m^*(p)$ for an ideal SBS copolymer at 298 K.

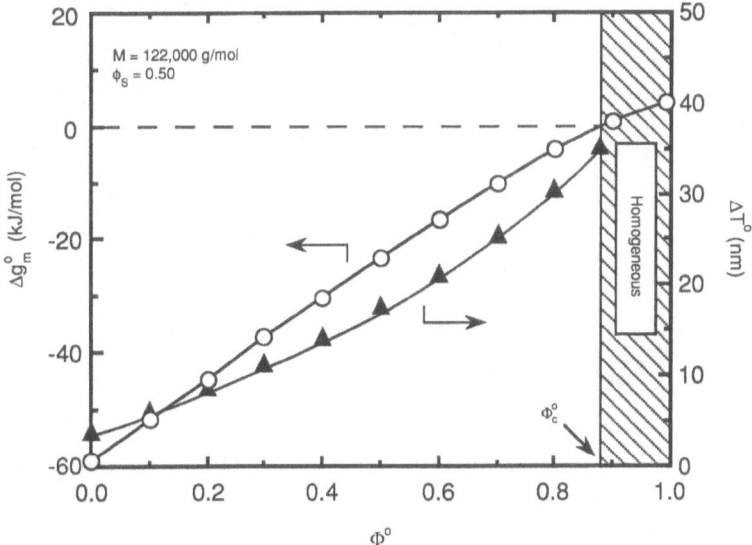

Fig. 2. Predicted $\Delta g_m{}^o(\Phi^o)$ and $\Delta T^o(\Phi^o)$ for a tapered copolymer.

p_c in Fig. 1) in a copolymer whose monodisperse analog would otherwise remain homogeneous ($\Delta g_m > 0$). Predictions of microstructural dimensions indicate that the domain and matrix cores enlarge in size as p increases due to the increasing fraction of long blocks included in the MWD. Increases in the domain repeat distance predicted with this model[9] are in agreement (±5%) with those obtained from mean-field theory.[11]

Effect of "Tapering"

The general version of the LHW theory also assumes that the monodisperse block copolymers are *ideal*, in which the composition exhibits a step change from one block to another. It has been shown[12-14] that the addition of a random copolymer (of equal composition to the block copolymer) between the blocks "tapers" the size (ΔT^o) and volume fraction (f^o) of the interphase region, thereby permitting better control of resulting bulk properties. Incorporation of the volume fraction of random copolymer (Φ^o) into the *diblock* (SB) formalism is facilitated by assuming that this added copolymer (a) is restricted to ΔT^o, which corresponds to the "domain-boundary mixing" effect,[12] and (b) functions thermodynamically as an extended "junction." As discussed in detail elsewhere,[15] the method followed here is to first determine the system energetics and microstructural dimensions in the absence of Φ^o and then, maintaining a constant molecular weight, perturb the system with the addition of Φ^o to the interphase. Predictions for the free-energy minimum ($\Delta g_m{}^o$) and interphase thickness (ΔT^o) as functions of added random copolymer are presented in Fig. 2. It is clear that ΔT^o increases dramatically with increasing Φ^o. Predicted values of $\Delta g_m{}^o$ are observed to become less negative as Φ^o increases, and significant additions of Φ^o can result in microscopic homogenization (at $\Phi_c{}^o$). A phase transition such as this has been reported[16] in an analogous poly(styrene-isoprene) (SI) copolymer, substantiating the trends predicted here. In addition, Fig. 3 indicates that predictions of $f^o(\Phi^o)$ agree very well with reported data up to about 20 wt% added random copolymer.

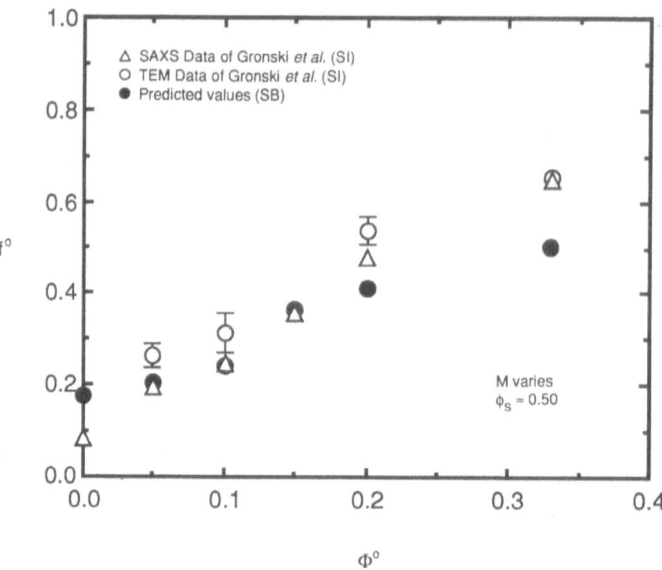

Fig. 3. Comparison of predicted and measured[13,14] $f^o(\Phi^o)$ at 298 K.

ACKNOWLEDGMENTS

This work was supported by the Director, Office of Energy Research, Office of Basic Energy Sciences, Materials Science Division of the U.S. Department of Energy under Contract No. DE-AC03-76SF00098.

REFERENCES

1. Meier, D.J., *J. Polym. Sci., Part C*, **26**, 81 (1969); *ACS Polym. Prepr.*, **15**(1), 171 (1974).

2. Helfand, E. and Wassermann, Z.R., *Macromolecules*, **9**, 879 (1976); **11**, 960 (1978); **13**, 994 (1980).

3. Leary, D.F. and Williams, M.C., *J. Polym. Sci., Polym. Phys. Ed.*, **11**, 345 (1973).

4. Henderson, C.P. and Williams, M.C., *J. Polym. Sci., Polym. Phys. Ed.*, **23**, 1001 (1985).

5. Spontak, R.J., Williams, M.C., and Agard, D.A., *Macromolecules*, **21**, 1377 (1988).

6. Leibler, L., *Macromolecules*, **13**, 1602 (1980).

7. Hadziioannou, G. and Skoulios, A., *Macromolecules*, **15**, 267 (1982).

8. Lin, S.B., Hwang, K.K.S., Wu, K.S., Tsay, S.Y, and Cooper, S.L., *Polym. Mats. Sci. Eng.*, **49**, 53 (1983).

9. Spontak, R.J. and Williams, M.C., *J. Polym. Sci., Polym. Phys. Ed.* (submitted).

10. Schultz, G.V., *Z. Phys. Chem.*, **B47**, 155 (1940).

11. Hong, K.M. and Noolandi, J., *Polym. Comm.*, **25**, 1671 (1984).

12. Hashimoto, T., Tsukahara, Y., Tachi, K., and Kawai, H., *Macromolecules*, **16**, 648 (1983).

13. Gronski, W., Annighöfer, F., and Stadler, R., *Makromol. Chem. Suppl.*, **6**, 141 (1984).

14. Annighöfer, F. and Gronski, W., *Makromol. Chem.*, **185**, 2213 (1984).

15. Spontak, R.J. and Williams, M.C., *J. Macromol. Sci.–Phys.*, **B28**(1), 1, (1989).

16. Annighöfer, F. and Gronski, W., *Colloid & Polym. Sci.*, **261**, 15 (1983).

^{2}H NMR LINE SHAPE IN POLYMER NETWORKS

Paul Sotta

Laboratoire de Physique des Solides (CNRS LA2)
Université Paris-Sud
91405 - Orsay, France

INTRODUCTION

Some properties of polydimethylsiloxane (PDMS) polymer networks, which have been put into evidence by use of Deuterium Nuclear Magnetic Resonance (^{2}H NMR), will be described in this paper. The anisotropy of molecular motions (in the time range 10^{-6} to 10^{-5}s) may be investigated with this technique. It has been already shown that deuteriated probes(solvent or free PDMS chains)dissolved in uniaxially deformed networks, acquire a uniaxial orientational order, which means that their motions, in the range 10^{-5}s, become uniaxial around the applied force direction [1-3]. This property and specifically the free chain orientation has been interpreted as the effect of orientational interactions between chain segments.

Properties of network chains themselves will be emphasized herein. The PDMS networks which are studied are model networks, obtained by end-linking deuteriated PDMS precursor chains of known average length (about 130 monomers) and polydispersity [4]. These networks are liquid-like at a monemeric scale, while at a larger scale (50 Å), chemical junctions prevent mutual diffusion and global reorientations of chains, resulting in a solid-like (amorphous) character. It will be pointed out how ^{2}H NMR allows to monitor motions in different time ranges, corresponding to different spatial scales. Then, the anisotropy induced upon deforming the network will be described and discussed.

MOTIONAL TIME SCALES AND ^{2}H NMR

The deuterium (^{2}H) nucleus (spin 1) may be described by a 3x3 density matrix ρ, whose relaxation is related to molecular reorientationl motions, modulating a second-order quadrupolar interaction tensor. The relaxation of the various matrix elements, which may be observed by specific pulse sequences, is sensitive to motions at different frequencies [5].

Longitudinal relaxation

This component, associated with diagonal elements of ρ, obeys an exponential relaxation (rate T_{1z}^{-1}), affected by motions in the range 10 MHz and faster. These are intramolecular rotations, involving short parts of the chains, which are not expected to depend much on constraints imposed on chains at larger scales (network structure). Indeed, it is found that T_{1Z} does not depend on wether the chains are cross-linked together or not.

Transverse relaxation and frequency spectrum

This component, associated with off-diagonal elements of ρ, is modulated by motions in an intermediate frequency range (10^5 to 10^6 Hz). Its structure reflects an eventual anisotropy of such motions, resulting in a non-zero average quadrupolar interaction. This anisotropy is related to the network structure, and to the constraints occasionally applied to it (see following section)

Quadrupolar polarization relaxation

In systems with non-zero average interactions, a 'Jeener echo' may be observed after a convenient pulse sequence, whose decay is sensitive to very slow motions (in the range 1 to 100 Mz)[6]. These motions should be related to slow eventual rearrangements of junction points, involving many chain collective motions. An exponential decay, with characteristic rate $T_{1Q}^{-1} = 3/5\ T_{1Z}^{-1}$ [5], is observed experimentally, up to 0.3 s, which means that no slow motions is present in that time range [7] : thus, the junction average locations are actually fixed in time.

TRANSVERSE RELAXATION

In isotropic liquid systems, this component obeys an exponential relaxation, corresponding when Fourier transformed to a unique Lorentzian resonance line (line width a few H_z). When molecular motions are anisotropic (in the range 10^5 to 10^6 Hz) the resulting average interaction leads to a structure in the resonance spectrum, reflecting the distribution of anisotropies in the system [8]. This non-zero average interaction may be checked clearly by the presence of a 'pseudo-solid echo' after a convenient pulse sequence[9]. The following points may be emphasized respectively in the absence & presence of an external constraint imposed on the network.

Relaxed network

Motions in the 10^{-6} s time range involve ensemble fluctuations of the whole chains. In networks, these fluctuations are highly restricted by cross-link junctions: specifically, an anisotropy is induced along network chains. The corresponding non-zero average interactions are indeed observed experimentally: the resonance line (line width of the order 100 Hz) is not a single narrow Lorentzian one (fig.a) and a well defined pseudo-solid echo is observed [10]. This result may be pictured by considering the network as an ensemble of chains with fixed ends: each chain is therefore anisotropic along its end-to-end results from adding the contributions of all the chains [10].

Uniaxially stretched network

When a uniaxial elongation or compression is applied to the network, the resonance spectrum suffers large alterations. A well resolved doublet appears, together with spectral wings expanding on both sides of the spectrum (fig b).

In order to interpret this result, a crucial experiment is to vary the angle Ω between the applied constraint $\vec{F}$ and the magnetic field $\vec{B}$ (which is the reference direction in NMR experiments). It has already been pointed out that the doublet splitting reproduces with great accuracy a $(3\cos^2\Omega-1)$ variation, [10-11] and that the wings themselves may be viewed as a distribution of doublets following the same Ω dependence.

The precise meaning of this results is that each average quadrupolar interaction is along the external force direction.

Thus, an information is obtained on both the local symmetry and the anisotropy (or orientation degree) of chain segment dynamics, in the uniaxially deformed network. First, the motions of each segment are uniaxial around the direction of the applied force, or in other words, the local symmetry axis for the segmental dynamics is along the force. On the other hand, the different doublet splittings contained in the spectral wings are related to different orientation degrees in the system.

DISCUSSION

Network properties at different scales emerge from the above results. Some constraint points at a semi-local scale (say, cross-link junctions) remains in average fixed, which reflects a solid-like property at this scale. This results in long-time coherence in the local average anisotropy directions.

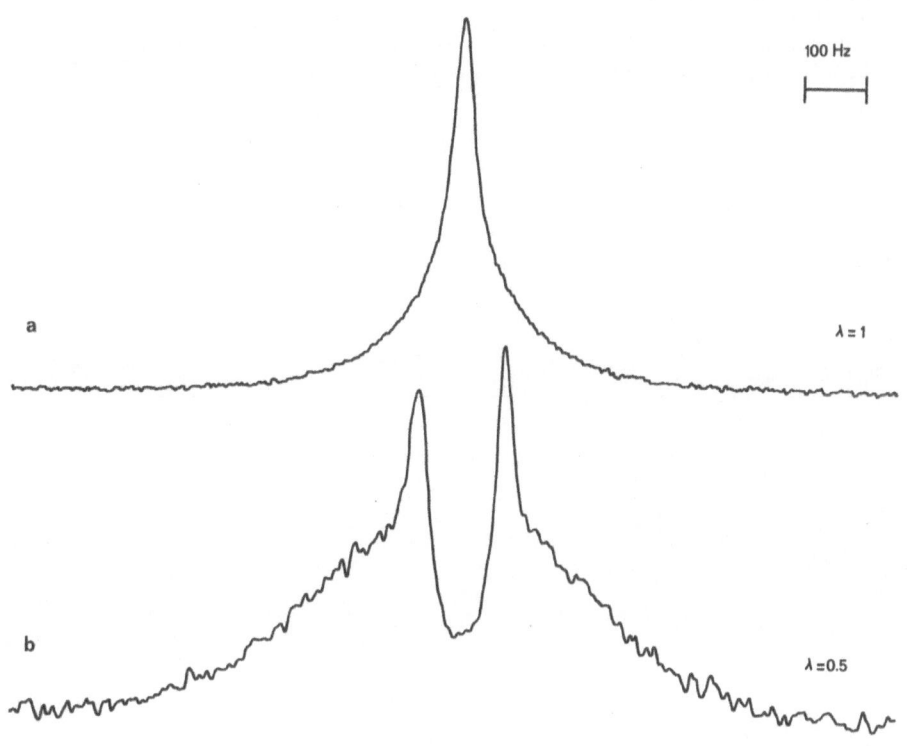

Figure 1.

^{2}H NMR spectra of a deuterated PDMS network:
a) in the absence of an central force
b) uniaxially compressed network; the deformation ratio λ is 0.5; the force is applied along the magnetic field $\vec{B}$.

Upon uniaxial deformation of the network, it has been demonstrated[10] that the observed induced uniaxiality is not accounted for by a mere deformation, homothetic to the macroscopic one, of this semi-local solid matrix. On the contrary, it emphasizes the collective response of chain segments, at a local (fluid)scale. This results in a polarization of the network which acquires the symmetry of a nematic (ie uniaxial) fluid, though the average orientation degree remains very weak (corresponding to an orientational order parameter $S \sim 10^{-3}$). Last, it may be pointed out that fluctuations in the magnitude of the orientation degree (reflected in the spectral wings upon network deformation) are important, whereas in liquid crystals – wherein the orientation degree is much stronger – fluctuations in the direction of orientation are predominent.

REFERENCES

1. P.Sotta, B.Deloche, J.Herz, A.Lapp, D.Durand and J.C.Rabadeux, Macromolecules 20: 2769 (1987).
2. P.Sotta, B.Deloche and J.Herz, Polymer 29: 1171 (1988)
3. B.Deloche and E.T.Samulski, Macromolecules 14: 575 (1981)
4. M.Beltzung, C.Picot, P.Rempp and J.Herz, Macromolecules 15: 1594 (1982).
5. J.H.Davis, Biochim.et Biophys.Acta 737: 117 (1983)
6. H.W.Spieso, Pure & Appl.Chem.57: 1617 (1985)
7. P.Sotta, to be published
8. E.T.Samulski, Polymer 26: 177 (1985)
9. J.P.Cohen-Addad, J.Chem.Phys.60: 2440 (1974)
10. P.Sotta and B.Deloche, submitted for publication in Macromolecules
11. B.Deloche, M.Beltzung and J.Herz, J.Phys.Lett.43: L-763 (1974)

GELATION AND ASSOCIATING POLYMERS

M. E. Cates
Cavendish Laboratory
Madingley Road
Cambridge CB3 0HE, UK

INTRODUCTION

The static theories of equilibrium gelation (percolation theory) and vulcanization (Flory-Stockmayer theory) are well known. But many of the interesting experiments on polymer sols and gels concern dynamics (*e.g.*, viscoelasticity), which are less universal and more difficult to describe theoretically than static properties. The dynamical response of a polymer sol near its gel-point obviously depends on whether (i) the reaction process has been quenched, or (ii) reversible exchange between one sol-molecule and another is taking place. For associating polymer systems, such as solutions of ionomers close to the overlap threshold, case (ii) is of interest. However, if the rates of exchange are not too large, clusters can move a certain amount before losing their identity: at short time-scales the system is effectively quenched. This means that case (i) must be understood first. Some old and some new ideas concerning the quenched problem are reviewed below, followed by a few comments on future prospects for understanding case (ii).

EQUILIBRIUM CROSS-LINKING

Suppose we have a set of polymer chains that can form reversible cross-links with one another. There are two known classes of static behaviour near the gelation threshold[1,2]. [Further classes arise for *irreversible* cross-linking[3].] The first is when the polymers barely overlap at the gel point (defined as the first occurrence of an infinite connected cluster). This occurs, for example, with polymers end-functionalized with carboxylate groups in the presence of metal counterions[4]. Each chain can be thought of as a spherical blob with a probability p of being connected to each of its neighbours; p is of order one at the overlap concentration C^*. The blobs can be thought of as sites on a lattice; the gelation transition is described by percolation on the lattice.

The second class is known as "classical percolation" or vulcanization, and arises when a small concentration of reversible cross-linking agents are added to a concentrated polymer system[5]. Gelation occurs at of order one cross-link per chain. Since each chain is overlapped with a huge number of others, loop-formation is suppressed and the resulting molecules are trees[5]. These may be thought of as sites (cross-links) connected by bonds (the original polymers).

Let us quickly derive two well-known exponents for the classical case. At the gel point, the number of trees of mass m scales as $m^{-\tau}$; each tree has a fractal dimension D (so its gyration radius $R(m) \sim m^{1/D}$). Actually it is easier to work with D', the fractal dimension of the tree in "chemical space". This is defined so that the length $L(m)$ of the longest connected path in a tree of mass m obeys $L(m) \sim m^{1/D'}$. Since such a path is a random walk in real space, $D' = D/2$.

Choose any point in the system at random, and ask what is the expected number $N(l)$ of l-neighbours (sites connected to this point by a path of l bonds in chemical space). Obviously $N(l) = N(l-1)ps$ where p is the bond probability and s the branching ratio ($s+1$ is the maximum functionality at a junction). The critical point is when $N(\infty)$ first becomes finite, so that $p_c s = 1$; at this point, $N(l)$ is a constant. But $N(l)$ may also be written as a sum over clusters:

$$N(l) \sim n(l) \int_{l^{D'}}^{\infty} m^{1-\tau} dm \tag{1}$$

where the integral is over clusters of chemical size larger than l, and $n(l)$ is the number of l-neighbours in each of these. Obviously $n(l) \sim l^{D'-1}$, so doing the integral and requiring the result to be independent of l, gives

$$D'(3 - \tau) - 1 = 0 \ . \tag{2}$$

Next choose a random site (i) and ask[6] the probability $p(L)$ that the longest chemical path (lcp) emanating from site i and passing through a neighbouring site j is exactly of length L. The value of L is simply $1 + \max[L']$, where L' are the various lcp's emanating from site j and passing through each of its bonds (other than the one connecting it to i). The calculation of $p(L)$ is simplest for the case $s = 2$, when site j has at most two other bonds, one of which must have $L' = L - 1$ and the other $L' \leq L - 1$. It follows that

$$p(L) = p(L - 1) \int_0^{L-1} p(L') dL' \ . \tag{3}$$

Taking the limit of continuous L, and differentiating, gives[6] $\frac{d}{dL}\left(\frac{1}{p}\frac{dp}{dL}\right) = p$, from which we infer the the asymptotic behaviour $p(L) \sim L^{-2}$. Clearly, $p(L)$ is (to within an order unity factor) the probability that a randomly chosen bond is on a cluster of chemical size L. Changing variable from chemical size to mass $m \sim L^{D'}$ must give the bond-averaged mass distribution $P(m) \sim m^{1-\tau}$. Hence

$$P(m) \sim m^{1-\tau} \sim L^{D'(1-\tau)}$$

$$= \frac{p(L)}{dm/dL} \sim L^{-2}/L^{D'-1} \sim L^{-1-D'}$$

$$\Rightarrow D'(2-\tau) = -1 . \qquad (4)$$

In combination with (2) this gives the well-known results

$$D \equiv 2D' = 4 ; \quad \tau = 2.5 . \qquad (6)$$

We now turn to the percolation model, for which the corresponding results[2] are less easy to derive. The best estimates in $d = 3$ are

$$D \simeq 2.5 , \tau \simeq 2.2 \qquad (7a)$$

which obey exactly a hyperscaling relation

$$D(\tau - 1) = d \qquad (7b) .$$

The clusters are no longer trees, but of a more complicated self-similar structure with many loops. The hyperscaling relation, as discussed further below, means that clusters of any chosen size are roughly at the overlap concentration with one another.

WHAT HAPPENS WHEN YOU QUENCH?

Imagine a polymer sol with a certain fixed density of mobile cross-links that are suddenly and instantaneously frozen. (We assume that this is the only effect of an "ideal quench".) The system is still free to change configuration but the *network structure* or "connectivity" of the clusters must remain fixed. Subject only to this, let us suppose that the system is ergodic. [This neglects any effects of permanent topological constraints arising from tangled loops, which could be important in the percolation case[7]. In principle such effects could be incorporated into the definition of the word "connectivity", but we ignore them here for simplicity.] If we wait a long time, clusters will have changed positions with one another, and changed shape, but neither broken nor formed any bonds.

We may ask how the equilibrium static properties of the final system differ from those we started with. A little thought shows that they do not differ at all. To see this, consider a rather formal object, the probability distribution $P[\zeta]$ for a complete specification ζ of all the connectivities of all the clusters in the system. Denoting also the complete specification of spatial arrangements for the clusters by ρ, we may write, before the quench,

$$P(\rho, \zeta) = P(\zeta)P(\rho|\zeta) \qquad (8)$$

where $P(\rho|\zeta)$ is a conditional probability. The basic point is that this conditional probability is unaffected by our idealized quenching operation; in both cases, it is simply proportional to the Boltzmann factor $e^{-F(\rho)/T}$.

Now, since we are below the gel point, all clusters are finite, and it is clear that $P[\zeta]$ is a self-averaging quantity in the usual sense. [Any large enough system will contain all possible cluster connectivities many times over.] Thus $P[\zeta]$ may be replaced by $\delta(\zeta - \bar{\zeta})$, where $\bar{\zeta}$ is a (rather complicated, but intensive) specification of the number density of clusters of each different possible connectivity. Using this self-averaging property, and the fact that $P(\rho|\zeta)$ is unaffected, we see that all volume-independent static properties are exactly preserved in our ideal quench; stirring the system makes no difference. What is more, if we add a few reversible cross-links, equilibrate, quench these (permanently), stir, and repeat *ad nauseam*, we should still get the same results. In view of this, we should not be surprised that many irreversible cross-linking processes yield the same static behaviour as equilibrium gelation [1-3,8-10].

INTERACTING QUENCHED CLUSTERS

The previous remarks imply that for static purposes, a polymer sol can always be thought of as a set of interacting quenched clusters with the appropriate $\bar{\zeta}$. In simple cases, the relevant interaction is excluded volume. Let us take the percolation model in $d = 3$ as an example, and try to calculate the fractal dimension $D \simeq 2.5$ of the interacting clusters, using only known information about $\bar{\zeta}$. As well as the size distribution exponent $\tau \simeq 2.2$, much is known[2,3,11] about the cluster connectivities; in particular, the clusters have spectral dimension

$$d_s \simeq 1.33 . \qquad (9)$$

There are many equivalent definitions of d_s; a convenient one is in terms of the average *resistance* $\Omega(m)$ between two randomly chosen points on a cluster of mass m. [The resistance of the network is obtained by placing a unit conductor along each bond, with connections at the cross-links. Accidental contacts, other than crosslinks, are not conducting; hence Ω only depends on the network structure of a cluster and not on its spatial arrangement.] In terms of Ω, d_s may be defined[3,11-13]

$$\Omega(m) \sim m^{2/d_s - 1} . \qquad (10)$$

The resistance properties of a cluster are interesting, because they determine its natural "gaussian" size – the one it would have with excluded volume interactions switched off. Specifically, there is an exact result for polymer networks (first used in this context in Ref.12) that

$$R_g^2 \propto \Omega \qquad (11a)$$

$$\Rightarrow D_g = \frac{2d_s}{2 - d_s} \qquad (11b)$$

where the subscript g denotes gaussian. We see that for percolation clusters D_g is close to 4 (it is exactly 4 in the vulcanization case, as was shown above). Using this, the usual Flory estimate[5,12] for the it swollen fractal dimension D_F of such a cluster (on its own in a good solvent) is found by minimizing $F(R) = (R/R_g)^2 + M^2/R^d$, giving

$$D_F = d_s(d + 2)/(d_s + 2) \simeq 2.0 . \qquad (12)$$

It is immediately noticeable that the actual value of $D \simeq 2.5$ (Eq.7a) in percolation lies in between $D_F = 2.0$ and $D_g = 4$.

According to the present viewpoint, the intermediate value $D \simeq 2.5$ must be understandable as arising from excluded volume interactions in a system having a quenched power law distribution $\sim m^{-\tau}$ of clusters with spectral dimension d_s. To explain the result in these terms[14], we introduce an overlap parameter for clusters of mass m, which should tell us the degree to which excluded volume interactions are screened:

$$\Phi(m) = m^{d/D-1} \int_m^\infty m'^{1-\tau} dm'. \qquad (13)$$

In words, we imagine cutting all clusters of mass $\geq m$ into blobs of mass m, draw a sphere round each blob, and add up the volume fractions of the spheres; this is $\Phi(m)$. As $m \to \infty$, there are three possibilities:

$$D > d/(\tau - 1) \Rightarrow \Phi \to 0 \qquad (14a)$$

$$D < d/(\tau - 1) \Rightarrow \Phi \to \infty \qquad (14b)$$

$$D = d/(\tau - 1) \Rightarrow \Phi \to \text{const. (hyperscaling).} \qquad (14c)$$

Now observe that for $\tau \simeq 2.2$ and $d_s \simeq 1.33$, setting $D = D_g$ gives case (a). Gaussian clusters would not overlap ($\Phi = 0$) and so should swell because excluded volume forces are unscreened. On the other hand, setting $D = D_F$ gives case (b): such clusters would strongly interpenetrate, and so the excluded volume interaction should be screened out, causing them to collapse. The only possible solution is (c), the hyperscaling result, in which clusters marginally interpenetrate, giving partial screening, and an intermediate fractal dimension. Hyperscaling arises whenever D_g and D_F are such as to obey (14a) and (14b) respectively[14].

DYNAMICS OF QUENCHED PERCOLATION-TYPE SOLS

For quenched systems in the percolation class (but not for vulcanized ones), the sol molecules are only marginally interpenetrating, and it makes sense, at least initially, to study the dynamics of cluster rearrangements while neglecting entanglement effects. For linear polymers (with or without excluded volume) there are two main types of behaviour in the unentangled case, known as the Rouse and Zimm regimes respectively[15]. In the Rouse model, random local bond-flips are made: the chain moves in a uniform frictional enviroment. In the Zimm model, the chain is strongly coupled to itself by hydrodynamic interactions, and it moves roughly like a solid sphere. Both models were originally set up for gaussian chains, but excluded volume can be treated either with a mode-coupling approach or at the level of scaling arguments[15,5]. There is no problem of principle in extending each of these models to the case of flexible fractal clusters, either in scaling terms[16-19,9] or with a more complete description that analyzes the dynamical relaxation modes in terms of the spectral dimension[13].

Let us start with the Rouse model (we follow the presentation in Ref.9). In this model, friction is local so that the force needed to drag a cluster through its surroundings at a certain speed is linear in its mass m. Thus the centre of mass diffusion constant $\mathcal{D} \propto 1/m$. (An equivalent argument is that, with local friction, the net Langevin random force on the centre-of-mass is the sum of m uncorrelated local terms. That $\mathcal{D} \propto 1/m$ follows immediately.) By scaling, we expect the longest relaxation time $\mathcal{T}_{rouse}$ for internal shape relaxation model to be comparable to the time taken for centre of mass to move of order the gyration radius $R \sim m^{1/D}$. Hence

$$\mathcal{T}_{rouse} \sim m^{1+2/D} . \tag{15}$$

This result also gives the relaxation time for a piece of size m that is attached to a larger cluster. Using it, we may for example work out a scaling law for the real part of the dynamic modulus $\mathrm{Re}(G(\omega)) \equiv G'(\omega)$. This is the in-phase stress response to an applied oscillatory shear deformation at frequency ω. There is a contribution of order $k_B T$ from each blob of size m_b, with the blob size chosen so that $\mathcal{T}_{rouse}(m_b) \sim 1/\omega$. [Hence $m_b \sim \omega^{-D/(2+D)}$.] We obtain

$$G'(\omega) \sim \int_{m_b}^{\infty} m^{-\tau} \frac{m}{m_b} \, dm \sim \omega^u \tag{16}$$

$$u = u_{rouse} = \frac{D(\tau - 1)}{D + 2} \simeq 0.67 . \tag{17}$$

The imaginary part $G''(\omega)$ is easily shown to have the same power law. Several other dynamic exponents can also be worked out in a similar fashion. Two recent experiments[9,10] give a measured exponent of $u = 0.69$, although this value is certainly not unique in the literature[1,20].

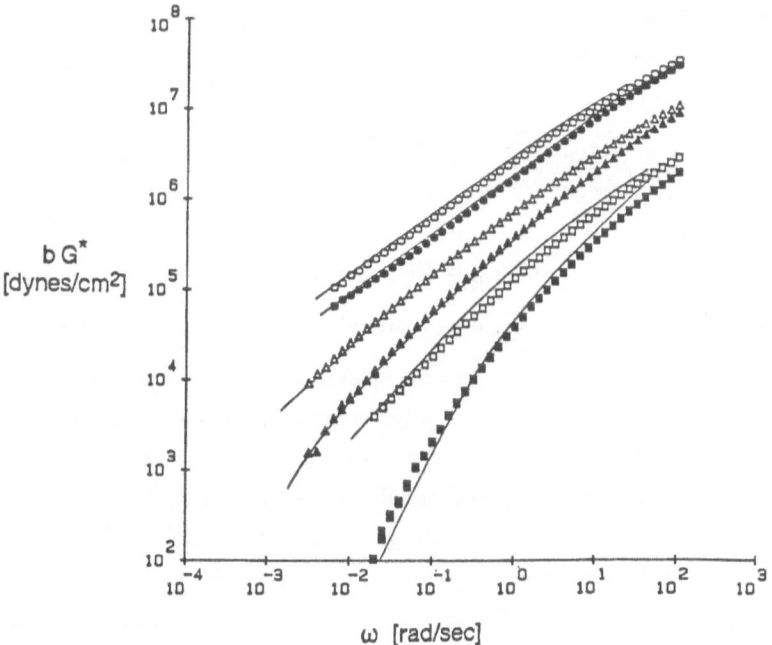

Fig. 1. Complex modulus of branched polyesters near a gel point. (Three samples.) Open and closed symbols refer to real and imaginary parts respectively. The lower curves are shifted by factors 3 and 10 to avoid data overlap. From Ref.9.

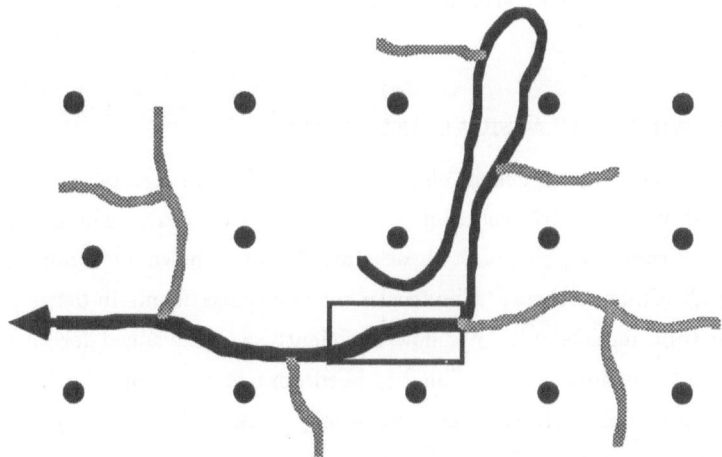

Fig. 2. The relaxation time for the bond indicated is the waiting time for the longest path in the smaller of the two trees rooted at the bond to return to the origin without enclosing any obstacles. In the case shown, the bond can now move to the left. The motion of less senior parts of the cluster is rapid.

A similar argument can be made in the case of Zimm dynamics[16-18], for which the relaxation time for a cluster of mass m scales with the rotational diffusion time of a sphere[5] of radius $m^{1/D}$:

$$T_{zimm} \sim R^d \sim m^{D/d} . \tag{18}$$

This corresponds to a frequency dependent blob size $m_b \sim \omega^{-d/D}$. Repeating the integral (16) we find

$$u_{zimm} = D(\tau - 1)/d = 1 \tag{19}$$

which is much further from the experimental values than the Rouse prediction.

In view of this, one would like to be able to show theoretically that the Rouse result is in fact the correct one for quenched polymer sols (always neglecting entanglement effects). Unfortunately this is far from clear. In general one expects the local friction picture to be correct only when hydrodynamic interactions are fully screened, and this requires strong interpenetration of the clusters[15]. But the hyperscaling argument tells us that interpenetration is only marginal; *a priori*, there is no reason to prefer the Rouse to the Zimm result[16-18]. There is a rather complicated discussion of this issue in Ref.13, accompanied by an attempt at a self-consistent calculation. As with static excluded volume, the marginal overlap condition can lead to partial screening, with a residual hydrodynamic interaction that is a power law. Unlike in the static case, there seems to be no simple self-consistency argument that fixes the exponent u uniquely. All that can be said is that the Zimm and Rouse results are upper and lower bounds respectively on u. The experiments[9,10] clearly suggest that the true answer is close to the lower bound, u_{rouse}, but we still don't really know why this is the case.

DYNAMICS NEAR THE CLASSICAL GEL POINT

While the neglect of entanglement effects on dynamics is reasonable in a quenched hyperscaling system, it is obviously not so in the classical case. There one has a tangled mass of branched polymers, whose statistics are known in complete detail from the Flory-Stockmayer theory. The treatment of entanglements in dense polymer systems (using tube models[15,20,21]) is now sufficiently well-advanced for this rather complicated case to be approached. This has been done very recently by Rubinstein, Zurek, McLeish and Ball[6] (RZMB); here we briefly outline their main results. (For simplicity we consider only $s = 2$.)

The first observation is that a bond between sites i and j (say) on a branched polymer can be ranked according to its "seniority"[6], which is defined as the minimum over i and j of the lcp emanating from one and passing through the other. This is essentially the chemical size of the smaller of the two trees that would be created by cutting the bond. RZMB show that the concentration of sites of seniority x scales as x^{-3} near the gel point. Treating the surrounding chains as a grid of topological constraints, their basic idea is that a given bond can only relax its orientation when the lcp in the smaller of these two trees folds back on itself without enclosing any obstacles (Fig.2). This situation is extremely rare; the waiting time depends on the seniority and can be calculated recursively[6]. For a branched polymer in a fixed network, the calculation is straightforward and gives an exponential increase of the relaxation time of a bond with its seniority x. In the present system, the increase is less dramatic, because those parts of the surroundings that have seniority less than x relax fast, and do not contribute to the topological grid for the purposes of calculating the relaxation of that bond. A self-consistent treatment of this "dynamic dilution effect"[22] allows the stress relaxation function $G(t)$ (whose fourier transform is $G(\omega)/i\omega$) to be calculated, with the result[6]

$$G(t) \propto \left(1 - \log[t/T_1]/c\bar{N}\right)^4 . \qquad (20)$$

Here c is a geometrical constant ($\simeq 15$ for $s = 2$) and $\bar{N}$ is the average number of entanglements per chain in the pre-crosslinked state; T_1 is the relaxation time (by reptation) of a chain in this state. As the gel point is approached, the longest relaxation time does not diverge, according this treatment, but approaches a constant value

$$T_\infty = T_1 \exp[c\bar{N}] . \qquad (21)$$

Thus, very close to the gel point, there must be a crossover to some form of unentangled motion. Careful experiments will be needed to test these various new predictions.

DYNAMICS OF REVERSIBLY CROSS-LINKED SYSTEMS

As mentioned in the introduction, it is very hard to treat the dynamics of branched polymer clusters that can reversibly exchange material at a slow rate, without first having understood the fully quenched case. We are now approaching the point where such modelling can be seriously attempted.

Let us give an oversimplified calculation, just to indicate the issues involved. Take a percolation sol, and assume that the correct dynamics for unbreakable clusters is Rouse motion. We now ask what happens when there is a small reaction rate for bond scission and formation. A simple proposal is that the clusters still move by Rouse motion, except that a large cluster will fall to pieces before its Rouse time is reached. In this case, we assume that the longest relaxation time in the system is that of a cluster whose Rouse time is equal to the time it takes to fall apart. The latter can be estimated from the scission rate (call this k) and the known behaviour of the density of "red bonds"[2,3,23]. The red bonds on a cluster are those which, if cut, would divide it into two part of roughly equal size (they are bottleneck bonds on the backbone of the cluster). Indeed, we may argue (crudely) that the time for a cluster to fall apart is none other than the time taken for one of its red bonds to break. Since for a cluster of mass m there are of order $m^{1/\nu D} \sim m^{0.47}$ red bonds[23] (where $\nu = \nu_{perc} \simeq 0.85$), we have a breaking time for a cluster of mass m

$$\mathcal{T}_{break} \sim k^{-1} m^{-0.47} . \tag{21}$$

Equating this to the Rouse time from Eq.15, $\mathcal{T}_{rouse} \sim \mathcal{T}_o m^{1.8}$ (with $\mathcal{T}_o$ a microscopic time for bond motion), we get a characteristic mass $\bar{m}$ for a cluster whose Rouse time and breaking time are equal:

$$\bar{m} \sim (\mathcal{T}_o k)^{-0.44} . \tag{22}$$

This corresponds to a longest relaxation time

$$\mathcal{T}^* \sim \mathcal{T}_{rouse}(\bar{m}) \sim \mathcal{T}_{break}(\bar{m}) \sim \mathcal{T}_0^{0.2} k^{-0.8} . \tag{24}$$

The above argument generalizes of one given by Faivre and Gardissat[24] for the motion of unentangled, breakable, linear polymers. We expect that $G(\omega)$ continues to behave as a power law with exponent u for high frequencies, but this behaviour is cutoff at $\omega \sim 1/\mathcal{T}^*$ which remains finite as the gel point is approached. Indeed, with breakage present, there is no reason for the viscosity of the system to diverge in the gel phase. The transition should be invisible to viscoelastic measurements, with the above calculation for $\mathcal{T}^*$ only breaking down on the gel side when the correlation mass ($m_\xi \sim \xi^D$) becomes smaller than $\bar{m}$ obeying (23).

Equation (24) for $\mathcal{T}^*$ is quite interesting since it has a fractional power law dependence on the dissociation rate constant k. For reversibly associating polymers[4] this rate constant is highly activated ($k \sim e^{-E/T}$ with E large). Neglecting the (much weaker) temperature dependence of $\mathcal{T}_o$, Eq.24 predicts Arrhenius behaviour for the terminal time, but with an effective activation energy $E' \simeq 0.8E$. So it may be dangerous[4,25] to presume that measurements of the temperature dependence of τ near the gel point allow a direct estimate of E for associating polymers. [This conclusion would stand even if our estimate of the power-law involved is not very accurate.]

The dynamics of *highly entangled* branched polymer motion in the presence of reversible cross-links is more complicated. In the case of entangled but *unbranched* polymers, it was recently shown how to include reversible scission effects within the framework of a tube model[26]. With the results of Ref.6 to hand, a similar development is now possible in the branched case, and will be described in a future publication[27].

Acknowledgements: Thanks are due to numerous colleagues for discussions on many of the topics covered here. I am particularly grateful to the authors of Ref.6 for allowing me to disclose their results before publication. In addition, I thank one of them (Tom McLeish) for his forbearance and helpfulness during the frantic last-minute preparation of these notes.

REFERENCES

1. D. Stauffer, A. Coniglio and M. Adam, Adv. Polym. Sci. 44:103 (1982).

2. D. Stauffer, *Introduction to Percolation Theory*, Taylor and Francis, London (1985).

3. H. Hermann, Phys. Reports 136:153 (1986).

4. G. Broze, R. Jerome, P. Teyssie, Macromolecules 14:224 (1981); *ibid.* 15:920,1305 (1982); 16:996 (1983).

5. P. G. de Gennes, *Scaling Concepts in Polymer Physics*, Cornell, Ithaca (1979).

6. M. Rubinstein, S. Zurek, T. C. B. McLeish and R. C. Ball, to be published.

7. Y. Kantor and G. Hassold, Phys. Rev. Lett. 60:1457 (1988).

8. F. Schosseler and L. Leibler, Macromolecules 18:398 (1985).

9. M. Rubinstein R. H. Colby and J. R. Gillmor, to be published.

10. D. Durand, M. Delsanti, M. Adam and J. M. Luck, Europhys. Lett. 3:297 (1987).

11. S. Alexander and R. Orbach, J. Phys. (Paris) Lett. 43:L625 (1982).

12. M. E. Cates, Phys. Rev. Lett. 53:926 (1984); *ibid.* 55:131C (1985).

13. M. E. Cates, J. Phys. (Paris) 46:1059 (1985).

14. M. E. Cates, J. Phys. (Paris) Lett. 46:L837 (1985).

15. M. Doi and S. F. Edwards, *The Theory of Polymer Dynamics*, Clarendon, Oxford (1986).

16. D. Sievers, J. Phys. (Paris) Lett. 41:L535 (1980).

17. P. G. de Gennes, J. Phys. (Paris) Lett. 40:L197 (1979).

18. M. Daoud, J. Phys. A 21:L237 (1988).

19. J. E. Martin and J. P. Wilcoxon, Phys. Rev. Lett. 61:373 (1988).

20. H. H. Winter and F. Chambon, J. Rheol. 31:683 (1987).

21. T. C. B. McLeish, Europhys. Lett. 6:511 (1988); Macromolecules 21:1062 (1988).

22. T. C. B. McLeish and R. C. Ball, Macromolecules 22:1911 (1989).

23. A. Coniglio, Phys. Rev. Lett. 46:250 (1981); J. Phys. A 15:3824 (1982).

24. G. Faivre and J. L. Gardissat, Macromolecules 19:1988 (1986).

25. T. A. Witten, J. Phys. (Paris) 49:1055 (1988).

26. M. E. Cates, Macromolecules 20: 2289 (1987); J. Phys. (Paris) 49:1593 (1988).

27. M. E. Cates, T. C. B. McLeish and M. Rubinstein, to be published.

SOL-GEL TRANSITION : AN EXPERIMENTAL STUDY

J.P.Munch([1,3]), M.Delsanti([1]), M.Adam([1]) and D.Durand([2])

([1]) Service de Physique du Solide et de Résonance Magnétique, CEN-Saclay, 91191 GIF SUR YVETTE Cedex, France
([2]) Laboratoire de Chimie et de Physico-Chimie Macromoléculaire, Université du Maine, route de Laval, 72017 LE MANS Cedex, France
([3]) Laboratoire de Spectrométrie et d'Imagerie Ultrasonores, Université Louis Pasteur, 4, rue Blaise Pascal, 67070 STRASBOURG Cedex, France

KEYWORDS/ABSTRACT : polymer / gelation / percolation / cluster growth process / relaxation in disordered system / diffusion / glasslike behavior

The experimental values of the exponents γ and τ, which characterize the divergence of a mean average mass distribution of clusters at the gelation threshold and the mass distribution of clusters respectively, show that gelation and percolation belong to the same class of universality. The dynamical structure factor, measured on solution of such percolating clusters elaborated near and below the gelation threshold, exhibits relaxation behavior similar to those observed in disordered media near the glass transition.

INTRODUCTION

The investigation of the sol-gel transition can be done by rheological studies which show the divergence of the viscosity in the sol phase and the increase from zero of the elastic modulus in the gel phase [1]. This behavior reflects the fact that, at the sol-gel transition (or gel point), appears a randomly ramified cluster whose size is that of the reacting bath considered as infinite. In the sol phase, only finite polymer clusters are present. Here we investigate this transition by scattering measurements which allow us to determine statistical properties of the population of clusters characteristic of the connectivity properties at large scale. We report and comment measurements obtained using the technique of dynamical light scattering [2], static light scattering [3], and small angle neutrons scattering [4].

EXPERIMENTAL CONDITIONS

The chemical procedure used to elaborate the sample is a polycondensation of triol monomers (trifunctional OH group) with diisocyanate monomers (difunctional NCO group) at different values of stoichiometric ratio p which is the ratio of NCO groups to OH groups. The chemical reaction is carried out in bulk, until complete reaction.

The gelation threshold p_c, which defines a stoichiometric gel point, is determined by solubility measurements and the different samples are characterized by : $\epsilon = |p-p_c|/p_c$. The studies were carried out using values of ϵ between 3×10^{-3} and 10^{-1}.

Light scattering experiments were performed on a spectrometer which is a home-built apparatus. Intensity and quasi-elastic light scattering experiments allow to measure the space-time dependence of the structure factor on a space scale length q^{-1} between 27 nm and 300 nm and within a time window which extends typically from 0.2 μs to 10 ms.

Small angle neutrons scattering experiments were performed on the PACE spectrometer in Saclay. The range of investigated scattering vector defines a space length q^{-1} between 1 nm and 20 nm.

CONNECTIVITY PROPERTIES IN THE SOL PHASE

<u>Intensity scattering measurements</u> In order to describe connectivity properties of polymer clusters, scattering measurements must be done on dilute solutions and the different characteristic parameters must be determined at zero concentration limit : $c \to 0$ (independent clusters).

For a wave transfer momentum q, the intensity $I(m,q)$ scattered by a cluster of mass m and characteristic size $R(m)$ is proportional to the Fourier transform of its pair correlation function − $(r^D/r^3)g(r/R(m))$ − where D is the fractal dimension, $R(m) \sim m^{1/D}$.

In the limit of infinite dilution the total intensity $I(q)$ is the averaged sum of the intensities scattered by all the clusters. The quantity which describes the degree of polymerization of the clusters is the intensity scattered per monomer:

$$I(q)/c \sim \int n(m) \; I(m,q) \; dm \; / \int n(m) \; m \; dm$$

where $n(m)$ is the number of clusters having a mass between m and m+dm.

The experimentally explored domain of q^{-1} values compared with the size of the clusters in solution defines the measurable quantities characteristic of clusters distribution.

<u>Guinier regime</u> The condition $q^{-1} > R(m)$ can be satisfied, in this case, with light scattering. The observation scale is larger than the clusters size (Guinier regime). The intensity scattered per monomer is :

$$I(q)/c \sim M_w (1 - q^2 x \; R_z^2/3)$$

where M_w is a mean average mass equal to $\int n(m) \; m^2 \; dm \; / \int n(m) \; m \; dm$ and R_z a mean radius of gyration $R_z^2 \sim \int n(m) \; m^2 \; R^2(m) \; dm \; / \int n(m) \; m^2 \; dm$. Through M_w and R_z is measured a mean degree of connectivity of clusters and a correlation lenght of connectivity respectively. We find that [3] :

$M_w = 8.1\times10^2 x \; \epsilon^{-\gamma}$ with $\gamma = 1.71 \pm 0.06$ and R_z (nm) $= 1.1 x \epsilon^{-\nu}$ with $\nu = 1.01 \pm 0.05$

<u>Fractal dimension : Neutrons scattering</u> If the magnitude of the scattering wave vector q is such that $q^{-1} < R(m)$ for all m, the scattered intensity per monomer is only q dependent and the fractal dimension D can be measured. This experimental condition was obtained with a sample, at a given ϵ, fractionated by size exclusion chromatography in order to select only clusters having a size larger than 20 nm. The result is [4] :

$$I(q)/c \sim q^{-D} \text{ with } D = 1.98 \pm 0.03$$

The experiment was repeated on the unfractioned sample. In this case only the condition $q^{-1} < R_z$ was realized and the result is [4] :

$$I(q)/c \sim q^{-D'} \text{ with } D' = 1.59 \pm 0.05.$$

<u>Discussion</u> : Some years ago [5,6] it was pointed out that the connectivity properties at the sol-gel transition can be described by a percolation model. The basic concept of this theory is that the mass distribution of clusters obeys a power law $m^{-\tau}$ below a mass m^* which diverges at the gel point as $\epsilon^{-1/\sigma}$. This leads to [4] : $D' = D \times (3-\tau)$ whence $\tau = 2.2 \pm 0.04$.

Thus if we analyse these results taking into account the swelling effect due to dilution, which modifies the fractal dimension [7] ($D = 2$ instead of 2.5), we can conclude that gelation is a critical phenomenon of connectivity described by the percolation theory.

The exponents γ and τ are, within experimental precision, identical to those determined by Monte-Carlo simulation following percolation model.

QUASI ELASTIC LIGHT SCATTERING MEASUREMENTS

Quasi elastic light scattering experiments allow the time correlation function of the intensity to be determined :

$$\langle I_q(t)I_q(0)\rangle / \langle I_q\rangle^2 = A \times G^2(q,t) + 1$$

where A depends on the geometry of the experiment and $G(q,t)$, the dynamical structure factor, is by definition the normalised correlation function of the concentration fluctuations, δc_q :

$$G^2(q,t) = [\langle\delta c_q(t)\delta c_{-q}(0)\rangle / \langle|\delta c_q|^2\rangle]^2$$

We report here experiments [2] which were performed on solutions of clusters obtained near the gelation threshold as a function of three independent variables ϵ ,c and q .

To comment the results we can say firstly that the q dependence, for given ϵ and c values, can be analysed as controlled by a diffusive process, and secondly that the dynamical structure factor is not in general a simple exponential function of time. This last point cannot be analysed as a trivial polydispersity problem, as at fixed ϵ the profile of the structure factor changes tremendously as a function of concentration. Thus a single change in time scale is not useful to analyse the concentration dependence of the measurements.

In order to characterize the evolution of the profile as a function of the concentration, two mean characteristic times are determined. One of them τ_0, is the harmonic mean relaxation time which is obtained from the slope at time origin :

$$\tau_0^{-1}(c,q) = \lim_{t\to 0}\left[-\frac{1}{2}\frac{d}{dt}(\text{Log }(G^2(q,t)))\right]$$

This mean relaxation time defines an usual cooperative diffusion coefficient, which increases with the concentration. This description is only useful for times shorter than $\tau_0(c,q)$.

A most important fact is that, if we exclude this short time scale, we observe a profile which goes from a stretched exponential function to a power law function as the concentration increases from the dilute state to concentration where largest clusters are penetrated by smaller clusters.

In more quantitative terms, using this chemical system, we observed that for concentrations lower than 6×10^{-2} g/cm^3, for various ϵ values, the major part ($t > \tau_0$) of the time dependence of $G^2(q,t)$ can be approximated by a stretched exponential :

$$G^2(q,t) \sim \exp(-(t/\tau)^\beta)$$

β is independent of q and ϵ, and decreases from a value close to 1, near zero concentration, to a value of 1/3 at the concentration where the power law behavior appears.

For a concentration larger than 8×10^{-2} g/cm^3 and for times larger than $\tau_0(q,c)$ the profile is well described by a power law :

$$G^2(q,t) \sim (t/\tau_0(q,c))^{-2\alpha}$$

α is independent of q and ϵ and depends slightly on concentration. For c = 0.11 g/cm^3 α = 0.25.

In the first concentration domain, c < 6×10^{-2} g/cm^3, it is possible to define a second mean characteristic time, the arithmetic mean relaxation time which is calculated through integration :

$$\tau = \int_0^\infty G(q,t) \, dt$$

We find that the corresponding diffusion coefficient decreases and the associated dynamical length increases when the size of the largest cluster, ϵ dependent, is increased. Moreover this dynamical length increases as the concentration increases. This means that all the clusters present are interacting dynamically for c > 7.5×10^{-2} g/cm^3 and that the concentration fluctuations cannot relax on the quasi elastic light scattering time scale. This is consistent with the fact that we do not observe any measurable cutoff function for the power law behavior.

In summary, for such solutions of percolating clusters – which are not entangled systems, as clusters of same size do not overlap the global overlap betweeen clusters rather proceeds through hierarchical filling – we observe relaxation behaviors similar to those observed in other disordered systems like spin glasses [8], metallic glasses and other amorphous systems. In our case the inverse of the concentration seems to play the same role as the temperature in the case of glasses. The transition between stretched exponential and power law behavior occurs for β equal to one third, this value was also obtained in a recent theoretical calculation on spin glasses [9]. The present result must be related to studies on dynamical properties observed during the gelation process [10], at fixed concentration. Some recent measurements on other chemical systems confirm these results.

References

1. M.Adam, M.Delsanti, D.Durand, G.Hild, J.P.Munch. Mechanical properties near gelation threshold, comparison with classical and 3d percolation theories. Pure & Appl. Chem. , 53, 1489 (1981)
2. M.Adam, M.Delsanti, J.P.Munch, D.Durand. Dynamical Studies of Polymeric Clusters Solutions Obtained near the Gelation Threshold : Glasslike Behavior Phys. Rev. Lett. 61:706 (1988)
3. M.Adam, M.Delsanti, J.P.Munch, D.Durand. Size and mass determination of clusters obtained by polycondensation near the gelation threshold. J.Phys. (Paris) 48:1809 (1987)
4. E.Bouchaud, M.Delsanti, M.Adam, M.Daoud, D.Durand. Gelation and Percolation: Swelling effect. J. Physique (Paris). 47:1273 (1986)
5. P.G. de Gennes."Scaling concepts in polymer physics". Cornell Univ. Press, Ithaca, NY (1979).
6. D.Stauffer. Scaling Theory of Percolation Clusters. Phys.Rep.54:1(1979)
7. M.Daoud, F.Family, G.Jannink. J.Physique Lett. 45:199 (1984)
8. M.Alba, J.Hamman, M.Ocio, P.Refregier, H.Bouchiat. J.Appl.Phys. 61:3683 (1987)
9. I.A.Campbell. J.Physique.Lett. 46:1159 (1985)
10. J.P.Munch, M.Akrim, G.Hild, S.Candau. Macromolecules 17:110 (1984)

STRUCTURE AND DYNAMICS OF AEROGELS

Eric Courtens

IBM Research Division

Zurich Research Laboratory

CH-8803 Rüschlikon, Switzerland

René Vacher

Lab. de Science des Matériaux Vitreux*

Univ. des Sciences et Techn. du Languedoc

F-34060 Montpellier, France

ABSTRACT

In these lectures, the structure and vibrations of random media that are self-similar at various length scales are described. Small-angle scattering demonstrates that silica aerogels are excellent fractals. Brillouin, Raman, and inelastic neutron scattering are used to investigate their vibrational dynamics. The existence of new localized collective excitations, called *fractons*, is clearly revealed. Values are obtained for the fractal dimension D, the spectral dimension $\bar{\bar{d}}$, an "internal length" exponent σ, as well as for the density of vibrational states $N(\omega)$. The picture that emerges is in remarkable agreement with fracton theory.

1. INTRODUCTION

The results presented in these lectures have been or are soon to be documented extensively elsewhere, including recent reviews.[1,2] The present account is an extended abstract to serve as a guide to the literature and to the original publications.

The collective vibrations of fractal structures have been the subject of many recent theoretical investigations.[3-8] The strongly localized vibrations, called fractons, have also been studied by computer simulations of random networks.[9-13] Fractons were proposed to explain the low-temperature thermal properties and low-frequency vibrations of various amorphous materials.[14-24] This assumes that these materials, although homogeneous in their mass, are fractal in some way, for example in their connectivity. Such a property has never been demonstrated experimentally. Hence, to investigate fracton properties, it is preferable to select materials whose fractality can be established, as for example is done in Refs. 25 to 28. Of these materials, aerogels have already been known to be fractal over restricted length scales.[29] With suitable preparation, we obtained silica aerogels that are fractal over more than two orders of magnitude in length.[30] These turned out to be excellent material for the investigation of fractons.

2. FRACTAL STRUCTURE OF AEROGELS

Small-angle neutron scattering experiments have been performed to determine the fractal structure of various series of aerogels.[30-32] It is generally possible to extract three parameters from a measurement: 1) the mean particle size a; 2) the correlation length ξ beyond which the material becomes homogeneous and 3) the fractal dimension D. One important result[30] is that series of materials can be prepared for which: 1) ξ is progressively changed; 2) D is constant and 3) ξ is related to the macroscopic density ρ by $\xi \propto \rho^{1/(D-3)}$, with the same value of D and a constant proportionality factor across the series. Such materials can be called *mutually self-similar*. They allow scaling to be investigated as a function of length in the fractal regime ($a < L < \xi$) without actually performing microscopic experiments at these lengths.[2] It is such a series of neutrally reacted aerogels with D $\simeq$ 2.4 that was used to obtain most of the dynamical results.[33-37]

3. THE PHONON-FRACTON CROSSOVER

The phonon-fracton crossover frequency, ω_{co1}, and wave vector, q_{co}, are accessible by Brillouin scattering experiments.[33,34] Spectra can be measured for various sample densities ρ, and for each ρ for various momentum exchanges q. Each spectrum is determined by three fitting parameters, namely ω_{co1}, q_{co}, and the absolute intensity. Using the former two, and assuming that mutual self-similarity also extends to dynamical properties, one obtains the fracton "dispersion curve",[34,1] $\omega \propto q^{D/\bar{\bar{d}}}$. Here, $\bar{\bar{d}}$ is the fracton spectral dimension. From q_{co} alone, one derives an acoustical correlation length, $\xi \propto 1/q_{co}$. This yields a value for D in agreement with the direct structural information.[34,2] From $D/\bar{\bar{d}}$ and D, one derives[34] $\bar{\bar{d}} = 1.3 \pm 0.1$. Such a value can be caused either by scalar elasticity, or by tensorial elasticity in structures that are more connected than percolation clusters. The latter appears more likely.[34]

4. RAMAN SCATTERING THROUGHOUT THE FRACTON REGIME

In addition to coherent contributions, there are incoherent contributions to light scattering in which fractons of characteristic size smaller than the optical wavelength scatter independently from each other. In that case, the scattered intensity is the sum of the individual scattering intensities, while for coherent contributions it is the scattered amplitude which is the sum of the amplitudes. The incoherent (Raman) spectrum of fractons falls typically in the range from 1 to 300 GHz.[35] Beyond this frequency, particle modes are observed. In the fracton range, the dependence of the Raman susceptibility χ on frequency ω is given by $\chi \propto \omega^{-2+2\sigma\bar{\bar{d}}/D}$ [35,8] Here, σ is an internal length dimension,[8] for which we find $\sigma \simeq 1.5$.

5. THE DENSITY OF VIBRATIONAL STATES

Of great interest is to obtain a measurement of $N(\omega)$ for *one* single sample. This should allow $\bar{\bar{d}}$ to be determined independently of scaling and mutual self-similarity, since for fractons $N(\omega) \propto \omega^{\bar{\bar{d}}-1}$. Of the various inelastic neutron scattering approaches[37] for determining $N(\omega)$, we chose a difference technique that allows us to extract the incoherent scattering from protons alone.[36-38] The initial results covered the region of particle modes, and probably the upper frequency fractons.[36] The particle density of states was also calculated and compared to experimental results, hence affording an approximate absolute calibration of $N(\omega)$. Combined with the Brillouin results, the overall behavior of $N(\omega)$ was thus obtained.[36] The thermal properties calculated therefrom agreed remarkably well with literature data.[39] The latter also yielded $\bar{\bar{d}} \simeq 1.4$.

More recently, measurements were repeated with a differently prepared sample, and in such a manner that the relative density of high-frequency modes should have increased.[38] Using higher resolution neutron spectroscopy, more than one order of magnitude in ω of the fracton density of states was observed. It indeed yields the higher value $\bar{\bar{d}} \simeq 1.8 \ldots 1.9$. This confirms that $\bar{\bar{d}}$ is not a universal quantity, but depends on the microstructure of the material.

[1]Laboratoire de Science des Matériaux Vitreux is associated with CNRS No. 1119.

REFERENCES

1. E. Courtens and R. Vacher, in *Random Fluctuations and Pattern Growth: Experiments and Models* H.E. Stanley and N. Ostrowsky, Eds. (Kluwer Academic Publishers, Dordrecht, 1988) p. 21.
2. E. Courtens and R. Vacher, Proc. R. Soc. Lond. A **423,** 55 (1989).
3. S. Alexander and R. Orbach, J. Phys. (Paris) Lett. **43,** L625 (1982).
4. Y. Gefen, A. Aharony, and S. Alexander, Phys. Rev. Lett. **50,** 77 (1983).
5. R. Rammal and G. Toulouse, J. Phys. (Paris) Lett. **44,** L13 (1983).
6. S. Alexander, O. Entin-Wohlman, and R. Orbach, Phys. Rev. B **34,** 2726 (1986); A. Jagannathan, R. Orbach, and O. Entin-Wohlman, Phys. Rev. B (to be published).

7. A. Aharony, S. Alexander, O. Entin-Wohlman, and R. Orbach, Phys. Rev. Lett. **58**, 132 (1987).

8. S. Alexander, Phys. Rev. B (to be published).

9. G.S. Grest and I. Webman, J. Phys. Lett. **45**, 1155 (1984).

10. I. Webman and G.S. Grest, Phys. Rev. B **31**, 1689 (1985).

11. K. Yakubo and T. Nakayama, Phys. Rev. B **36**, 8933 (1987).

12. T. Nakayama, K. Yakubo, and R. Orbach, J. Phys. Soc. Japan **58**, 1891 (1989).

13. K. Yakubo and T. Nakayama, Phys. Rev. B **40** (1 July 1989).

14. S. Alexander, C. Laermans, R. Orbach, and H.M. Rosenberg, Phys. Rev. B **28**, 4615 (1983).

15. H.M. Rosenberg, Phys. Rev. Lett. **54**, 704 (1985).

16. A.J. Dianoux, J.N. Page, and H.M. Rosenberg, Phys. Rev. Lett. **58**, 886 (1987).

17. J.E. de Oliveira, J.N. Page, and H.M. Rosenberg, Phys. Rev. Lett. **62**, 780 (1989).

18. A. Boukenter, E. Duval, and H.M. Rosenberg, J. Phys. C **21**, L541 (1988).

19. M. Arai and J.E. Jörgensen, Phys. Lett. A **133**, 70 (1988).

20. A. Avogadro, S. Aldrovandi, and F. Borsa, Phys. Rev. B **33**, 5637 (1986).

21. A. Fontana, F. Rocca, and M.P. Fontana, Phys. Rev. Lett. **58**, 503 (1987).

22. E. Duval, G. Mariotto, M. Montagna, O. Pilla, G. Viliani, and M. Barland, Europhys. Lett. **3**, 333 (1987).

23. G. Mariotto, M. Montagna, G. Viliani, R. Campostrini, and G. Carturan, J. Phys. C **21**, L797 (1988).

24. A. Boukenter, B. Champagnon, E. Duval, J.L. Rousset, J. Dumas, and J. Serughetti, J. Phys. C **21**, L1097 (1988); J.L. Rousset, E. Duval, A. Boukenter, B. Champagnon, A. Monteil, J. Serughetti, and J. Dumas, J. Non-Cryst. Solids **107**, 27 (1988).

25. D. Deptuck, J.P. Harrison, and P. Zawadski, Phys. Rev. Lett. **54**, 913 (1985).

26. J.H. Page and R.D. McCulloch, Phys. Rev. Lett. **57**, 1324 (1986).

27. Y.J. Uemura and R.J. Birgeneau, Phys. Rev. Lett. **57**, 1947 (1986) and Phys. Rev. B. **36**, 7024 (1987).

28. T. Freltoft, J. Kjems, and D. Richter, Phys. Rev. Lett. **59**, 1212 (1987).

29. D.W. Schaefer and K.D. Keefer, Phys. Rev. Lett. **56**, 2199 (1986).

30. R. Vacher, T. Woignier, J. Pelous, and E. Courtens, Phys. Rev. B **37**, 6500 (1988).

31. R. Vacher, T. Woignier, J. Phalippou, J. Pelous, and E. Courtens, J. Non-Cryst. Solids **106**, 161 (1988).

32. R. Vacher, T. Woignier, J. Phalippou, J. Pelous, E. Courtens, Rev. Phys. Appl. **24**, C4-127 (1989).

33. E. Courtens, J. Pelous, J. Phalippou, R. Vacher, and T. Woignier, Phys. Rev. Lett. **58**, 128 (1987).

34. E. Courtens, R. Vacher, J. Pelous, and T. Woignier, Europhys. Lett. **6**, 245 (1988).

35. Y. Tsujimi, E. Courtens, J. Pelous, and R. Vacher, Phys. Rev. Lett. **60**, 2757 (1988).

36. R. Vacher, T. Woignier, J. Pelous, G. Coddens, and E. Courtens, Europhys. Lett. **8**, 161 (1989).

37. G. Coddens, R. Vacher, T. Woignier, J. Pelous, and E. Courtens, Rev. Phys. Appl. **24**, C4-151 (1989).

38. R. Vacher, E. Courtens, G. Coddens, J. Pelous, and T. Woignier, Phys. Rev. B **39**, 7384 (1989).

39. R. Maynard, R. Calemczuk, A.M. de Goër, B. Salce, J. Bon, E. Bonjour, and A. Bourret, Rev. Phys. Appl. **24**, C4-107 (1989).

NEW CRITICAL EXPONENTS FOR SPATIAL AND TEMPORAL

FLUCTUATIONS IN STOCHASTIC GROWTH PHENOMENA

Preben Alstrøm, Paul Trunfio, and H. Eugene Stanley

Center for Polymer Studies
Boston University
Boston, MA 02215, USA

INTRODUCTION

Scale-invariant structures originating from growth processes have been found to be extremely widespread in nature.[1] This observation have led to a number of careful experiments, and various growth models have been suggested to describe the fractal *outcome*; but why did they become fractal in the first place? To answer this question we must understand the spatio-temporal *evolution*. Dynamically, the interface is observed to be unstable, and the system eventually reaches a statistically stationary state where a rich ramified pattern is created. A major observation is that this state can be described by power laws - the pattern becomes scale invariant.

In this paper, it is argued that growth processes *naturally* develop critical states. The growth is viewed as a branching process: every perturbation may either 'die' or give rise to further branching. The dynamics stabilizes at the critical point where branching *precisely* balances extinction. This suggests a natural way to divide a cluster into sub-clusters that exhibit spatio-temporal fluctuations. These fluctuations are found to be scale invariant for invasion percolation and diffusion-limited aggregation (DLA), which both have a variety of applications such as fluid displacement in a porous medium. The critical exponents characterizing these fluctuations are surprisingly close to the values we obtain in the 'mean-field' limit in which growth in different regions of the cluster is statistically independent. For invasion percolation the fluctuations exhibit $1/f^\varphi$ noise with φ close to unity. Also, we identify the associated order parameter j, and we determine its scaling exponent β.

MEAN-FIELD THEORY

Consider first the formation of viscous fingers when one fluid displaces another fluid with higher viscosity. To understand why viscous fingers become scale-invariant, one must follow the dynamical process that created them. Basically (Fig. 1), (i) the flow can stop, (ii) the flow can continue, or (iii) the flow can branch, creating a new finger. However, eventually every finger can *not* branch, since this would imply a *persistent* decrease of the average flow rate, and the system would never reach stationarity. Thus, some of the fingers *must* stop growing. *The system reaches stationarity exactly when the branching has been broken down to the level where the flow barely survives.* At this point extinction balances branching, and the growth process is stable with respect to fluctuations. It is in this sense the dynamical stationary states for growth phenomena become critical.

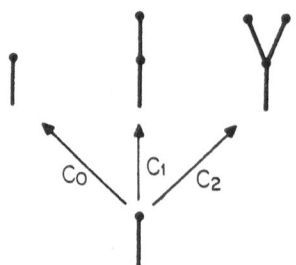

Fig. 1. Formation of spatial scaling structure described by branching processes. Criticality appears for $C_0 = C_2$.

To be more specific, the flow above is dynamically modeled by the branching process where in each generation an 'individual' is replaced by zero, one, or two descendants with probabilities C_0, C_1, and C_2, respectively (Fig. 1). On the average, the number of descendants increases by a factor of $C_1 + 2C_2 = 1 + C_2 - C_0$ from one generation to the next. At criticality where the 'family' barely survives, $C_0 = C_2$, and the structure of branches become scale invariant.[2,3] The probability $D(s)$ that a branching process creates a (total) family with *exactly* s individuals is for large s

$$D(s) \propto s^{1-\tau}, \tag{1}$$

where $\tau = 5/2$.

Also, we shall consider the lifetime distribution $D(t)$, where we relate the lifetime t to the number of generations ℓ through a dynamical exponent z, $t \propto \ell^z$. The probability $D(\ell)$ that a process stops *exactly* at the generation ℓ is for large ℓ, $D(\ell) \propto \ell^{-2}$. Hence, if z takes on the simple diffusion value 2,

$$D(t) \propto t^{\varphi-2}. \tag{2a}$$

where $\varphi = 1$. We shall see that the branching process may be interpreted as a 'noise'. Since the distribution of lifetimes (2b) translates directly into the power frequency spectrum,

$$P(f) \equiv \int_0^{1/f} tD(t)dt \propto f^{-\varphi}, \tag{2b}$$

the result $\varphi = 1$ implies that the noise is $1/f$ noise.

INVASION PERCOLATION AND DLA

For invasion percolation, the growth is started from a seed on a lattice with a random distribution of site probabilities, and at each time unit growth occurs at the perimeter site with the lowest probability. Figure 2a shows a cluster of 'mass' $M = 4000$. In the infinite mass limit there exists a critical probability p_c at which the normalized distribution of probabilities $\mathcal{N}(p)$ along the perimeter jumps from *zero* to the constant value $1/(1 - p_c)$. In this limit the average probability becomes

$$\langle p \rangle \equiv \int_0^1 p\mathcal{N}(p)dp = (1 + p_c)/2. \tag{3a}$$

Hence, p_c can be calculated as the limit value of p^* defined by

$$p^* \equiv 2\langle p \rangle - 1. \tag{3b}$$

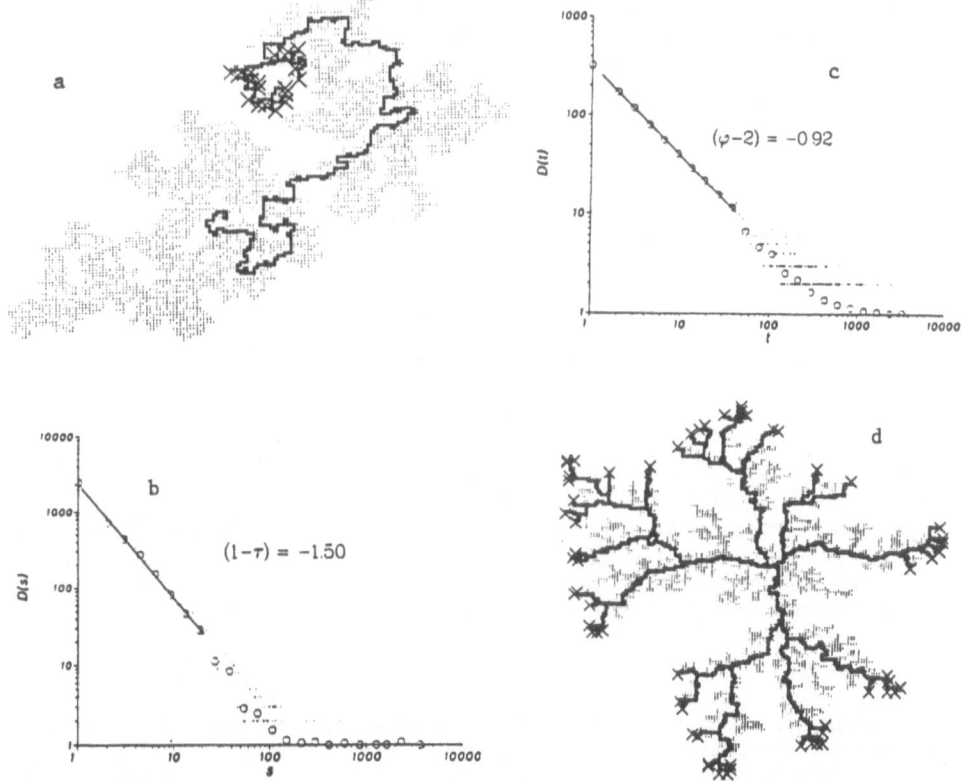

Fig. 2. (a) Invasion percolation cluster of mass $M = 4000$. The surviving paths are shown as heavy lines. The rest is a disconnected set of extinct branches. (b) The size distribution $D(s)$ of extinct branches. (c) The distribution $D(t)$ of lifetimes t. The slopes of the straight lines are -1.50 and -0.92, respectively. The distributions are based on 40 clusters of mass $M = 4000$. The circles represent binned data. (d) DLA of mass $M = 4000$. The heavy lines are surviving paths. The rest is extinct branches.

Perimeter sites can be divided into *surviving* sites with $p \leq p^*$ and *extinct* sites with $p > p^*$. Each surviving site is traced *back* to the original seed, choosing at each step the neighbor site which was grown most recently. We call the paths thereby obtained *surviving paths*. Removing all the surviving paths from the cluster, the cluster splits into 'sub-clusters' (Fig. 2a), denoted *extinct branches*. Each extinct branch has a number s of sites and a lifetime t, which is the difference between the first and the last site grown on that branch. The extinct branches may be considered as noise on the surviving paths. Figures 2b,c show the distribution $D(s)$ of sizes and $D(t)$ of lifetimes for invasion percolation. We find $\tau = 2.50 \pm 0.05$ and $\varphi = 1.08 \pm 0.03$, which are close to the 'mean-field' predictions $\tau = 5/2$ and $\varphi = 1$ above.

In DLA, the growth is initiated by launching a random walker from a random point on a circle surrounding the cluster. Starting from a seed, growth takes place when the random walker touches a perimeter site. Figure 2d shows a cluster of mass $M = 4000$. The growth probabilities p_i along the perimeter are determined by solving Laplace's equation $\nabla^2 \varphi = 0$ for the probability field $p \propto |\nabla \varphi|$, where $\varphi = 0$ on the cluster and $\varphi = 1$ on the surrounding circle. Then, at the perimeter, $p_i \propto |\nabla \varphi_i|$. Since DLA is created through a *multiplicative* process where all p_i converge to zero, the sites where the growth eventually takes place have probability p^* given by the geometrical mean,[2]

$$\ln p^* = \sum_i p_i \ln p_i. \tag{4}$$

341

We now identify the surviving sites as those with $p \geq p^*$. These sites are traced back to the seed to find the surviving paths (Fig. 2d). Removing these, we obtain the extinct branches. From the size distribution $D(s)$ we find $\tau = 2.48 \pm 0.10$, close to the mean-field result. For DLA, we have not been able to reliably calculate the exponent φ.

ORDER PARAMETER

We may introduce the deviation from criticality

$$\epsilon \equiv |p^* - p_c|, \tag{5a}$$

where $p_c = 0$ for DLA. As the cluster grows, $\epsilon \to 0$ and

$$j(\epsilon) \equiv N_s(\epsilon)/M \propto \epsilon^\beta, \tag{5b}$$

where N_s is the number of surviving sites. The 'flow' j is an order parameter. We find $\beta = 0.78 \pm 0.05$ for DLA, and $\beta = 0.7 \pm 0.15$ for invasion percolation.

For DLA, we can relate β to the fractal (mass) dimension D, $M \propto L^D$, where $2L$ is the diameter of the cluster. To this end, we use that $p^* \propto L^{-D_I}$ and $N_s \propto (p^*)^{-\mathcal{R}}$, where D_I is the information dimension and $\mathcal{R} \to 1$ since p^* dominates the growth in the infinite mass limit. For DLA, $D_I = 1$. Hence,

$$\beta = D - 1. \tag{6}$$

For finite clusters $\mathcal{R}$ is always smaller than 1, and $\mathcal{R}$ expresses how well p^* dominates the growth. A direct study of N_s versus p^* yields $\mathcal{R} = 0.9$, consistent with a dimension $D \simeq 1.7$. Note that a numerical determination of the length scale L of the clusters is *not* necessary to obtain β, D and $\mathcal{R}$.

CONCLUSIONS

In conclusion, branching processes have been used to understand the underlying mechanism for fractal growth. Our conceptual framework has practical utility in quantifying the spatio-temporal fluctuations in dynamic growth phenomena. In particular, we have found new critical exponents — τ, φ, and β — describing these fluctuations. For DLA, $\beta + 1 = D$. Our results for the exponents τ and φ are close to the mean-field values derived from a general *non-interacting* branching process. For invasion percolation we find $1/f^\varphi$ noise with φ surprisingly close to 1. We emphasize that the critical exponents are experimentally measurable. The surviving sites can be found by observing the interface over a small time period. For DLA type growth, p^* can be estimated since the small growth probabilities do *not* contribute to p^*, and the large probabilities are well approximated by the local growth.

REFERENCES

1. See e.g. B. B. Mandelbrot, "The Fractal Geometry of Nature" (Freeman, San Fransisco, 1982); R. Pynn and T. Riste, eds., "Time-Dependent Effects in Disordered Materials" (Plenum, New York, 1987); H. E. Stanley and N. Ostrowsky, eds., "Random Fluctuations and Pattern Growth: Experiments and Models" (Kluwer, Dordrecht, 1988); J. Feder, "Fractals" (Plenum, New York, 1988).

2. P. Alstrøm, Phys. Rev. A **38**, 4905 (1988). P. Alstrøm, preprint.

3. T. E. Harris, "The Theory of Branching Processes" (Springer, Berlin, 1963).

EFFECTS OF PHASE TRANSITIONS AND FLUCTUATIONS
ON MASS TRANSPORT

Jørgen Vitting Andersen and **Ole G. Mouritsen**
Department of Structural Properties of Materials
The Technical University of Denmark, Building 307
DK-2800 Lyngby, Denmark

I. INTRODUCTION

It is well known[1] that transport phenomena in condensed matter are strongly dependent on cooperative phenomena and that transport coefficients display anomaleous behavior near phase transitions. The general physical reason for this behavior is that transport coefficients can be expressed in terms of correlations of fluctuations in appropriate fields.[2] Since these correlations decay slowly in time and get long ranged close to phase transitions it is expected that the the transport capacity will be strongly affected as the transition point is approached and the thermal density fluctuations blow up.

Calculations based on microscopic models of fluctuation effects on transport properties are sparce.[3-8] Some of these calculations are based on computer-simulation methods applied to statistical mechanical lattice models of closed systems. Most of these deal with models of spin self-diffusion. Recently, a very interesting Monte Carlo simulation of the thermal conductivity of a kinetic Ising model has been reported by Harris and Grant.[7] Building upon an idea due to Creutz[8] involving microcanonical Monte Carlo sampling by the demon technique, these authors were able to calculate the heat flux and hence the temperature dependence of the heat transport coefficient, $\kappa(T)$ of the two-dimensional Ising model. They found that $\kappa(T)$ is strongly influenced by thermal fluctuations and that it displays a peak close to and above the critical temperature. At the critical temperature, $\kappa(T)$ vanishes due to critical slowing down. In the simulations by Harris and Grant,[7] the temperature profile and the heat flux through a sample system was calculated by assigning different temperatures (via heat baths administered by an ordinary Metropolis Monte Carlo sampling procedure) to the two opposite ends of the sample.

In this paper we report on the first results of a calculation by Monte Carlo computer-simulation techniques of the mass transport coefficient in a lattice gas model. We calculate in a rectangular geometry the mass flux, $Q(T)$, and the mass density profile, $\rho(x,T)$, for an open system which is subject to an externally imposed chemical potential gradient. x is a spatial coordinate. From these quantities, the mass transport coefficient, $D(T)$, can be derived using the equivalent of Ohm's law,

$$Q(T) = D(T)\frac{\mathrm{d}\rho(x,T)}{\mathrm{d}x}. \tag{1}$$

$D(T)$ is a measure of the system's capacity for transporting mass.

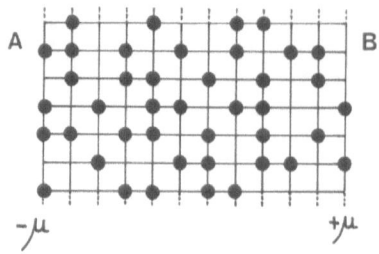

Fig. 1. Two-dimensional lattice gas model on a square lattice with open boundaries A and B in one dimension and periodic boundary conditions in the other dimension. Chemical potentials, $\pm\mu$, are applied at the two ends. Black dots denote particles on the lattice.

II. MODEL AND COMPUTATIONAL TECHNIQUES

We consider a two-dimensional lattice gas on a quadratic net which is finite in one dimension and periodically infinite in the other dimension, cf. Fig. 1. Chemical potentials, $\pm\mu$, are applied at the two edges A and B. The interactions between the particles in the lattice gas are governed by the Hamiltonian

$$\mathcal{H} = -J_{NN} \sum_{i,j}^{NN} n_i n_j - J_{NNN} \sum_{i,j}^{NNN} n_i n_j - \mu \sum_{i \in A} n_i + \mu \sum_{i \in B} n_i, \qquad (2)$$

where $n_i = 0, 1$ is the occupation variable of the ith site. The two first summations of Eq. (2) are over respectively nearest-neighbor (NN) and next-nearest neighbor (NNN) sites. In the present work we have chosen the interaction parameters as $J_{NN} = -J_{NNN} = J < 0$ which in thermodynamic equilibrium at low temperatures lead to a simple two-fold degenerate (2×1)-ordering.

The applied external chemical potentials act like particle baths and the system is therefore an open one. However, particles can only enter or leave the system via the edges A and B. This chemical potential step-profile is the prerequisite for setting up a net particle flow through the system and it hence permits the transport capacity of the model to be probed. In the steady state, the net flux of particles into the system through edge A will equal the net flux out of edge B. The mass flux leads to an energy gain whenever a particle has dropped through the chemical potential. In order to keep the potential energy of the system constant this gain is compensated by adding 2μ to the system energy when a particle leaves the system.

The steady-state dynamics is assumed to be governed by a Kawasaki-type particle-vacancy NN pair-exchange mechanism inside the system combined with a Glauber-type particle creation/annihilation mechanism at the two edges A and B. Hence, neither the particle number nor the total energy are conserved quantities. This implementation corresponds to a canonical ensemble inside the lattice and a grand canonical ensemble at the edges. The total density is hence a dependent variable which has to be calculated. The dynamical processes are subject to the conventional Monte Carlo Metropolis criterion.[9]

The calculations are carried out on different lattice sizes (mostly with 50^2 and 100^2 sites) in order to estimate finite-size effects. As an operational criterion for the non-equilibrium steady-state condition we have required that the incoming and outgoing fluxes are the same within less than 0.5 %. The statistics correspond to 10^5 Monte Carlo steps per site.

III. RESULTS

In Fig. 2a are shown selected results for the density profile, $\rho(x)$, at temperatures close to and above the critical temperature for different values of the chemical potential. Except for some special effects at the edges, $\rho(x)$ is as expected a linear function.

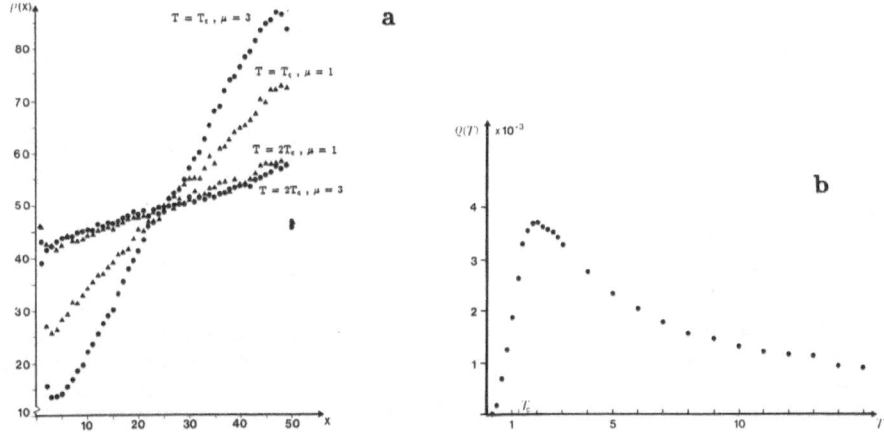

Fig. 2.(a): Selected mass density profiles, $\rho(x)$, for s system with 50^2 sites calculated at different temperatures and chemical potentials. T_c ($\simeq 1.2 J/k_B$) is the critical temperature. $\bigcirc$: $\mu = 3$, $\triangle$: $\mu = 1$. The steep curves correspond to $T = T_c = 1.2$ and the flatter curves to $T = 2T_c$. (b): Temperature dependence of the mass flux, $Q(T)$, for a system with 50^2 sites and $\mu = 3$. $Q(T)$ is in units of inverse Monte Carlo time.

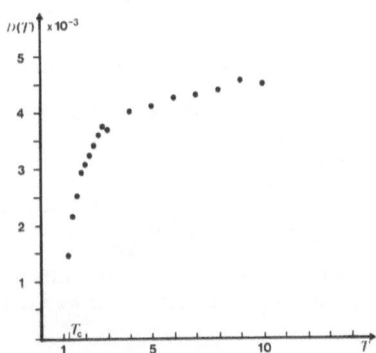

Fig. 3. Mass transport coefficient $D(T)$, Eq. (1), in units of inverse Monte Carlo time.

The edge effects reflect the temperature-dependent correlation length of the symmetry-breaking chemical potentials. In all of the calculations it is found that the total average density is $\langle \rho \rangle \simeq \frac{1}{2}$.

The mass flux, $Q(T)$, as a function of temperature is shown in Fig. 2b for $\mu = 3$. $Q(T)$ has a distinct maximum around $2T_c$ and decays to zero at high and low temperatures. At any given value of T, $Q(T)$ increases with μ. At the same time larger values of μ lead to steeper density profiles in a way which makes the mass transport coefficient, $D(T)$ of Eq. (1), independent of μ.

The final results for the mass transport coefficient, $D(T)$, are given in Fig. 3. This figure, which is the main result of the present work, shows that $D(T)$ has a strong temperature dependence and in particular that $D(T)$ is suppressed by thermal density fluctuations and vanishes at the phase transition. At high temperatures, $D(T)$ approaches a constant. With the present data we are unable to bring the system into a steady-state at low temperatures with respect to the density profile. Due to the ordering phenomenon, the system is very stiff in this regime and not even equilibration times of up to 10^5 Monte Carlo steps per site are sufficient to stabilize $\rho(x)$, whereas $Q(T)$ within this time has approached a stable value. Similar difficulties were encountered in Harris and Grant's[7] study of the thermal conductivity at low temperatures. The reason for these difficulties is that kinetically activated processes limit the transport in the ordered phase.

IV. FUTURE PERSPECTIVE

The methods described in this paper and the preliminary results reported for a simple lattice model hold a promise for the feasibility of calculating mass transport and flow in more material-specific situations. Immediate applications include surface diffusion in chemisorbed overlayers, e.g. O/W(110) [which has a four-fold degenerate (2×1) ordering symmetry produced by Eq. (2) with $2J_{NN} < J_{NNN} < 0$] and O/W(211) [which has the ordering symmetry studied in the present work]. It is of obvious interest to investigate the influence on the ordering symmetry on the mass transport. Furthermore, we are currently exploring the potential for studying oxygen diffusion in the CuO-planes of high-T_c superconductors during the annealing process as well as flow in random, fractal, and porous media.

ACKNOWLEDGEMENTS

This work was supported by the Danish Natural Science Research Council under grant No. 5.21.99.72 and the Danish Technical Research Council under grant No. 16-4296.K.

REFERENCES

1. H. E. Stanley, *Introduction to Phase Transitions and Critical Phenomena* (Oxford University Press, New York, 1971) Chap. 13.
2. D. Forster, *Hydrodynamic Fluctuations, Broken Symmetry, and Correlation Functions* (Benjamin, Reading, MA, 1975).
3. A. Sadiq, Phys. Rev. B **9**, 2299 (1974).
4. D. A. Reed and G. Erlich, Surf. Sci. **105**, 603 (1981).
5. R. Kutner, K. Binder, and K. W. Kehr, Phys. Rev. B **26**, 2967 (1982).
6. A. Sadiq and K. Binder, Surf. Sci. **128**, 350 (1983).
7. R. Harris and M. Grant, Phys. Rev. B **38**, 9323 (1988).
8. M. Creutz, Ann. Phys. **167**, 62 (1986); Phys. Rev. Lett. **50**, 1411 (1983).
9. O. G. Mouritsen, *Computer Studies of Phase Transitions and Critical Phenomena* (Springer-Verlag, Heidelberg, 1984).

THE STRETCHED EXPONENTIAL, THE VOGEL LAW, AND ALL THAT

Michael F. Shlesinger

Office of Naval Research, Physics Division
800 North Quincy Street, Arlington, VA 2217 USA

John T. Bendler

Polymer Physics and Engineering Branch
General Electric Corporate Research and Development
Schenectady, NY 12301 USA

ABSTRACT

The nature of relaxation in glassy materials is investigated via
a defect transport model. Relaxation is assumed to occur when a mobile
defect reaches a frozen-in region of the material. A many defect calculation leads to the ubiquitous stretched exponential law as a probability limit distribution for the relaxation. The time scale for this
relaxation varies inversely with the concentration of uncorrelated
defects. The fast disappearance of these defects with decreasing
temperature leads to a Vogel-like law for the relaxation time scale.

1. INTRODUCTION

Three questions stand out in the behavior of many glassy
materials:

i)   Perturbations relax back to equilibrium according to the
stretched exponential law[1]

$$\Phi(t) = \exp[-(t/\tau)^{\beta}] \ , \ \beta < 1 \tag{1}$$

How does this universal form arise?

ii)   The relaxation time τ in Eqn (1) in many glasses follows the
Vogel-Fulcher-Tammann[2-4] law

$$\tau \sim \exp\left(\frac{\text{const}}{T-T_0}\right) \tag{2}$$

What is the mechanism for this law?

Exceptions to this law do exist, e.g., several materials do follow the more familiar Arrhenius law[5].

iii) The third question is what is the meaning of T_0 in equation (2). It is known that

$$T_0 < T_g \tag{3}$$

where T_g is the glass transition temperature. So the relaxation time τ is keyed to a temperature lower than T_g.

We will present a model which provides a self-consistent answer to the above questions based on the transport and aggregation of defects in glassy systems.

2. A DEFECT TRANSPORT MODEL OF RELAXATION

We assume that a frozen-in part of a glass can be relaxed if it is reached by a defect[6]. The defect may encapsulate some free volume which would provide room for molecules to freely move. We start with a system with N defects which can move between V sites[7]. The frozen-in part of our system is placed at one of these sites which we label as an origin. The relaxation function $\Phi(t)$ is the probability that none of the N defects (whose positions we randomize at time t = 0) has reached the origin by time t. The expression for $\Phi(t)$ is

$$\Phi(t) = \left[1 - \frac{1}{V} \sum_{l_0} \int_0^t F(l_0,\tau)d\tau \right]^N \tag{4}$$

The term in brackets is one minus the probability that a particular defect did reach the origin by time t. If the defect started at site l_0 at time t=0, $F(l_0,\tau)$ is the probability density that the defect reaches the origin for the first time at time t = τ. The bracket is raised to the N-th power to obtain the probability that none of the N defects has reached the origin at time t. In the limit N,V → ∞ with N/V = c, a constant concentration of defects Eqn (4)

348

becomes

$$\Phi(t) = \exp\left[-c \sum_{l_0} \int_0^t F(l_0,\tau)d\tau \right] \qquad (5)$$

The term in bracktes is -c times the flux of defects in to the origin in a time t. Eqn (5) can be rewritten as [7]

$$\Phi(t) = \exp\left[-c\, S(t)\right] \qquad (6)$$

where $S(t)$ is the number of distinct sites that a single defect starting from the origin visits in a time t. To proceed further we must specify how a defect moves.

3. DEFECT MOTION

We introduce a probability density function $\psi(t)$ which governs the duration of a defect waiting at a site before it jumps to the next site. If the first moment of $\psi(t)$ exists (we denote it by $\langle t \rangle$) then [7]

$$S(t) \sim \text{const. } t/\langle t \rangle \qquad (7)$$

and $\Phi(t)$ decays exponentially. If, however, at long times

$$\psi(t) \sim t^{-1-\beta}, \quad \beta < 1 \qquad (8)$$

then $\langle t \rangle = \infty$ and no characteristic waiting time at a site exists. For this case[7]

$$S(t) \sim t^\beta \qquad (9)$$

and

$$\Phi(t) = \exp(-\text{const. } c t^\beta) \qquad (10)$$

For the parallel reaction scheme we have analyzed (i.e. any one of N defects can cause the relaxation) we find two relaxation laws, i) The Poisson law, and ii) The Stretched exponential law. Both are probability limit distributions. The universality of the stretched exponential law is tied to this fact, i.e. specific details of the defect motion do not matter, only that $\langle t \rangle = \infty$. This case can arise from activated hopping over, a distribution $f(\Delta)$ of, potential barriers. If t, the time to overcome a barrier of height Δ is

Arrhenius so $t \sim e^{\Delta/kT}$, then $f(\Delta) = e^{-\Delta/\Delta_0}$ leads to Eqn (8) with $\beta = T/k\Delta_0$ for $T < k\Delta_0$. Note that the distribution $f(\Delta)$ needs only to lead to Eqn (8), i.e. we do not need to find an $f(\Delta)$ which would yield $\Phi(t)$ directly.

We also note that a random walk on a fractal has $S(t) \sim t^{d_s/2}$ where d_s is the spectral dimension of the fractal lattice. So defect motion on fractal structures also leads to Eqn (1), with β independent of temperature.

Finally, in one dimension, for a random walk with $\langle t \rangle$ finite $S(t) \sim \sqrt{t}$, and $\Phi(t) \sim \exp(-\sqrt{t})$, a result found for several 1D non-random walk models. Fluctuations, in these models, must, however, move around in a random walk fashion.

4. A VOGEL-LIKE LAW

Writing Eqn (10) in the form of Eqn (1) we obtain

$$\tau \sim \frac{1}{c^{1/\beta}} \tag{11}$$

i.e., the time constant in the stretched exponential depends on the concentration of mobile defects.

If we lower the temperature, in our model, the only way we can lower the entropy is to have our defects cluster as they are the only mobile entities. We further assume that as the defects become correlated only single (uncorrelated) defects contribute to the relaxation, so Eqn (10) becomes

$$\Phi(t) = \exp(-\text{const. } c_1 t^\beta) ,$$

$$= \exp(-[t/\tau]^\beta) . \tag{12}$$

Likewise we now have $\tau \sim 1/c_1^{1/\beta}$, where c_1 is the concentration of uncorrelated single defects, which is given by

$$c_1 = c(1-c)^Z \tag{13}$$

The above equation says that a site has an uncorrelated defect if there is a defect at the site (the c factor) and none of the sites it interacts with has a defect (the $(1-c)^Z$ factor). Z is the number of sites within the defect's sphere of influence.

In a lattice gas model[8] of the defect's interaction, an interaction correlation lenght ξ grows as the temperature is lowered (in the mean field approximation) as

$$\xi \sim (T - T_0)^{-1/2} \qquad (14)$$

The mean field approximation is valid because experiments are performed above T_g and $T_0 < T_g$.

A correlation volume $Z = \xi^3 \sim (T - T_0)^{-3/2}$

as the temperature is lowered, so

$$c_1 \sim c\, e^{\ln(1-c)/(T-T_0)^{3/2}} \qquad (15)$$

(note $c < 1$),

and thus, $\tau \sim \exp\left(\dfrac{1}{(T-T_0)^{3/2}} \right)$ $\qquad (16)$

This relaxation time is keyed to T_0, the temperature where all uncorrelated defects disappear. In our model the defects embody the flexibility in the viscous liquid above the glass transition. As the temperature is lowered the mobile single defects disappear until at some critical temperature, T_g, rigidity percolates, but some uncorrelated defects still exist, so $T_0 < T_g$. We have had success[9] in describing much experimental data with Eqn (16). Note our model does not make any predictions about the nature of the glass dynamics at T_g.

If the defects repel each other (as in $S_i O_2$) or the defect concentration is too small to begin with, then the phase transition of defect clustering need not occur and a Vogel like law will not hold. Thus finding the stretched exponential does not imply that the Vogel law will follow.

One example of a defect is the high energy CIS conformation in polycarbonate which can propagate along the more predominant trans-chains. NMR studies by Li et al.[11] find static and mobile defects in polycarbonate with the mobile species disappearing according to a Vogel-like law. Aging in polycarbonate can be related to the loss of defects off of chain ends[10]. All of this suggests that the introduction of defects during processing of polymeric material may be advantageous in enhancing their flexibility and extending their lifetime.

REFERENCES

1. G. Williams and D. C. Watts, Trans. Faraday Soc., 66, 80 (1970).
2. H. Vogel, Z. Phys. 22, 645 (1921).
3. G. Fulcher, J. Amer. Ceram. Soc. 8, 339 (1925).
4. G. Tammann and H. Hesse, Z. Anorg. Chem. 156, 245 (1926).
5. C. A. Angell, in "Relaxation in Complex Systems," ed. K. Ngai and G. Wright, U.S. Govt. Printing Office (1985).
6. S. H. Glarum, J. Chem. Phys. 33, 1371 (1960).
7. M. F. Shlesinger and E. W. Montroll, Proc. Nat. Acad. Sci (USA) 81, 1280 (1984).
8. M. E. Fisher, J. Math. Phys. 5, 944 (1964).
9. J. T. Bendler and M. F. Shlesinger, J. Stat. Phys. 53, 531 (1988).
10. J. T. Bendler and M. F. Shlesinger, J. Molec. Liq. 36, 37 (1987).
11. K. Li, T. Inglefield, A. A. Jones, J. T. Bendler and A. D. English, Macromolecules, 21, 2940 (1988).

SOME INDUSTRIAL APPLICATIONS OF SOFT CONDENSED MATTER

IN HIGH-PERFORMANCE POLYMERS PROCESSING

Paul Smith

Materials Department
University of California
Santa Barbara, California 93106

ABSTRACT

Soft condensed matter is of paramount importance both as final product and processing intermediate. This paper describes the role of this unusual state of matter, in the form of polymeric liquid crystals and gels, as processing intermediates, or precursors, in manufacturing technologies of high-performance polymer materials. The unique structure of soft condensed matter intermediates allows particular conformational changes, rearrangement and packing of long chain molecules, yielding materials with exceptional mechanical properties. Of practical importance and scientific interest is that the structure of these soft intermediates can be controlled in a detailed manner, which is dramatically reflected in their processing characteristics and properties of the final materials. The correlation between the structure of the soft condensed matter and the properties of the resulting high-performance materials will be elucidated. In addition, some recent uses and an unusual synthesis of this unique state of matter will be presented.

INTRODUCTION

The past two decades have witnessed a dramatic development in polymer materials science and technology. Only 25 years ago, polymers were regarded primarily as cheap replacements for many existing, frequently naturally occurring, materials. The low cost generally was associated with poor or, at best moderate, properties. Today, polymers are manufactured that exhibit extraordinary mechanical properties, thermal stability, high electrical conductivity, tailor-made optical properties, etc.

Illustrative in this respect is the graph in Figure 1, which displays the progression in time of the tensile strength (stress at break) of some industrial polymer fibers. Prior to the 70's, polymer fibers were manufactured primarily from polyesters and polyamides and, of course, cellulose. The 1970 step-wise increment in strength of about 300 % was the result of the discovery, and subsequent commercialization by Du Pont, of lyotropic liquid crystals of rigid chain molecules. The second, still continuing, surge in mechanical properties found its origin in the discovery of [at the time] unusual tensile deformation behavior of thermo-reversible gels of certain flexible macromolecules.

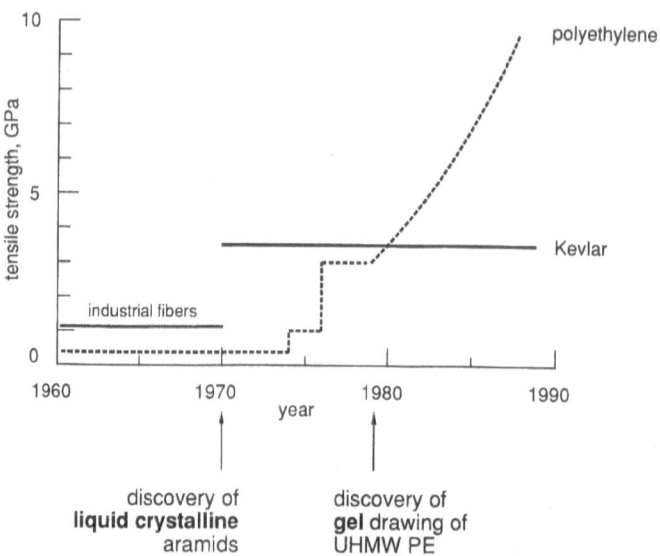

Figure 1. Tensile strength (stress at break) *versus* year of some man-made industrial fibers. Recent highlights are the discovery of high-performance fibers derived from liquid-crystalline aramids, first commercialized by Du Pont under the registered trade name Kevlar; and the discovery of the unusual tensile drawing characteristics of thermo-reversible gels of high molecular weight polyethylene, which resulted in ultra-high strength fibers sold by DSM/Toyobo (Dyneema) and, under licence from DSM, by Allied (Spectra).

Soft condensed matter has played, and is expected to continue to play, an important role in this remarkable development. This paper focuses on the distinctive role that soft condensed matter, in the form of macromolecular liquid crystals and gels, plays in the manufacturing technologies of these strong polymer materials. In addition, a more specialized and recent example of the utilization of soft condensed matter in processing of intractable materials will be described, as well as an unusual method of synthesizing this remarkable class of matter.

2. LIQUID CRYSTALLINE POLYMERS

2.1 Processing Intermediate for High-Performance Fibers

Stephanie L. Kwolek, at Du Pont, prepared in the late 1960's the first high molecular weight aromatic polyamides (1, 2). The chemical structures of two of the more important polymers of this class, which later became known as aramids, are presented in Figure 2.

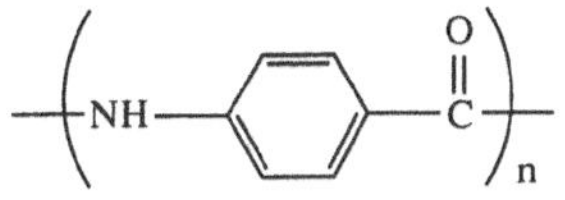

I

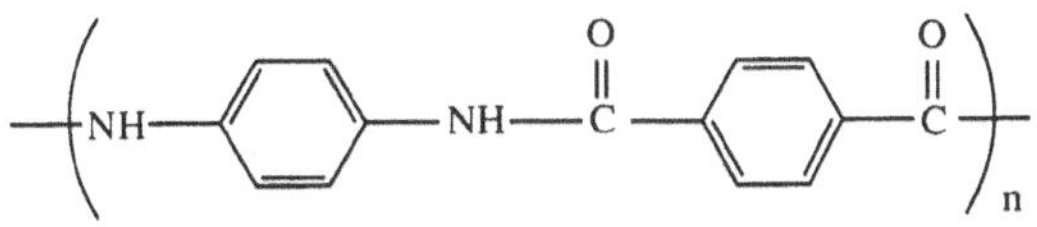

II

Figure 2. Chemical structures of the aromatic polyamides [aramids] poly-(1,4-benzamide) (I) and poly(1,4-phenylene terephthalamide) (II). The latter macromolecule constitutes Du Pont's high-performance Kevlar fibers.

Poly(1,4-benzamide) (I, PBA) is most commonly prepared by low-temperature solution polycondensation of 4-aminobenzoyl chloride hydrochloride. Poly(1,4-phenylene terephthalamide) (II, PPTA), which is the aramid that has become of prime commercial importance, typically is synthesized by solution polycondensation of terephthaloyl chloride and 1,4-phenylene diamine (1-3; see also references 4 and 5 for excellent reviews of the polymerization and processing of aramids).

Both poly(1,4-benzamide) and poly(1,4-phenylene terephthalamide) are macromolecules that are characterized by a remarkable rigidity, unlike the flexible nature of the more common polymers, such as the polyamides nylon 6, 6.6, etc. The characteristic ratio, C_n, (6) of these aramids was estimated to be as high as 124 (7). This quantity is defined as

$$C_n = <r^2>_0/nl^2 \tag{1}$$

where nl^2 is the mean square end-to-end length for a freely jointed chain of n segments of length l and $<r^2>_0$ is the corresponding value of the unperturbed, actual polymer chain. The value of 124 should be contrasted to a characteristic ratio C_n= 6-7 of most flexible chain molecules (6). The rigidity of the above aramids, which is believed to originate in a partial double-bond character of the C-N liaison, causes these macromolecules to exist in a near fully extended conformation. The latter is of fundamental importance for the formation of high-modulus/high-strength materials, as will become apparent below.

A direct consequence of the intrinsic rigidity of the aramids is that solutions of these macromolecules, above a minimum polymer volume fraction ϕ, separate into an ordered phase and an isotropic solution (8) of respective volume fractions $\phi^{*'}$ and ϕ^*, with $\phi^{*'} > \phi > \phi^*$. As long ago as the 1950's, this interesting (lyotropic) phase behavior of solutions of rigid rodlike particles was theoretically investigated, among others, by Onsager (9), Ishihara (10) and Flory (11). Using his lattice model for polymer solutions, Flory showed that the rod volume fraction ϕ^* may be approximated by the relation

$$\phi^* \sim (8/x)(1-2x) \sim 8/x \tag{2}$$

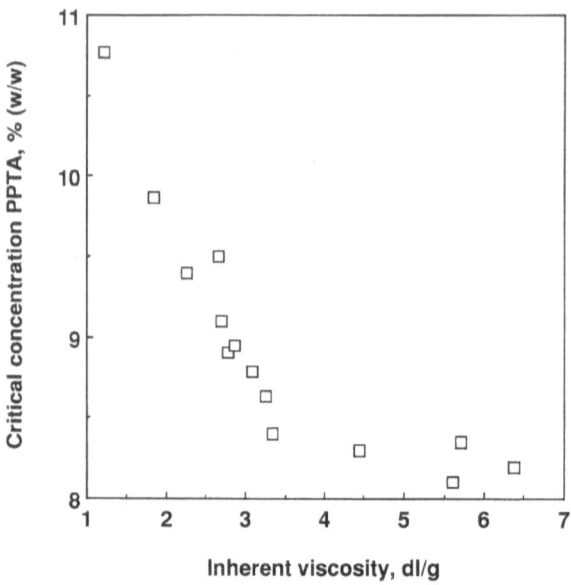

Figure 3. The minimum polymer concentration, in (fuming) sulfuric acid, for the onset of the formation of an anisotropic, liquid crystalline phase *versus* inherent viscosity of PPTA (II). Data from Bair et al. (3).

for large values of x. Here x is the axial ratio of the rods. At concentrations above $\phi^{*\prime}$ the biphasic nature of the solutions disappears. The solutions are again homogeneous, but ordered.

Indeed, PBA and PPTA, which are soluble in concentrated sulfuric acid, display simple lyotropic behavior, closely obeying the simple relation (2). Figure 3 shows a plot of the minimum polymer concentration for the onset of the formation of an anisotropic, liquid crystalline phase *versus* the inherent viscosity of PPTA. The data in Figure 3, taken from ref. 3, reveal the systematic, inverse relation between the minimum concentration and the polymer viscosity. The latter quantity scales with the molecular weight M approximately as $M_w^{1+\delta}$ (12). The ordered phase of the present lyotropic aramids is nematic, and consist of arrays of approximately parallel chains, presumably without regular arrangement of the end groups or chain units.

A most conspicuous and, as it turns out, a technologically most important property of the ordered phase is its relatively low viscosity. This characteristic is illustrated in Figure 4, which displays the low-shear rate viscosity against concentration of PPTA solutions in 100% sulfuric acid (13). Clearly, a sharp drop in viscosity is observed at the onset of the formation of the anisotropic phase. It is essential to point out that this maximum in the viscosity is sensitive to the applied shear rate, and, in fact, fully disappears at high shear rates. This observation can be construed as to indicate that at high shear rates the critical concentration for the onset of the formation of the anisotropic phase shifts to increasingly lower values (see also Figure 6).

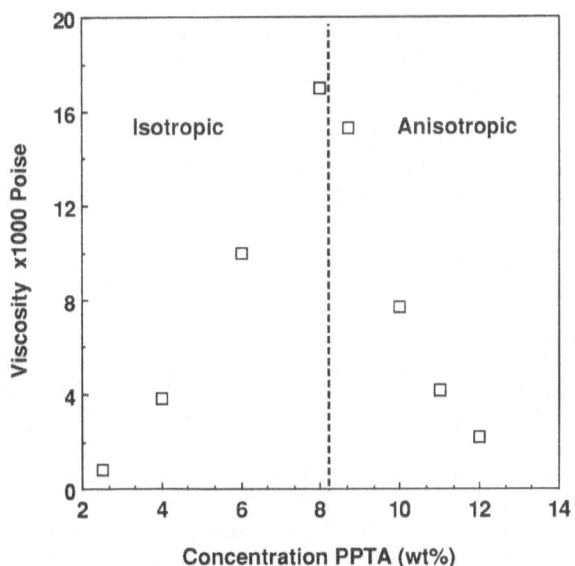

Figure 4. Low-shear viscosity of PPTA solutions in 100% sulfuric acid *vs* polymer concentration. Note the conspicuous maximum in the viscosity prior to the onset of the formation of the anisotropic phase. Polymer inherent viscosity was 2.7 dl/g . Data from Jingsheng et al. (13).

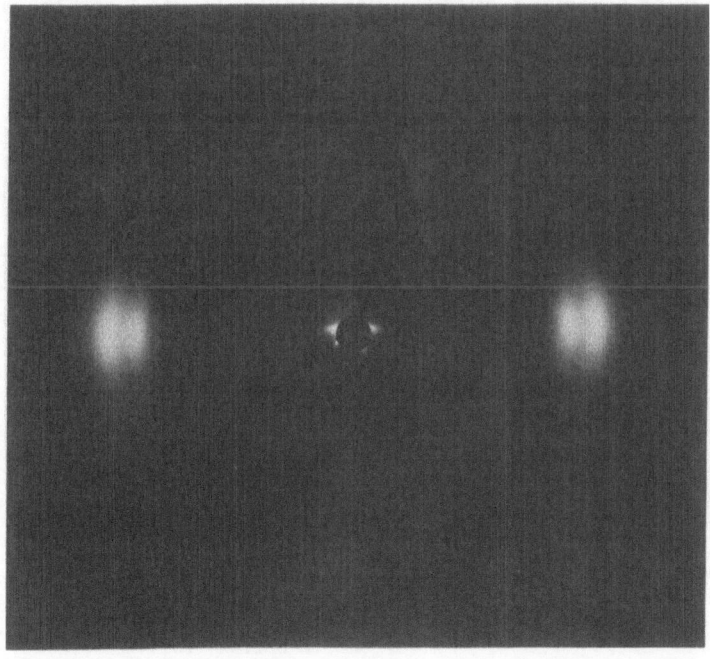

Figure 5. Wide-angle X-ray pattern of PPTA fiber spun from the optically anisotropic, liquid-crystalline phase. Fiber axis is vertical.

Typically, the anisotropic, liquid-crystalline phase is transformed into the high-strength/high-modulus fibers referred to in Figure 1, employing a fiber extrusion process that has become known as dry-jet wet, or air-gap spinning. In this technique, the importance of which was first recognized by Herbert Blades of Du Pont (14), the extruded fluid, liquid-crystalline, spin line is attenuated immediately below the spinneret, prior to coagulation of the fiber in, e.g., water. The latter operation consolidates the orientation of the rigid chain molecules that is generated in the elongational-flow field generated during attenuation in the air gap. Figure 5 shows a wide-angle X-ray pattern of an as-spun PPTA fiber. This pattern reveals the high degree of crystallinity and preferred alignment of the macromolecules along the fiber axis. It is from this particular molecular arrangement that the fiber derives it exceptional strength and stiffness, which typically are of the order of, resp., 3 and 80 GPa (1 GPa = 10^9 N/m^2. These values compare favorably with the mechanical properties of high-quality glass fibers and steel, and, due to a specific weight of only ~1.45 g/cc, exceed them by many times on an equal weight basis. The latter combination is the origin of the successful application of these unusual materials in high-strength, light-weight structures.

The importance of the anisotropic phase, as observed under conditions of low shear, should not be overestimated. Of course, its existence allows processing of aramids at convenient pressures and rates into highly oriented materials. But, high-strength/high-modulus materials may also be derived, although with some difficulty, from isotropic solutions of rigid chain molecules, provided that fiber extrusion is carried out at sufficiently high shear rates to induce uniaxial orientation, and that rapid coagulation is ensured to prevent relaxation of the order. The latter statements are particularly well-illustrated by the schematic, after Hermans (15), of Figure 6 and the results in Figure 7.

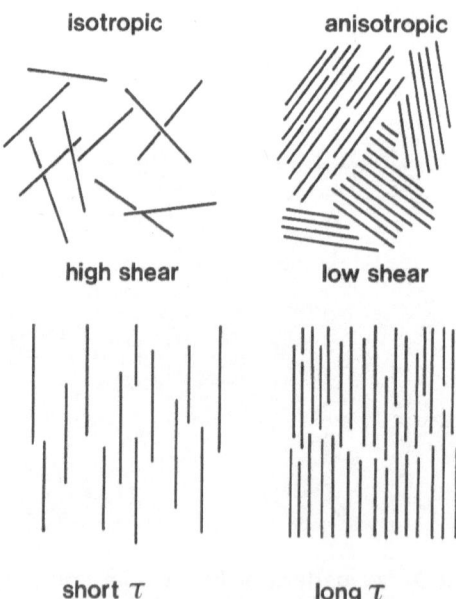

Figure 6. Schematic of alignment of rigid chain molecules, and the role of the anisotropic phase. τ is the relaxation time. See also Hermans (15).

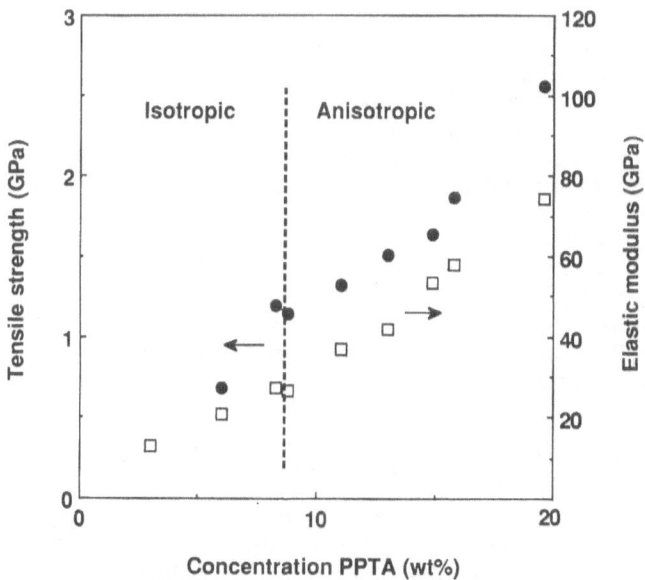

Figure 7. Mechanical properties (Young's modulus and tensile strength) *versus* solution concentration of poly(1,4-phenylene terephthalamide) (in sulfuric acid) in the "spin-dope". In the fiber spinning process, the draw down ratio was increased to maintain constant filament diameter at increased concentration. Also indicated is the isotropic/anisotropic phase boundary. Note the absence of any discontinuity at the transition. The inherent viscosity of the polymer was 4.2 dl/g. Data from taken from Weyland (16).

The latter figure shows the mechanical properties, measured at room temperature, of as-spun poly(1,4-phenylene terephthalamide) fibers plotted against the polymer concentration of the solutions from which the fibers were produced. The draw-down ratio, or spin-stretch factor (wind-up speed/ extrusion rate) was adjusted for each concentration to maintain a constant fiber diameter. Remarkably, these reported results display no discontinuity in the fiber tensile strength or stiffness at the onset of the formation of the anisotropic phase.

Also, it is of importance to point out that the appearance of liquid crystallinity should not be interpreted as an insurance for high strength or high stiffness of the materials obtained from such phases. The aramids described above derive their extraordinary mechanical properties from their intrinsic stiffness, which causes these macromolecules to readily orient in flow fields, to yield uniaxially oriented structures, *and* their ability to form strong intermolecular secondary bonds, such as the hydrogen and π/π bonding. These bonds overcome the stress concentrations caused by the large number of chain ends resulting from the modest molecular weights of these polymers. Deviation from chain rigidity and secondary bond strengths (at constant chain length) will invariably be associated with deterioration of mechanical properties.

The monumental discoveries by Kwolek of the synthesis of high molecular weight, rigid aromatic polyamides and by Blades of the processing of theses materials into a, then, novel class of high-performance polymers has led to an explosion of research activity. This enormous endeavor in what has become the field of liquid-crystalline polymers, has resulted in a host of other rigid chain molecules that form lyotropic liquid crystalline phases (4). Currently, the most important of these new materials are considered to be the polybenzothiazoles and oxazoles families (17), which are still more rigid than the aramids. Also, in addition to new lyotropic macromolecules, a large number of *thermo*tropic polymers (see the overview of ref. 4) have been synthesized, which generally are chain molecules of reduced rigidity. These materials, unlike the aramids, can be processed from the melt due to their lower, accessible melting temperatures, but, not surprisingly, exhibit lower stiffness and strength.

2.2 Ordered Dispersions of Polymer Whiskers

Liquid crystallinity, as described in the previous section for aromatic polyamides, is a phenomenon not *per se* limited to molecular species. It has been recognized for some time that liquid crystalline order is dominated by geometrical factors (9-11). Therefore, mesomorphic order can be expected in dispersions of any kind of asymmetric objects, provided that flocculation does not occur. Well-known biological systems displaying supramolecular liquid crystallinity are suspensions of the tobacco mosaic virus (18) and Hemoglobin S (19). Supramolecular liquid crystalline order in synthetic polymer systems has received little attention, and liquid crystallinity is commonly associated with the intrinsically rigid or semi-rigid macromolecules discussed in the previous section, or macromolecules with mesogenic side groups (4, 20, 21).

The formation of supramolecular liquid crystalline order was recently reported in suspensions of (extended-chain) crystalline whiskers of the *flexible* macromolecule poly(tetrafluoroethylene) (PTFE) (22). This finding will be briefly reviewed, because it once more illustrates that spontaneous ordering is not the exclusive property of solutions or melts of (semi) rigid molecules. Flexible macromolecules, when assembled into asymmetric objects, such as extended-chain whiskers, also may exhibit the fascinating and useful self-alignment phenomenon that leads to liquid crystalline order. It is believed that this class of soft condensed matter will provide new avenues for processing (flexible) macromolecules into highly ordered materials with anticipated superior performance.

PTFE is commercially manufactured in an aqueous emulsion polymerization process. Typically, the polymerization is carried out at about 80 °C, and is initiated with $K_2S_2O_8$. The polymer is insoluble in the reaction medium, and, therefore, precipitates during the polymerization. Normally, about 0.1 % w/w of a perfluorinated surfactant, such as $(C_7F_{15}COONH_4)$, is used to improve the dissolution of the monomer, tetrafluoroethylene (TFE) gas, and to prevent flocculation of the growing particles.

Figure 8 displays a transmission electron micrograph of a dried dispersion of fully polymerized PTFE. This micrograph reveals the globular nature of common, commercial PTFE particles. More detailed studies on these particles revealed that these entities are not polycrystalline, but, in fact, are ribbon, or rodlike extended chain crystals that are wrapped around themselves in a more or less random manner (23).

Recently, it was reconfirmed (22), in agreement with earlier findings (24), that the shape and size of the PTFE particles is markedly dependent on

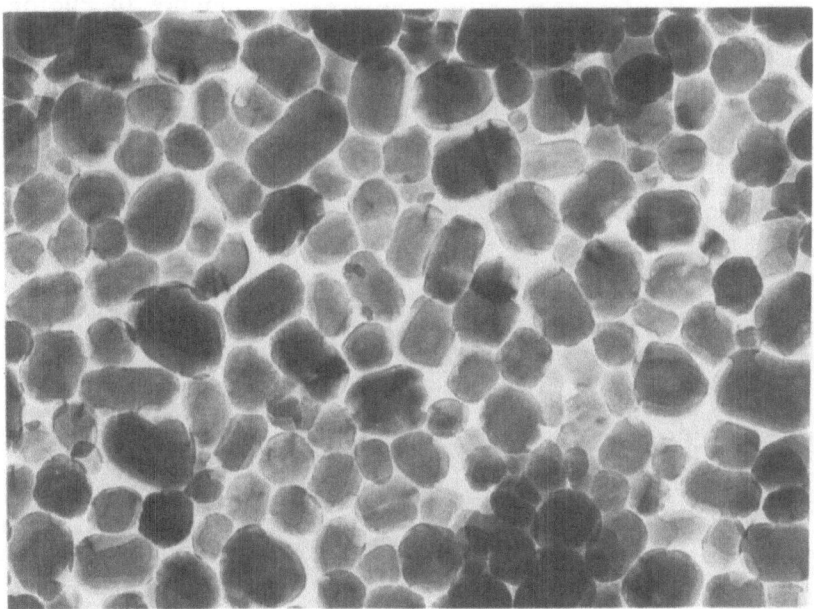

Figure 8. Transmission electron micrograph of [dried] commercial aqueous dispersion of as-polymerized poly(tetrafluoroethylene) [Teflon, Du Pont]. PTFE is formed as near-spherical particles of ~ 0.2 μm diameter.

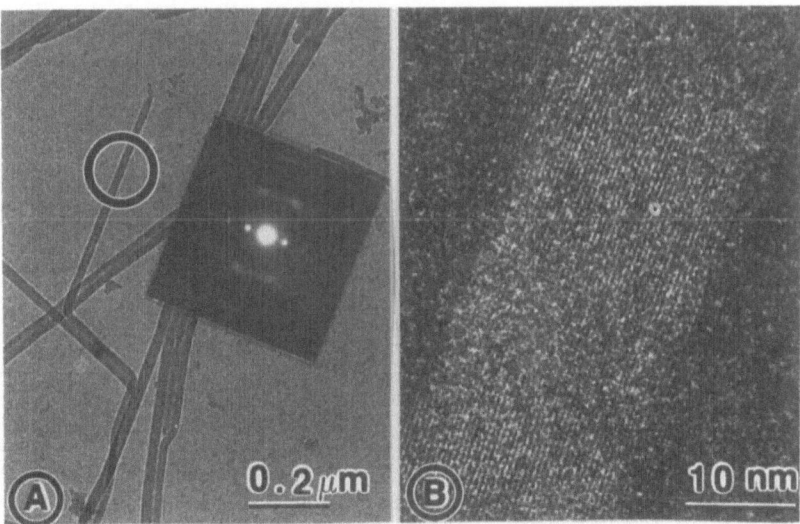

Figure 9. PTFE whiskers, synthesized as the material of Figure 8, but at increased levels of surfactant of slightly greater length (1% w/w of $C_9F_{19}COONH_4$, instead of 0.1% w/w of $C_7F_{15}COONH_4$). A: Transmission electron micrograph of PTFE whiskers. Insert: electron diffraction pattern of single whisker, revealing the perfect orientation of the macromolecules along the whisker-axis. B: High-resolution electron micrograph (22) showing the unperturbed (100)-lattice planes across the PTFE whisker.

Figure 10. Photograph (polarized light, crossed polarizers) of a glass vial containing as-polymerized, aqueous dispersion of PTFE whiskers. Note the separation into a non-birefringent, iso-tropic upper layer, and highly birefringent, ordered bottom layer.

Figure 11. Optical micrograph of aniso-tropic dispersion of PTFE whiskers (bottom layer of Figure 10), showing large ordered domains. Magnification 100X, polarized light, crossed Nicols.

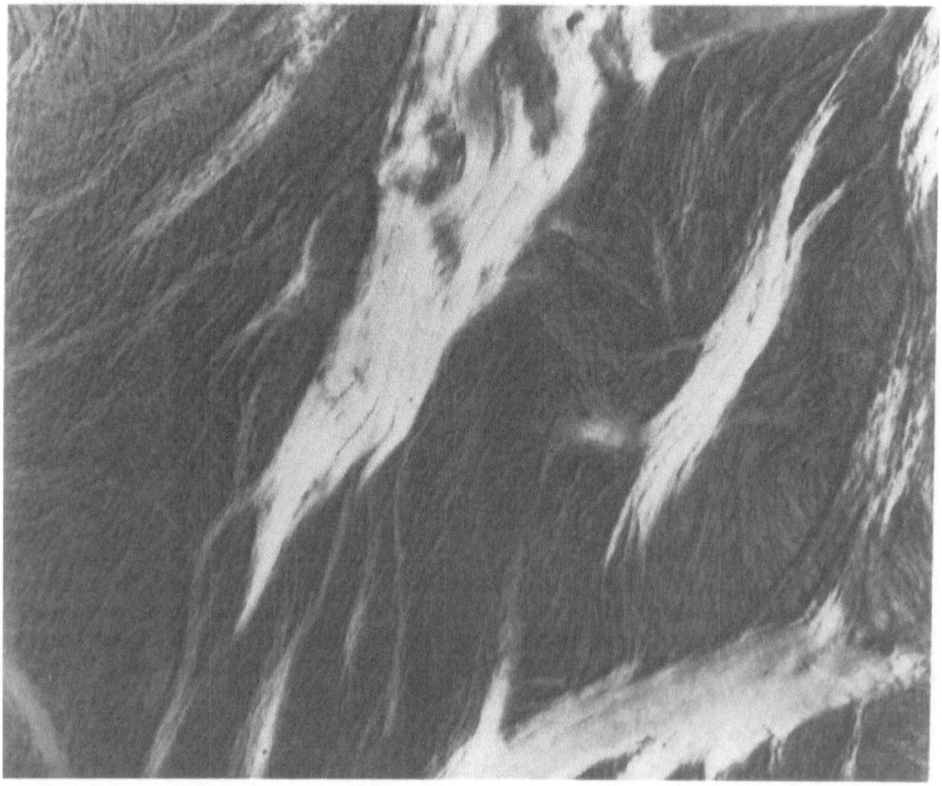

both the surfactant concentration and length of its non-polar, perfluorinated alkane part. Rod-like poly(tetrafluoroethylene) particles were recovered, for example, in a polymerization that was carried out with 1 % by weight (based on water) of the surfactant ammonium perfluorodecanoate ($C_9F_{19}COONH_4$), under otherwise identical conditions as those described above. Electron diffraction studies and high-resolution electron microscopy revealed that the rods, in fact, were fully extended chain whiskers of PTFE, as evidenced by the transmission electron micrographs in Figure 9. Figure 9A displays a few polymer whiskers, and a selected area electron diffraction pattern (insert) of a single rod, in the proper orientation with respect to the rod. Once indexed, this pattern indicated that the molecular axis of the polymer was parallel to that of the whisker. Figure 9B shows a high-resolution, (100)-lattice image of a part of a whisker; the uninterrupted lattice fringes illustrate the perfect parallel packing of the polymer molecules across the entire whisker. The poly(tetra-fluoroethylene) whiskers were of a rather uniform width of about 20 nm and a of length that ranged from about one micron to several tens of microns. In average, their axial ratio x (length/diameter) was estimated to be of the order of 10^3.

Upon standing at room temperature for about 4 hrs the PTFE suspension had separated into two phases. Figure 10 shows an optical photograph, taken between crossed polarizers, of a glass vial containing the phase-separated polymer suspension. The upper phase was isotropic and displayed no optical birefringence. Birefringence could readily be induced in this phase by shearing the suspension, or, notably, by placing it in a magnetic field. The polymer weight fraction in this upper phase was about 0.02. The bottom layer displayed a remarkably high birefringence (see optical micrograph in Figure 11), which is indicative of the liquid crystalline order in this phase. The PTFE concentration in the bottom layer was 6% w/w. According to Flory's approximative relation for stable anisotropy (11, eq. 2), a solution of rods having an axial ratio of 1000 should exhibit an anisotropic phase at a volume fraction exceeding about 0.008. This critical value corresponds to approximately 1.9 % by weight in the present PTFE system. The suspension described above had an overall rod concentration of 2.7% w/w, which, gratifyingly, satisfies the predicted requirement for the existence of an anisotropic phase.

Remarkably, if the surfactant $C_8F_{17}SO_3K$ was employed in the polymerization, performed under otherwise identical conditions, an entirely liquid crystalline suspension was obtained. On the other hand, if any of the two surfactants mentioned above was used at relatively low concentrations, e.g. 0.1 % w/w,, no birefringent phase was observed; at such low surfactant concentrations, only spherical PTFE particles were formed (as in Figure 8) and few, if any, whiskers were found.

It is clear that the formation of liquid crystalline suspensions of poly(tetrafluoroethylene) whiskers is dominated by the action of the surfactant and by the tendency of crystallizable macromolecules to polymerize in extended chain form when synthesized far below their melting or dissolution temperature (25). Strong indications exists that the presence of surfactant in the polymerization at a level above its critical concentration for rod-like micelle formation is a necessary requirement for the production of the liquid crystalline suspensions.

From the above considerations it would appear that the conditions for the synthesis of polymer whiskers and the formation of ordered phases are rather general, and applicable to a great many systems. The expected use of such whiskers is manyfold, since these materials are bound to exhibit properties, such as stiffness, strength, and optical, electric and magnetic characteristics, not far from their theoretical limits.

3. (THERMO-REVERSIBLE) GELS OF FLEXIBLE CHAIN MOLECULES

Macromolecular gels can be produced with a great variety of techniques. Most common, perhaps, is the formation of gels through chemical cross-linking of macromolecules in solution, or swelling of cross-linked polymers. Other mechanisms of gel formation include association of chain molecules on a segmental level, the formation of entities of high aspect ratios, such as the fibrils formed upon precipitation of rigid chain molecules from isotropic solutions, etc. (for an outstanding, brief review see ref. 26).

This section is confined to thermo-reversible gels of flexible macromolecules, because of their importance in processing of high-performance polymers (Section 3.1). These particular gels consist of a macroscopic, mechanically coherent network that is formed and stabilized through interconnected crystals. They can readily be obtained upon cooling of solutions of crystallizable polymers, provided that the polymer concentration is sufficiently high. A pictorial view of the formation of the gels of interest, and the relevance of the structure of the liquid from which they are formed, is given in Figure 12. Commonly, precipitation of crystallizable macromolecules from dilute solutions results in the formation of a dispersion of single crystals. On the other hand, crystallization from the melt mostly yields dense, semi-crystalline solids. Cooling of solutions of crystallizable polymers with concentrations above the onset of coil overlap, ϕ^*, however, normally generates thermo-reversible gels (27). (ϕ^* is approximately given by $\phi^* \sim 1/[\eta]$, where $[\eta]$ is the limiting viscosity number of the polymer).

Already from this very brief description it is clear that thermo-reversible gels may capture much of the arrangement of the macromolecules in solution, such as for example the spacing between chain entanglements, provided that during the formation of the connecting crystals no severe chain disentanglement occurs. The molecular weight between entanglements, M_{ent},

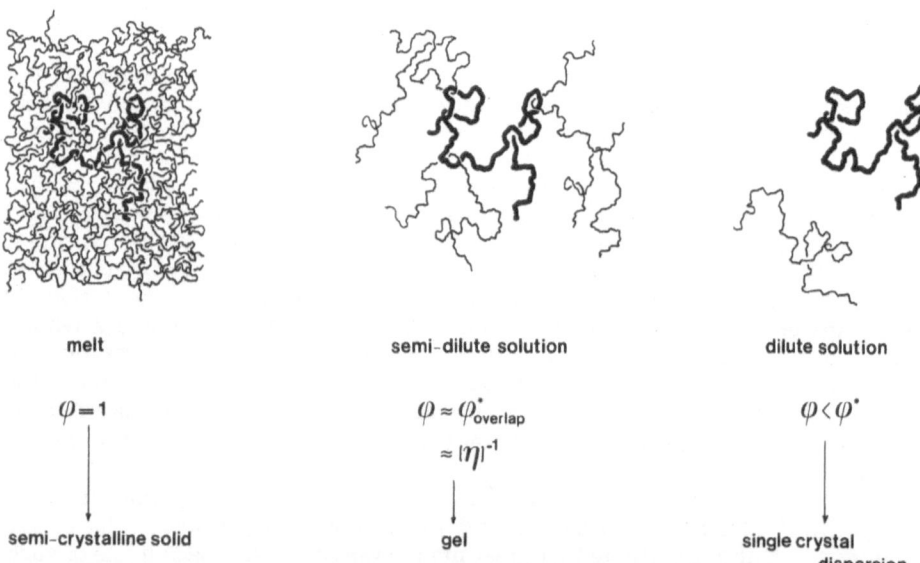

Figure 12. Schematic of the correlation between the polymer volume fraction and the nature of the derived solids. Note the importance of the chain overlap concentration ϕ^*, and, therewith, the chain length.

is known (28) to vary with the polymer volume fraction, ϕ, in semi-dilute and concentrated solutions approximately as

$$M_{ent}{}^{soln} \approx M_{ent}{}^{melt}/\phi \qquad (3)$$

The structure of thermo-reversible gels of high molecular weight polyethylene, discussed in the following sections, is revealed by the self-explanatory wide-angle X-ray pattern and micrographs in Figures 13 and 14.

It should be pointed out that thermo-reversible gels can be obtained from solutions of relatively low molecular weight species, and at concentrations well below ϕ^*, if these molecules precipitate as high-aspect ratio entities, such as fibrillar crystals or whiskers. The latter gels, however, invariably are brittle and not suitable for the applications discussed below.

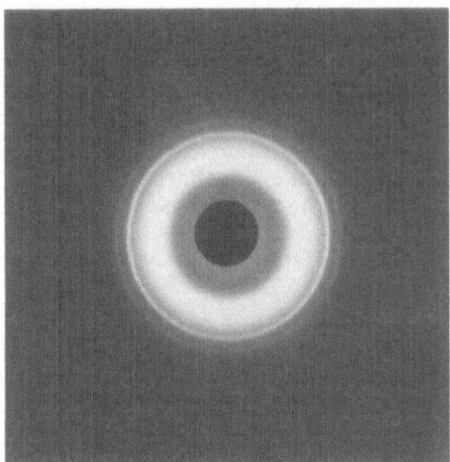

Figure 13. Wide-angle X-ray (WAXS) pattern of gel of 2 % v/v of ultra-high molecular weight polyethylene ($M_w = 1,500,000$ kg/kmol) in decalin. The amorphous halo is due to the solvent; the sharp circular reflections result from the 100 and 200 crystal planes of the crystalline polymer junctions.

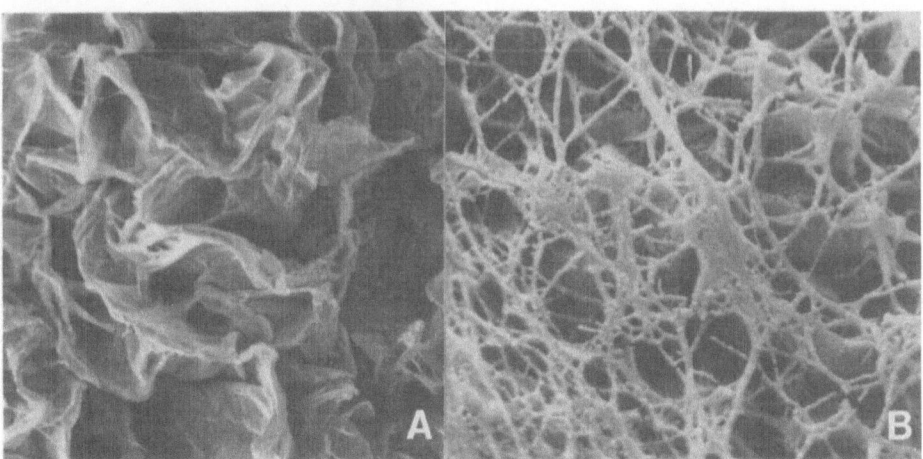

Figure 14. Scanning electron micrographs of critical-point dried gels of UHMW PE, obtained from 2 % solutions in decalin. A) quiescent conditions; B) agitated solutions (1000X). Note the lamellar nature of the polyethylene crystals in A, and the fibrillar crystals reminiscent of the flow that occurred during agitation prior to gelation, in B.

365

3.1 Precursor for Ultra-High Strength Fibers

Much like the saying "failure is an orphan; success has many fathers", the discovery of ultra-high strength polyethylene fibers is said to have been made by many inventors. It is clear from the graph in Figure 1, that this development features a number of milestones, selected few of which will be elaborated upon below. Initial studies by, among others, Peterlin et al. (29) on the deformation of polyethylenes paved the path for the developing notion that flexible polymers may be elongated beyond what was frequently believed to be their natural draw ratio (final/original length) of 4-6, that was found for many macromolecular, semi-crystalline materials. Subsequently, systematic studies by Ian Ward and his co-workers, at Leeds, focused on the optimization of various process parameters, such as molecular weight distribution, temperature and rate, on the maximum draw ratio of melt-crystallized, linear polyethylenes (see review in 30). Most importantly, these authors carried out extensive investigations of the mechanical properties of the drawn materials. They arrived at the crucial conclusion that the final stiffness of the drawn polyethylenes was dictated uniquely by the absolute draw, and not by the molecular weight. The efforts of Ward et al. resulted in a first significant increase in tensile strength of macroscopic polyethylene samples (Figure 1). An important result of the Leeds group was to re-establish the fact that melt-crystallized linear polyethylenes of increasing molecular weight display a systematically decrease in maximum draw ratio (Figure 15). This observation is, of course, in sharp contrast to expectations based on the maximum draw ratio, λ_{max}, of single, isolated chains, which is predicted (31) to vary as

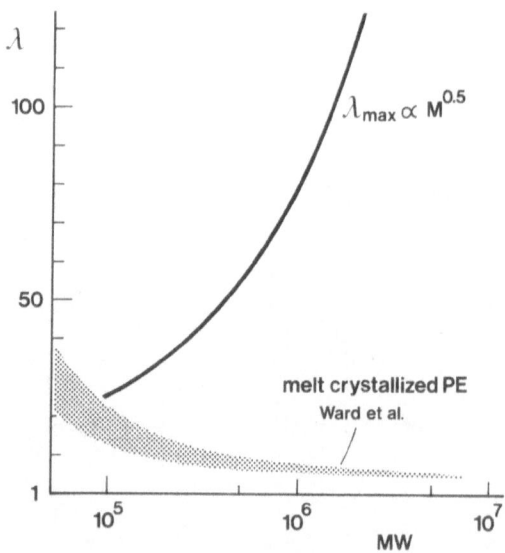

Figure 15. Maximum draw *versus* molecular weight of melt-crystallized polyethylene, under optimized conditions. Data from Ward et al. (30). Also plotted are the calculated average maximum draw ratios for isolated polyethylene chains. Note the dramatic discrepancy between the calculated and experimentally observed maximum draw ratios.

$$\lambda_{max} = l_p/l. \; (n/C_n)^{1/2} \qquad\qquad (4)$$

and thus should drastically increase with chain length. In equation (4), l is the length of a bond, l_p its projected length along the chain axis, and n the number of segments.

Results similar to those obtained by Ward et al., were produced, in the same period, by Porter and co-workers, using a technique known as solid-state extrusion (see their review in 32).

Also at that time, Albert Pennings and his group in Groningen investigated the crystallization behavior of ultra-high molecular weight, linear polyethylene (UHMW PE) in various flow fields. Among their many discoveries was the first continuous growth of strong fibers of this interesting polymer (33). This pioneering fiber, produced with a technique referred to as surface growth, exhibited mechanical properties not unlike those of the aramids discussed above (cf. the step-wise increase in tensile strength at 1976 in Figure 1). Although Pennings' findings attracted much attention, among others, of Keller et al. (review in 34), the surface growth process remained only of academic interest due to very low production rates and the (at the time) confusing growth mechanism.

The surge in tensile properties commencing at DSM in the late 70's finds it origin in the discovery of the unusual tensile drawing behavior of gels of UHMW PE, in comparison with the deformation of melt-crystallized samples of the same polymer (35, 36). This observation is most clearly presented in Figure 16, which displays a graph of the maximum draw ratio at 130 °C of these gels, from which all solvent was removed (to avoid misleading plasticizing effects). Indeed, UHMW PE, when crystallized from semi-dilute solutions, is readily elongated, *in the solid state,* to the draw ratios predicted by equation (4). This result, and particularly the systematic dependence of λ_{max} on the initial polymer volume fraction in the gel, was interpreted to indicate that the deformation behavior of weakly bonded, flexible macromolecules is controlled by the number of entanglements in the (semi-crystalline) polymer solid, and not by the crystal morphology (37). The latter, originally frequently questioned, conclusion was substantiated theoretically by the prediction that the maximum draw ratios of entanglement networks should scale as the square root of the molecular weight between entanglements. The latter quantity approximately varies as the inverse polymer volume fraction (eq. 3). Therefore, the maximum draw ratio of (dried) gels of ultra-high molecular weight polyethylene was predicted (37) to vary with the initial polymer volume fraction ϕ in the solutions, from which they were derived, as

$$\lambda_{max} \propto (\phi)^{-1/2} \qquad\qquad (5)$$

for ϕ exceeding the coil-overlap concentration. This dependence was verified and confirmed experimentally (37). Clearly, here the unique role of soft condensed matter in the form of thermo-reversible gels is the unprecedented control of chain entanglement densities in polymer solids.

On the basis of Ward's results, it follows that the extravagantly high draw ratios of UHMW PE gels were associated with marked increases in stiffness and tensile strength. The enhancement of the mechanical properties is shown in Figure 17. It should be noted that, unlike the Young's modulus, the tensile strength is dependent on the molecular weight (roughly as $M_w^{0.4}$ over the range of interest (38)). Accordingly, the use of ever higher molecular weights, and reduced chain length distribution, combined with optimized initial polymer solution concentrations, has resulted in the continuing increase in tensile strength shown in Figure 1. Currently, UHMW PE fibers are made with a reported strength of 10 GPa (39), which is within 80 % of the theoretical value for the molecular weights used (40).

The very high draw ratios cause drastic morphological and structural changes, which ultimately are responsible for the major improvement in mechanical properties. The results of these transformations are illustrated by the scanning electron micrograph and the wide-angle X-ray pattern of an ultra-oriented, dried gel of UHMW PE (respectively, Figure 18 and 19). The gel was prepared from a 0.6 % v/v solution in decalin, subsequently dried and drawn to 130 times its original length at 130 °C. Particularly interesting is the exceptional alignment of the polyethylene chains evidenced by the X-ray pattern; this high degree of order is to be contrasted with the orientation obtained in processing liquid crystalline materials (see Figure 5, for PPTA).

The above solution-spinning/drawing process, incorrectly referred to as gel-spinning, has now been commercialized by DSM/Toyobo (41) and, under DSM licence, by Allied. The commercially available fibers typically have a tensile strength of 3.5 GPa, and a modulus of 100-150 GPa. The major drawback of these materials is their low upper use temperature (~ 130 °C) in comparison with aramids; the polyethylene fibers out-perform aramids, however, in abrasion resistance and impact strength.

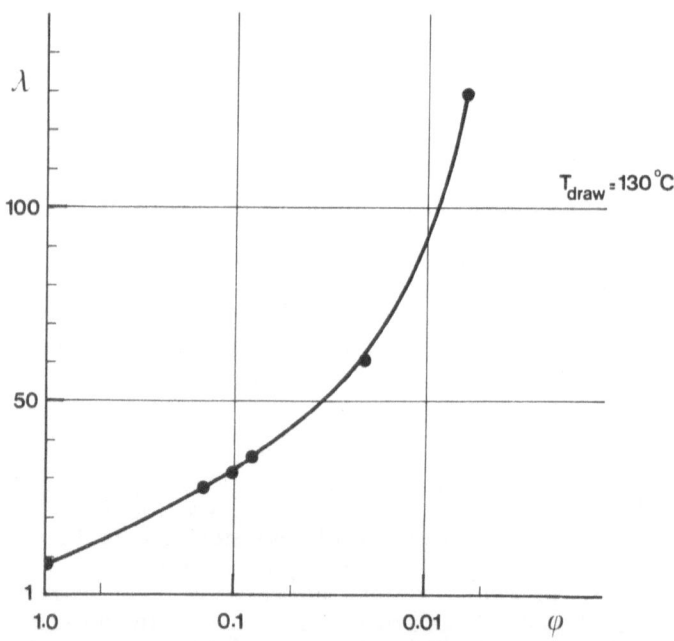

Figure 16. Maximum draw ratio of dried gels of UHMW PE versus initial polymer volume fraction (ϕ). The draw temperature was 130 °C. This graph illustrates that upon simple reduction of the entanglement density in otherwise identical macromolecular solids, the draw ratio can be increased by more than an order of magnitude.

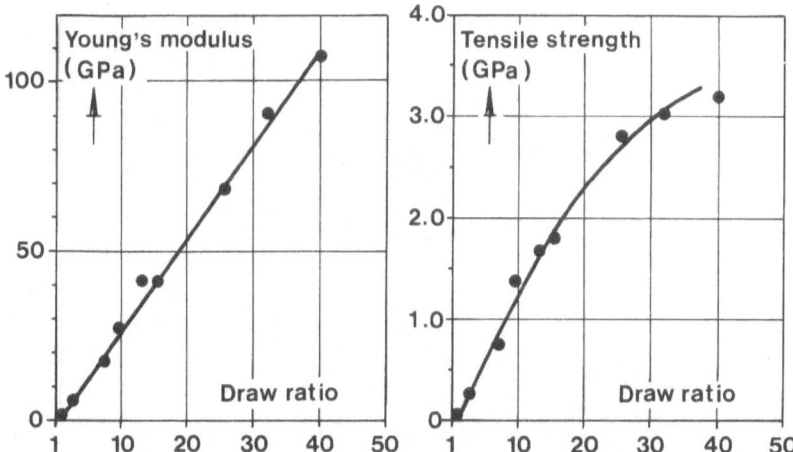

Figure 17. Mechanical properties (measured at room temperature) of gel fibers, spun from a 2 % solution of UHMW PE in decalin, as a function of the draw ratio. Drawing was performed at 120 °C using dried gel filaments. Note the dramatic increase (over 100-fold) in stiffness and strength.

Figure 18. Scanning electron micrograph of ultra-oriented UHMW PE, displaying defect free, fibrillar structure. Draw direction is diagonal.

Figure 19. Wide-angle X-ray pattern of ultra-oriented, dried gel of UHMW PE. The gel was prepared from a 0.6 % v/v solution in decalin. The dried gel was drawn to 130 times its original length at 130 °C. Draw direction is vertical. Note the exceptional alignment of the polyethylene chains; to be contrasted with the orientation obtained in processing of liquid crystalline materials (see Figure 5, for PPTA).

3.2 Gels as a Tool in Processing of Intractable Materials

A great many species, polymeric and small molecular, with intrinsically interesting and useful properties, are intractable, i.e. non-meltable and insoluble. Well-known examples are ceramics, rigid chain polymers and the conjugated polymers to be discussed hereafter. As a result, progress towards the development of useful objects of such materials, such as fibers and films, has been severely hampered. Elaborate techniques have been designed to process intractable substances. Such technologies involve, for instance, the synthesis of chemically related "precursor" species, which are processed and subsequently transformed, e.g. be heat, into the final object of the intractable material (e.g. 42-44). In many cases precursor routes are not available or feasible, or of use to only very few systems. Alternate methods include derivatization of the intractable systems. An example of the latter technique is the synthesis of alkyl derivatives of the conjugated polythiophenes (45-47). This relatively new class of polymers is soluble in a variety of common organic solvents and water (48) Unfortunately, the derivatization route is consistently accompanied by a reduction in the desired properties (in this case electrical conductivity), in comparison with the parent unsubstituted substance. As a result, it remains an objective to develop alternative, more general processing techniques for intractable materials.

It will be shown that soft condensed matter, in the form of polymeric gels, can play a unique role as a general tool in the processing of intractable materials. As an illustrative example we choose the manufacturing of fibers of the intractable polyacetylene $[-(CH=CH)_n-]$. The latter conjugated polymer in its "doped" form has an exceptionally high intrinsic electrical conductivity (> 100,000 S/cm) (49, 50); but, as is typical of many conductive polymers, it is insoluble and decomposes prior to melting because of its very same high level of conjugation.

Key to the concepts underlying this general technique (51), is to provide a pre-shaped reaction medium, in e. g. fiber, rod or film form, for the *in-situ* synthesis of the intractable material. This is achieved by producing a mechanically coherent and stable gel fiber or film comprised of a minor amount of a "carrier" polymer and a major amount of a suitable solvent. Ingredients for the synthesis of the final intractable material, such as catalysts, and/or initiators, monomers, etc., are introduced into the gel 1) through diffusion after the carrier gel is formed, or 2) prior to the gel formation process by dispersing or dissolution in the polymer solution that is used to generate the gel. Subsequently, these ingredients are transformed into the final product by the proper chemical reactions.

For the system of present interest, polyacetylene, the gels of ultra-high molecular weight polyethylene (discussed in Section 3.1) are exceptionally suitable because of their chemical "compatibility" with the acetylene synthesis. Polyethylene gel monofilaments were spun according to standard procedures (36) from a 2 % w/w solution of UHMW PE (Hostalen GUR 412, Hoechst, M_w = 4×10^6 g/mol) in mineral oil. The diameter of the polyethylene gel fiber was approximately 0.6 mm. An $Al(C_2H_5)_3/Ti(OC_4H_9)_4$ catalyst was employed for the synthesis of the intractable polyacetylene (PAc). It was prepared and dissolved in mineral oil also, following the method of Naarmann and Theophilou (49). The as-spun polyethylene gel fibers were transferred into an Argon filled dry-box and soaked in the catalyst solution for 1 hr at room temperature. The polyethylene/catalyst gel fibers were inserted in a tube-shaped container; it was degassed and subsequently exposed to 510 mm Hg pressure of acetylene monomer gas for a period of 20 min during which the polymerization proceeded to yield polyacetylene. The catalyst and mineral oil were extracted from the resulting polyacetylene/polyethylene fibers with toluene. The filaments were washed with a methanol/5 % HCl solution and then with methanol. Finally, the fibers were dried under a flow of Ar and then under vacuum.

Figure 20 displays an optical photograph (reflected light) of an as-polymerized polyacetylene/polyethylene (PAc/PE) fiber. The fiber was split longitudinally to expose the inner surface. This photograph clearly reveals that the polyacetylene was synthesized throughout the polyethylene gel fiber, and had not merely formed a skin. The diameter of the PAc/PE monofilaments was about 100 μm. The polyacetylene content in the composite fiber was calculated to be 82 % w/w from the weight up-take during the polymerization.

The electrical conductivity of Iodine-doped PAc/PE fibers was measured using the standard 4-probe method (52). The PAc/PE fibers were found to reach a maximum conductivity of 1,200 S/cm after 40 min of doping. This value is in good agreement with those previously reported for as-polymerized films of

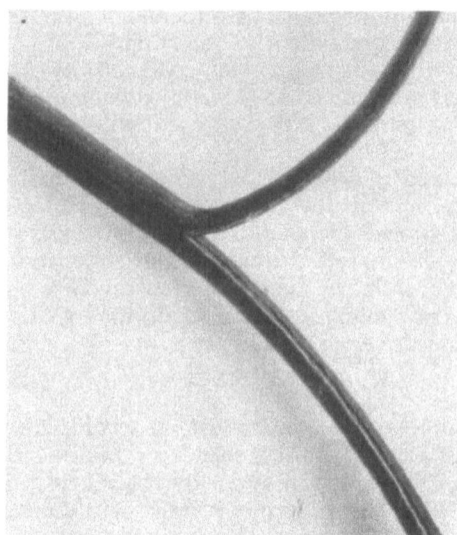

Figure 20. Optical photomicrograph (reflected light) of an as-polymerized polyacetylene/polyethylene (PAc/PE) fiber. Note that the polyacetylene was formed throughout the polyethylene gel fiber. The diameter of the monofilament is approximately 100 μm.

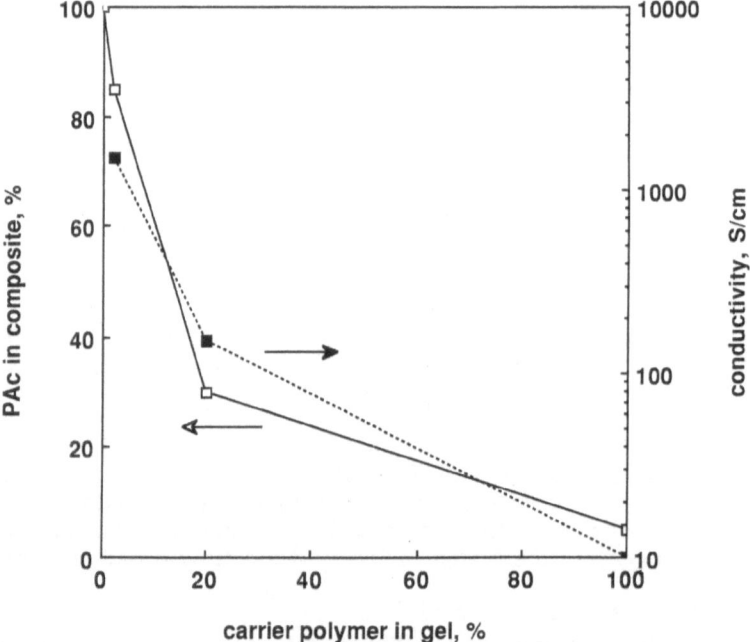

Figure 21. Yield, in % w/w polyacetylene, *versus* weight percent polymer in carrier gel [left axis], and electrical conductivity of the (Iodine doped) resulting PAc/PE composite [right axis]. This graph clearly illustrates the need for carrier gels that contain a very small amount of polymer, but have sufficient mechanical strength for handling.

polyacetylene (49), which indicates that the presence of a minor amount of UHMW PE did not have a detrimental effect on the conductivity of the polyblend fibers, which removes, of course, the necessity to extract the carrier polymer after the acetylene polymerization.

The specific role of the structure of the carrier gel in the manufacturing of fibers of intractable substances is most clearly illustrated in the graph of Figure 21. This graph shows the yield of polyacetylene *versus* weight percent of carrier polymer in the gel. This graph reveals that a sharply declining amount of polyacetylene is formed at increasing carrier polymer content. The electrical conductivities of the resulting composite materials, likewise, are strongly reduced when more concentrated carriers gels are employed (see also 53). It is clear, therefore, that the carrier gels should contain a minimal amount of polymer, just sufficient to yield adequate mechanical properties for handling. As demonstrated in this section, gels of ultra-high molecular weight polymers provide an outstanding source for such carrier materials.

This section briefly described a relatively simple, general, route to manufacture monofilaments, fibers, and film through *in-situ* synthesis of otherwise "intractable" materials in pre-shaped carrier gels. The technique is versatile and applicable to a wide variety of systems, including systems with desirable properties other than electrical conductivity, such as mechanical strength and stiffness, or high-temperature stability.

CONCLUDING REMARK

Soft condensed matter plays a vital and extremely versatile role in the narrow field that is the subject of this paper. The unusual fluidity of liquid crystals and the unique control of the spatial arrangement of long chain molecules in gels are just a few examples of the many ways in which soft condensed matter has impacted processing of high-performance materials. As the need for advanced materials with superior properties continues to grow, its role is likely to become even more important. The development of a detailed understanding of its formation, and new routes to manufacture and utilize this state of matter, no doubt, will prove to be rewarding.

Acknowledgement

Kevlar and Teflon are registered trade marks of E. I. du Pont de Nemours & Company, Inc.. This author was privileged to collaborate with Thomas Folda (now at BASF), who discovered the ordered dispersions of PTFE (section 2.2). He is indebted to Henri Chanzy, who took, and interpreted, most of the electron micrographs, Jin Chiang, Fred Wudl and Alan Heeger (section 3.2), Mike Jaffe, whose paper in ref. 5 has been frequently consulted, and Aaldrik Postema and Alejandro Andreatta for their invaluable help in the preparation of this manuscript.

References

1. S. L. Kwolek, Fr. Pat. 1,526,745 (1968).
2. S. L. Kwolek, U.S. Pat. 3, 600,350 (1971).
3. T. I. Bair, P. W. Morgan and F. L. Killian, *Macromolecules*, **10**, 1396 (1977).
4. S. L. Kwolek, P. W. Morgan and J. R. Schaefgen, in *Encyclopedia of Polymer Science and Engineering*, Vol. **9**, Wiley (New York), 1987, p. 1.

5. M. Jaffe and R. Sidney Jones, in *Handbook of Fiber Science and Technology*, Vol. **3A**, M. Lewin, J. Preston, Eds., Dekker (New York) 1985, 349.

6. P. J. Flory, *Statistical Mechanics of Chain Molecules*, Interscience (New York) 1969, p.11.

7. B. Erman, P. J. Flory, and J. P. Hummel, *Macromolecules*, **13**, 484 (1980).

8. S. L. Kwolek, U.S. Pat. 3,671,542 (1972)

9. L. Onsager, *Ann. N.Y. Acad. Sci.*, **51**, 627 (1949).

10. A. Isihara, *J. Chem. Phys.*, **18**, 1446 (1950); ibid **19**, 1142 (1951).

11. P. J. Flory, *Proc. Royal Soc. London*, Series A **234**, 73 (1956).

12. D. G. Baird and J.K. Smith, *J. Polym. Sci. Polym. Chem. Ed.*, **16**, 61 (1978).

13. B. Jingsheng, Y. Anji, Z. Shegging, Z. Shufan and H. Chang, *J. Appl. Polym. Sci.*, **26**, 1211 (1981).

14. H. Blades, U.S. Pat. 3,767,756 (1973), 3,869,429 (1975), 3,869,430 (1975).

15. J. Hermans, Jr., *Colloid. Sci.*, **17**, 638 (1962).

16. H. G. Weyland, *Polym. Bulletin*, **3**, 331 (1980).

17. J. F. Wolfe, P. D. Sybert and J. R. Sybert, U.S. Pat. 4,533,692 (1985).

18. M.F. Perutz, A.M. Lignori and S.Eirich, *Nature* (London), **167**, 929 (1951).

19.]. F.C. Bawden and N.W. Pirie, *Proc. Royal Soc. London* B **123**, 274 (1937).

20. *Polymer Liquid Crystals*, A. Ciferri, W. R. Krigbaum and R. B. Meyer, Eds. Academic Press (New York) 1982.

21. *Liquid Crystalline Order in Polymers*, A. Blumstein, Ed., Academic Press (New York), 1978.

22. T. Folda, H. Hoffmann, H. D. Chanzy and P. Smith, *Nature* (London), **24**, 123 (1989).

23. H. D. Chanzy, P. Smith and J.-F. Revol, *J. Polym. Sci. Polym. Lett. Ed.*, **24**, 557 (1986).

24. K. L. Berry, U.S. Pat. 2,559,750 (1951).

25. B. Wunderlich, *Adv. Polym. Sci.* **5**, 568 (1968).

26. T. Tanaka, in *Encyclopedia of Polymer Science and Engineering*, Vol. **7**, Wiley (New York), 1987, p. 514.

27. P. Smith, P. J. Lemstra, J. P. L. Pijpers and A. M. Kiel, *Colloid & Polym. Sci.*, **259**, 1070 (1981).

28. W. W. Graessley, *Adv. Polym. Sci.*, **16**, 4 (1974).

29. A. Peterlin and G. Meinel, *J. Polym. Sci. B*, **3**, 783 (1965).

30. G. Capaccio, A. G. Gibson and I.M. Ward, in *Ultra-High Modulus Polymers*, A. Ciferri and I. M. Ward, Eds., Appl. Sci. Publ (London) 1979, p. 1.

31. P. Smith, R. R. Matheson, Jr. and P. A. Irvine, *Polymer Communications*, **25**, 294 (1984).

32. A. E. Zachariades, W. T. Mead and R. S. Porter, ref. 30, p. 77.

33. A. Zwijnenburg and A. J. Pennings, *J. Polym. Sci. Polym. Lett. Ed.*, **14**, 339 (1976).

34. A. Keller, ref. 30, p. 325.

35. P. Smith, P. J. Lemstra, B. Kalb and A. J. Pennings, *Polym. Bull.*, **1**, 733, (1979).

36. P. Smith and P. J. Lemstra, U. S. Pat. 4,344,908 (1982); 4,411,854 (1983); 4,430,383 (1984).

37. P. Smith, P. J. Lemstra and H. C. Booij, *J. Polym. Sci. Polym. Phys.*, **19**, 877 (1981).

38. P. Smith, P. J. Lemstra and J. P. L. Pijpers, *J. Polym. Sci. Polym. Phys.*, **20**, 2229 (1982).

39. L. Myasnikova, presented at 20th Europhysics Conf. Macromol. Physics, Lausanne, Switzerland, Sept. 26-30, 1988.

40. Y. Termonia, P. Meakin and P. Smith, *Macromolecules*, **18**, 2246 (1985).

41. R. Kirschbaum, H. Yasuda and E. H. M. van Gorp, Int. Man-Made Fibers Congr., Dornbirn, Austria, Sept. 24, 1986.

42. J. H. Edwards and W. J. Feast, *Polymer Comm.*, **21**, 595 (1980).

43. H. Kahlert and G. Leising, *Mol. Cryst. Liq. Cryst.*, **117**, 1 (1985).

44. D. R. Gagnon, F. E. Karasz, E. L. Thomas and R. W. Lenz, *Synth. Met.*, **20**, 85 (1987).

45. R. L. Elsenbaumer, K. Y. Jen and R. Oboodi, *Synth. Met.*, **15**, 169 (1986).

46. M.-A. Sato, S. Tanaka and K. Kaeriyama, *J. Chem. Soc. Chem. Commun.*, 873 (1986).

47. K. Yoshino, S. Nakjima, M. Fujii and R. Sugimoto, *Polymer Comm.*, **28**, 309 (1987).

48. A.O. Patil, Y. Ikenoue, N. Basescu, N. Colaneri, J. Chen, F. Wudl and A. J. Heeger, *Synth. Met.*, **20**, 151 (1987).

49. H. Naarmann and N. Theophilou, *Synth. Met.*, **22**, 1 (1987).

50. S. Kivelson and A. J. Heeger, *Synth. Met.*, **23**, 45 (1988).

51. J. C. Chiang, P. Smith, A. J. Heeger and F. Wudl, *Polymer Comm.*, **29**, 161 (1988).

52. C. K. Chiang, M. A. Druy, S. C. Gau, A. J. Heeger, E. J. Louis, A. G. MacDiarmid, Y.W. Park and H. Shirakawa, *J. Am. Chem. Soc.*, **100**, 1014 (1980).

53. M. Galvin and G. E. Wnek, *Polymer Comm.*, **23**, 795 (1982).

WATER-PROCESSABLE CONDUCTING POLYMERS

Steven P. Armes and Mahmoud Aldissi

Materials Science and Technology Division
Los Alamos National Laboratory, P. O. Box 1663
Los Alamos, NM   87545, USA

INTRODUCTION

Until recently, most organic conducting polymers suffered from the lack of processability which is a prerequisite for their use in most technological applications. Furthermore, processability is of importance for the study of the polymers' intrinsic properties. Because of the rigid nature of their backbones, processing of most of the initial undoped systems was unsuccessful. For doped polymers, the difficulties were even greater because doping causes further aggregation due to inc-reased interchain interaction and certain local geometry changes of the chains. Therefore, we have utilized various methods to achieve processability, directly or indirectly. The discussion is limited to the use of small or polymeric surfactant molecules grafted or physically adsorbed onto the conducting polymer chain or particle. Both, solutions and colloidal dispersions are the subject of this brief discussion.

SOLUTION-PROCESSABILITY OF SELF-DOPED POLYHETEROCYCLES

Polypyrroles and polythiophenes are known to be highly conducting materials, particularly, when they are synthesized electrochemically. Their synthesis is accompanied by a simultaneous doping of the polymer by the anions of the supporting electrolyte such as tetrafluoroborate or toluene sulfonate anions that are incorporated in the structure as counterions to the polycations which are segments of the conjugated chains. Regardless of the anions size and nature, the materials are insoluble in any solvent.

Recently[1,2], we have found that grafting a surfactant molecule of a similar structure to the dopant onto the b-carbon atom of thiophene or pyrrole (Fig. 1) results in a water-soluble specie that polymerizes readily in the same medium by anodic oxidation. The surfactant molecule is preferably an alkylsulfonate in which the alkyl segment can be one or several carbon atoms. In addition to introducing solubility to the polymer, the surfactant molecule acts as an intrinsic dopant yielding a self-doped polymer. Such doping occurs when a charge is ejected from the p system, leading to the formation of a charge transfer complex between the anion and the defect (radical cation) on the conjugated chain.

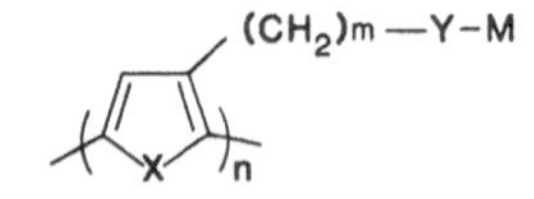

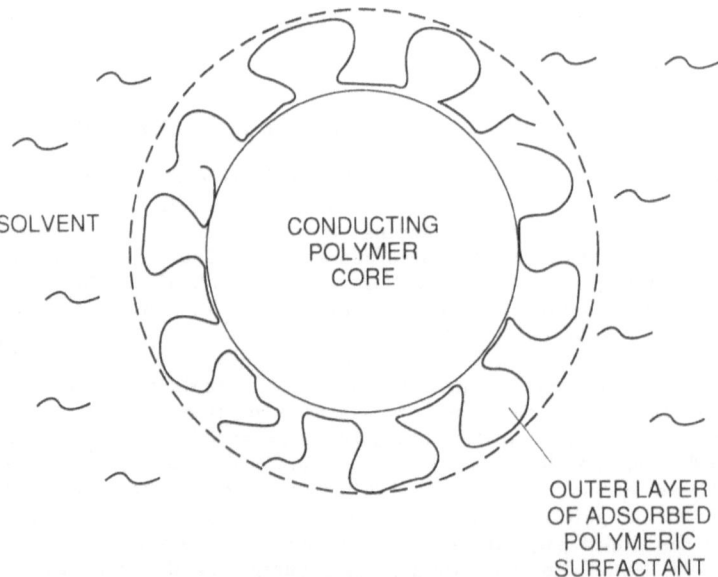

$(CH_2)m—Y—M$

X = S, NH

Y = SO_3, SO_4, CO_2

M = H, Li, Na, K, etc.

Fig. 1. General formula of water-soluble,
self-doped polyheterocycles.

Oxidation of the monomer on the anode can take place in the presence or in the absence of a supporting electrolyte. In the latter case, only self-doping occurs resulting in polymers with conductivities of 10^{-2} $(X.cm)^{-1}$. Additional doping of these polymers by sulfuric acid yields conductivities similar to those obtained when the polymerization is carried out in the presence of a supporting electrolyte, (1-100 $[X.cm]^{-1}$ depending on the length of the alkyl segment). The water-soluble polymers are characterized by highly conjugated chains similar to those of polythiophene or poly (3-alkylthiophene)s as suggested by their optical absorption characteristics.

COLLOIDAL DISPERSIONS OF CONDUCTING POLYMERS

As an indirect method toward processability, conducting colloids of polypyrrole[3-6] and polyaniline[7] have been prepared via a dispersion polymerization route. In this approach, the conducting polymer is che-mically synthesized in the presence of a suitable polymeric surfactant. The surfactant adsorbs or is chemically grafted onto the conducting polymer particle or chain and prevents macroscopic precipitation by a steric stabilization mechanism. The result is a stable dispersion of submicronic conducting polymer particles which consist of a conducting core and a thin outer layer of the non conducting surfactant as shown schematically in Fig. 2.

SOLVENT

CONDUCTING
POLYMER
CORE

OUTER LAYER
OF ADSORBED
POLYMERIC
SURFACTANT

Fig. 2. Schematic representation of a colloidal conducting
polymer particle.

Polypyrrole Colloids

Polymerization of pyrrole in the presence of an appropriate surfactant is achieved using $FeCl_3$ which is commonly used for the chemical synthesis of bulk polypyrrole. The polymer is obtained in its conducting form and requires no additional doping. Aggregation of polypyrrole colloids is prevented by adsorption of the polymeric surfactant onto the conducting particle. Spherical, monodisperse particles (Fig. 3) can be prepared in the size range of 50-250 nm diameter depending on the experimental conditions such as surfactant type, concentration and molecular weight. Compressed pellets or solution-cast films of colloidal polypyrrole exhibit conductivities in the range of 0.1-5.0 $(X.cm)^{-1}$. The chemical and physical properties of the dispersions are largely determined by the choice of adsorbing polymeric surfactant. For example, the use of poly(vinyl alcohol-co-vinyl acetate) results in colloids with good film-forming characteristics[4], while poly(vinyl pyridine)-based stabilizers produce colloids that exhibit reversible, pH-dependent aggregation-stabilization behavior[5,6]. Under favorable conditions, we can estimate the adsorbed mass of surfactant per unit mass of polypyrrole by various indirect spectroscopic techniques. Surfactant/conducting polymer mass compositions vary from 10/90% to 25/75%, depending on the type and initial concentration of the surfactant. The solid-state conductivities of these materials seem to be relatively insensitive to the presence of insulating surfactant layer, with values as high as 2 $(X.cm)^{-1}$ being obtained even for composites containing the highest levels of surfactant.

Polyaniline Colloids

Colloidal dispersions of polyaniline are prepared using KIO_3 as the oxidant in an aqueous medium which contains the polymeric surfactant[7]. Like polypyrrole, polyaniline in its bulk form or in its colloidal form is obtained in the conducting state. Unlike polypyrrole colloids, polyaniline colloids are being chemically grafted onto the polymeric surfactant chain forming a graft copolymer. The surfactant is a tailor-made poly(2-vinyl pyridine-co-p-aminostyrene) in which the p-aminostyrene component is 5.4 mol%. Compressed pellets or solution-cast films have conductivities in the range of 0.5-2.0 $(X.cm)^{-1}$. Polyaniline particles have a rice-grain morphology with an average length of 120 nm and an average width of 60 nm (Fig. 4).

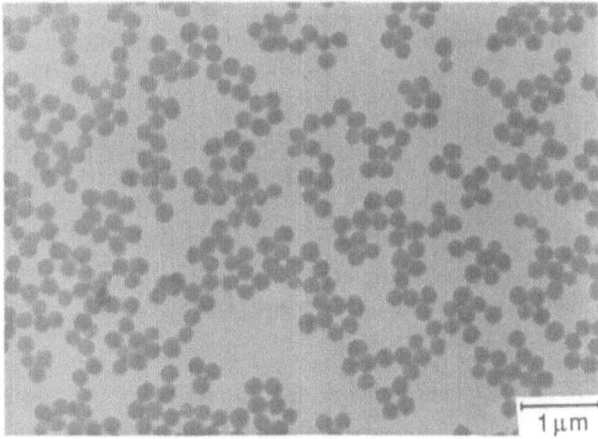

Fig. 3. Transmission electron micrograph of a
polypyrrole latex

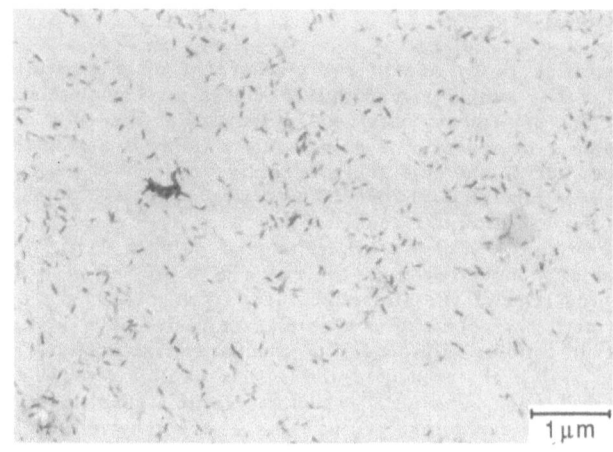

Fig. 4. Transmission electron micrograph of a
polyaniline latex

CONCLUSION

Quite often, processability of conducting polymers, achieved gene-rally by substitution of the monomers with flexible components, resulted in a decrease of their electrical activity. Furthermore, materials processability had been achieved in organic media. In this present work, we have demonstrated that these same conducting polymers can be processed in aqueous media by grafting or physical adsorption of small or large surfactant molecules onto the conducting polymer. Solutions or colloidal dispersions of the conducting polymers synthesized in this work yield materials with conductivities comparable to those of the bulk materials.

REFERENCES

1. M. Aldissi, Recent Advances in Inherently Conducting Polymers and Multicomponent Systems, J. Mat. Educ. 9(4):334 (1987)
2. M. Aldissi, Processability and Order in Conducting Polymers, Mol. Cryst. Liq. Cryst. 160:121 (1988).
3. S. P. Armes and B. Vincent, Dispersions of Electrically Conducting Polypyrrole Particles in Aqueous Media, J. Chem. Soc., Chem. Commun. 288 (1987).
4. S. P. Armes, J. F. Miller and B. Vincent, Aqueous Dispersions of Electrically Conducting Monodisperse Polypyrrole Particles, J. Coll. Inter. Sci. 118(2):410 (1987).
5. S. P. Armes, M. Aldissi and S. F. Agnew, Poly(vinyl pyridine)-Based Stabilizers for Aqueous Polypyrrole Latices, Synth. Met. 28:C837 (1989).
6. S. P. Armes and M. Aldissi, The Preparation and Characterization of Colloidal Dispersions of Polypyrrole Using Poly(2-vinyl pyridine)-Based Stabilizers, Submitted to Polymer.
7. S. P. Armes and M. Aldissi, Novel Colloidal Dispersions of Polyaniline, accepted for publication in J. Chem. Soc., Chem. Commun..

ORGANIZING COMMITTEE

Tormod Riste, director
Institutt for energiteknikk
POB 40, N-2007 Kjeller, Norway

David Sherrington, co-director
Dept. of Physics, Imperial College
London SW7, UK

Arne T. Skjeltorp
Institutt for energiteknikk
POB 40, N-2007 Kjeller, Norway

Gerd Jarrett, secretary
Institutt for energiteknikk
POB 40, N-2007 Kjeller, Norway

PARTICIPANTS

Aertsens, Marc
L.U.C. Universitaire Campus
B-3610 Diepenbeek, Belgium

Aharony, Amnon
School of Physics & Astronomy
Tel Aviv University, Tel Aviv 69978, Israel

Aldissi, Mahmoud
Los Alamos National Laboratory, POB 1663, MS: K764
Los Alamos, NM 87545, USA

Als-Nielsen, Jens
Dept. of Physics, Risø National Laboratory
DK-4000 Roskilde , Denmark

Alstrøm, Preben
Center for Polymer Studies, University of Boston
590 Commonwealth Ave., Boston, MA 02215, USA

Andelman, David
School of physics & Astronomy
Tel Aviv University, Tel Aviv 69978, Israel

Andresen, Arne F.
Institutt for energiteknikk
POB 40, N-2007 Kjeller, Norway

Aukrust, Trond
IBM Bergen Scientific Center
Thormøhlers gt. 55, N-5008 Bergen, Norway

Bahadiroglu, Asiye Gülay
Dept. of Physics, Yildiz University
Sisli-Istanbul, Turkey

Bechhoefer, John
Lab. de Physique des Solides, Bat. 510
F-91405 Orsay Cedex, France

Berg, Cecilie
SINTEF, Avd. 19
N-7034 Trondheim-NTH, Norway

Canessa, Enrique
School of Mathematics & Physics
University of East Anglia, Norwich NR4 7IJ, UK

Cates, Michael
Cavendish Laboratory, Madingley Road
Cambridge, CB3, 0HE, UK

Charvolin, Jean
Lab de Physique des Solides
Bat. 510, F-91504, Orsay Cedex, France

Ciccariello, Salvino
Dept. of Physics, University of Padova
I-35131 Padova, Italy

Cladis, Patricia E.
AT&T Bell Laboratories
Murray Hill, N.J. 07974, USA

Courtens, Eric
IBM Laboratory, Säumerstrasse 4
CH-8803 Rüschlikon, Switzerland

Craven, Jeremy
Dept. of Physics, University of Edinburgh
Mayfield Rd., Edinburgh EH9 3JZ, UK

Dhont, Jan K.G.
Van 't Hoff Laboaotium, University of Utrecht
Postbus 80.501, NL-3508 TB Utrecht, The Netherlands

Erbölükbas, Aysen
Dept. of Physics, Istanbul University
Vezneciler-Istanbul, Turkey

Farago, Bela
Institut Laue-Langevin, 156X
F-38042 Grenoble Cedex, France

Feder, Jens
Dept. of Physics, University of Oslo
POB 1048, N-0316 Oslo 3, Norway

Fogedby, Hans
Institute of Physics, University of Aarhus
DK-8000 aarhus, Denmark

Fossum, Jon
Risø Nationaanl Laboratory
DK-4000 Roskilde, Denmark

Frette, Vidar
Dept. of Physics, University of Oslo
POB 1048 Blindern, 0316 Oslo 3, Norway

Frisken, Barbara
Dept. of Physic, University of California
Santa Barbra, CA 93102, USA

Furuberg, Liv
Dept. of Physics, University of Oslo
POB 1048, Blindern, 9316 Oslo 3, Norway

Galam, Serge
Dept. de Recherches Physiques, T22-E3
University P. & M. Curie
F-75252 Paris Cedex 05, France

Gobron, Thierry
Lab. PMC, Ecole Polytechnique
F-91128 Palaiseau Cedex, France

Gorti, Sridhar
Dept. of Physics, R,-13-2033
MIT, Cambridge, MA 02139, USA

Guillaume, Francois
Lab. de Spectroscopie Moléculaire & Cristalline
351 cours de la Liberation, F-33405 Talence Cedex, France

Harden, James Leroy
Materials Dept., University of California
Santa Barbara, CA, 93106, USA

Helfrich, Wolfgang
Fachbereich Physik, Freie Universität Berlin
WE 5, Arnimallee 14, 1000 Berlin 33, W-Germany

Helgesen, Geir
Dept. of Physics, University of Oslo
POB 1048, Blindern, N-0316 OsLo 3, Norway

Hilfer, Rudolf
University of Mainz, p.t. Dept. of Physics
University of Oslo, POB 1048 Blindern
N-0316 OSLO 3, Norway

Hu, Chin-Kun
Inst. of Physics, Academia Sinica
Taipei, Taiwan, R.O.C.

Inganäs, Olle
IFM, University of Linköping
S-58183 Linköping, Sweden

Ipsen, John Hjort
Dept. of Struc. Properties, Technical
University of Denmark, DK-2800 Lyngby, Denmark

Itri, Rosangela
Inst. of Physics, University of Sao Paulo
CP 20516, Sao Paulo, Brazil CEP 01498

Jagodzinski, Otto
FB Physik, Universitat-GHS-Essen
PF 103764, D-4300 Essen 1, W-Germany

Janik, Janina M.
Inst. of Chemistry, Jagellonian University
ul. Karasia 3, Pl-30-060 Krakow, Poland

Janik, Jerzy A.
Institute of Nuclear Physics
ul. Radzikowskiego 152
PL-31-342 Krakow, Poland

Joanny, Jean-Francois
Ecole Normale Superieure, 46, Allée d'Italia
F-69364 Lyon, France

Jøssang, Torstein
Dept. of Physics, University of Oslo
POB 1048 Blindern, N-0316 OSLo 3, Norway

Kim, Mahn W.
Exxon Research & Engineering Co
22 Route E, Annandale, N.J. 08801, USA

Kjær, Kristian
Risø National Laboratory
DK-4000 Roskilde, Denmark

Kohanoff, Jorge Jose
SISSA, Strade Costiera 11, I-34014 Trieste, Italy

Langie, Greet
Lab. v. Molekuulfysika, Kath. Universite
Celestijnenlaan 200 D, B-3030 Leuven, Belgium

Lekkerkerker, Henk, N.W.
Van 't Hoff Laboratorium. Transitorium 3.
Rijksuniversiteit te Utrecht, Postbus 80.051
NL-3508 TB Utrecht, The Netherlands

Lothe, Jens
Dept. of Physics, University of Oslo
POB 1048 Blindern, N-0316 OsLo 3, Norway

Lundqvist, Bengt I.
Inst. of Theoretical Physics, Chalmers University
of Technology, S-412 96 Göteborg, Sweden

Maher, James V.
Physics Dept., University of Pittsburgh
Pittsburgh, PA 15260, USA

Marques Serra, Carlos M.
E.N.S.L., 46 Allée d'Italia
F-69364 Lyon, France

Martinez-Mekler, Gustavo
Instituto de Fusica, UNAM
Apdo. Postal 20-364Aretri), I-50125 Firenze, Italy
Mexico 01000 D.F. Mexico

McCauley, Joseph
Dept. of Physics, University of Houston
Houston, Tx 77004, USA

Mortensen, Kell
Risø National Laboratory
DK-4000 Roskilde, Denmark

Muller, Jiri
Institutt for energiteknikk, POB 40
N-2007 Kjeller, Norway

Munch, Jean-Pierre
Lab. für Spectrometrie, Univ. Lous Pasteur
4, rue Blaise Pascal, F-67070 Strasbourg Cedex, France

Möhwald, Helmuth
Inst. f. Physikalische Chemie, Johannes Gutenberg
Universität, Welder Weg 11, D-6500 Mainz, W-Germany

Naudts, Jan
Dept. of Physics, University of Antwerp
Universiteitsplein 1, B-2610 Antwerp, Belgium

Otnes, Kaare
Institutt for energiteknikk
POB 40, Kjeller, Norway

Oxaal, Unni
Dept. of Physics, University of Oslo
POB 1048, Blindern, N-0316 OSLO 3

Pedersen, Jan Skov
Risø National Laboratory
DK-4000 Roskilde, Denmark

Perrot, Francoise
SPSRM, CEN-Saclay
F-91191 Gif sur Yvette, France

Piazza, Roberti
Dipt. di Elettronica
via Abbiategrasso 209, I-27100 Pavia, Italy

Pieranski, Pawel
Lab. de Physique des Solidess
Bat. 510, F-91405 Orsay Cedex, France

Pieranski, Piotr
Inst. Fizyki Molekularnej Polskiej Akademi Nauk
Smoluchowskiego 17/19, Pl-60-179 Poznan, Poland

Pincus, Philip
Materials Dept., University of California
Santa Barbara, CA 93106, USA

Pleiner, Harald
FB Physik, Universität-Essen
PF 103764,D-4300 Essen, W-Germany

Posselt, Dorthe
Physics Dept., Risø National Laboratory
DK-4000 Roskilde, Denmark

Pynn, Roger
LANSCE, MS-H805, Los Alamos National Laboratory
Los Alamos, NM 87545, USA

Rakotomalala, Nicole
Dept. of Physics, University of Oslo
POB 1048, Blindern, N-0316 OsLo 3, Norway

Raphael, Elie
Matiere Condense, College de France
F-75231 Paris Cedex 05, France

Rieutord, Francois
Institut Laue-Lagevin, 156X
F-38042 Grenoble Cedex, France

Root, John
Neutron & Solid State Physics Branch
Chalk River Nuclear Laboratories
Chalk River, Ontario, Canada KOJ IJO

Safinya, Cyrus
Exxon Research and Engineering Co.
Route 33E, Annandale, N.J. 08801, USA

Samseth, Jon
Institutt for energiteknikk, POB 40
N-2007 Kjeller, Norway

Scheunders, Paul
Dept. of Physics, University of Antwerp
Universiteitsplein 1, B-2610 Antwerp, Belgium

Schwartz, Moshe
School of Physics & Astronomy
Tel Aviv University, Tel Aviv 69978, Israel

Shlesinger, Michael F.
Dept. of the Navy, ONR/1112
800 N Quincy St., Arlington VA 22217, USA

Smith, Paul
Materials Dept., University of California
Santa Barbara, CA 93106, USA

Sotta, Paul
Lab. de Physique des Solides, Bat. 510
F-91405 Orsay Cedex, France

Spontak, Richard John
Dept. of Mat. Science, University of Cambridge
Pembroke St., Cambridge CB2 3QZ, UK

Steinsvoll, Olav
Institutt for energiteknikk, POB 40
N-2007 Kjeller, Norway

Steitz, Roland
Inst. f. Physikalische Chemie
Johannes-Gutenberg-Universität
Jacob Welder Weg 11, D-6500 Mainz, West-Germany

Swift, Michael Robert
Inst. for Theoretical Physics, Oxford University
1. Keble Rd., Oxford OX1 3NP, UK

Tartaglia, Piero
Dept. of Physics, University of Rome
P. le A. Moro 2, I-00185 Rome, Italy

Teixeira, Paulo I.C.
CFMC, Av. Prof. Gama Pinto 2
P-1699 Lisboa Codex, Portugal

Tippmann-Krayer, Petra
Inst. f. Physikalische Chemie
Johannes-Gutenberg-Universität
Jacob Welder Weg 11, D-6500 Mainz, W-Germany

Thomas, Harry
Dept. of Physics, University of Basel
Klingelbergerstrasse. 82, CH-4056 Basel, Switzerland

Troian, Sandra
Exxon Research & Engineering Co
Route 22 E, Annandale, N.J. 08801, USA

Vanweert, Frans
Lab. v. Molekuulfysika, Kath. Universiteit
Celestijnenlaan 200D, B-3030 Leuven, Belgium

Velarde, Manuel G.
UNED-CIENCIAS, Ap. 60.141
E-28.071 Madrid, Spain

Vitting Andersen, Jørgen
Struc. Properties of Materiaals, Technical University
of Denmark, DK-2800 Lyngby, Denmark

Vives, Eduard
Dept. E.C.M., Diagonal 647
E-08028 Barcelona, Catalonia, Spain

Wang, Litian
Dept. of Physics, University of Oslo
POB 1048 Blindern, 0316 Oslo 3, Norway

Webman, Itzhak
IBM Bergen Scientific Center
Thormøhlersgt. 55, 5007 Bergen, Norway

Wilding, Nigel
Dept. of Physics, University of Edinburg
Mayfield Rd., Edinbrugh EH9 3JZ, UK

Winter, Anatol
Geological Survey of Denmark
Thoravej 8, DK-2400 Copenhagen, Denmark

INDEX

Rouse motion, 327

Saffman-Taylor instability, 248
Schmidt number, 141
Schulman-Montague condition, 253
Schulz function, 312
Scriven and Sterling waves, 142
Seam defect, 180
Second sound, 89
Seniority, 326
Snell's law, 114
Sol-gel transition, 331
Spectral dimension, 335
Specular reflection, 113
Sponge state, 186
Steric hindrance, 12
Stretched exponential, 334, 347
Surface tension, 125
Surfactant, 13, 139, 179, 195, 211,
    245, 249, 363, 377
Synchrotron X-ray, 249

Tanner law, 224
Tensile strength, 353
Tethered surface, 191
Tetracritical, 55
Thermal roughness, 123
Theta solvent, 297
Tiling, 103
Tilted molecules, 134
Topological changes, 109
Topological defects, 60
Transition, 52, 53, 61, 71
    anchoring, 71
    defect-mediated, 52
    hexatic-liquid, 61
    hexatic-solid, 61
    insulator-conductor, 203
    liquid condensed-extended, 147
    sphere-to-rod, 195
Transverse sound, 89
Triblock polymers, 311
Two-dimensional crystal, 45, 53, 113

Ultrasound, 88
Unbinding transition, 274
Undulation force, 15, 25, 251, 271

Van der Waals force, 7, 8, 28, 250
Van der Waals liquid, 8, 223
Virial coefficient, 9
Viscosity, 139
Viscous finger, 339
Volterra process, 103
Vogel law, 347
Vortex, 23
Vulcanization, 319

Wettability, 237
Wetting, 161, 221, 245

x,y-model, 55
X-ray scattering, 113

Young-Depré equation, 221
Young-Laplace equation, 239

Zimm model, 324